Chemical Arts and Technologies of Indigenous Americans

Burton L. Gordon

BAR International Series 1907
2009

Published in 2016 by
BAR Publishing, Oxford

BAR International Series 1907

Chemical Arts and Technologies of Indigenous Americans

ISBN 978 1 4073 0390 1

BAR Publishing is the trading name of British Archaeological Reports (Oxford) Ltd. British Archaeological Reports was first incorporated in 1974 to publish the BAR Series, International and British. In 1992 Hadrian Books Ltd became part of the BAR group. This volume was originally published by Archaeopress in conjunction with British Archaeological Reports (Oxford) Ltd / Hadrian Books Ltd, the Series principal publisher, in 2009. This present volume is published by BAR Publishing, 2016.

Printed in England

BAR
PUBLISHING

BAR titles are available from:

BAR Publishing
122 Banbury Rd, Oxford, OX2 7BP, UK
EMAIL info@barpublishing.com
PHONE +44 (0)1865 310431
FAX +44 (0)1865 316916
www.barpublishing.com

TO MYRA

ACKNOWLEDGEMENTS

I am particularly indebted to Miriam Beames and her husband, Dr. Paul Tutwiler, for thoroughly editing the present version of the manuscript. Without their help and invaluable advice this book would not have been completed.

My lasting thanks to Dr. Nicholas T. Mirov who taught me a little chemistry when I was an unskilled worker in his laboratory. As a graduate student, I learned much from the late Professors Sherborne F. Cook, Walter A. Hacker and Aden E. Treganza on several archaeological excavations in California.

I am grateful to Dr. Hanns Ahrens, a retired chemist from Berlin, who has read the manuscript, corrected several errors, and made numerous helpful suggestions. However, since many bits of text have been added after his reading, I am responsible for any mistakes that remain.

My thanks also to the following individuals: Anna L. Gordon who edited most of a preliminary manuscript and gave me frequent encouragement; Janice von Arnim for help in the interpretation of Bernardino de Sahagún's comments; Dr. Robin A. Gordon for assistance with translating several German sources; and Dr. Karen Gordon-Grube for advice on the subject of cookery.

Table of Contents

LIST OF ILLUSTRATIONS

INTRODUCTION

Written records of knowledge in the pre-Columbian New World are virtually non-existent (in contrast to an abundance of such records for ancient China and the Near East). Consequently chemical knowledge in the Americas, prior to the arrival of Europeans, is poorly documented. This knowledge is sometimes described as "rudimentary," as in the following passage:

> ...as far as known, aboriginal man in Neotropica did not process any animal product except possibly red dye from cochineal insects; i.e., he did not tan hides into leather by chemical means (tannic acid, etc.), prepare perfumes from animal musk or ambergris, or make glue from animal gelatin, or soap from fats (Gilmore 1963, 346).

Though mineral and metal artifacts are comparatively durable, organic artifacts tend to decompose quickly. Thus, archaeological evidence occasionally bears out the assessment given in the above quotation. Nevertheless, this is by no means a fair assessment of the chemical skills possessed by "aboriginal man in Neotropica." The discussion below--which includes the pertinent historical records, archaeological evidence, and ethnographic notes (as well as chemical observations)--tells a different story.

The arts and technologies discussed here are those known to have existed in pre-Columbian times--as well as those thought to have been developed by native peoples (independently of foreign influence) during the five centuries since the European conquest of the Americas began.

In the following discussion, the term "chemical arts" makes reference to those deliberate changes in the composition of substances made without the rational explanations provided by modern science; that is, changes that are based on practical experience with slight understanding of the chemical principles involved. Thus, the *chemical arts* are distinguished here from the *chemical sciences*. Even so, the differences between the two are not always clear cut. This is made evident historically in the Old World by the transition from alchemy (the "divine art") to scientific chemistry. Despite their mystical interpretation of procedures, the alchemists provided modern chemistry with practical tools, processes, and materials (for instance, the still, the process of distillation, and relatively pure alcohol). At the end of the 15th century alchemical concepts still reigned in Europe. Mystical explanations and religious ritual played a part in chemical proceedings in both the Old World and the New. Thus in the Americas at the time of the Conquest, these arts may have been no more burdened by irrational concepts than those of Europe.

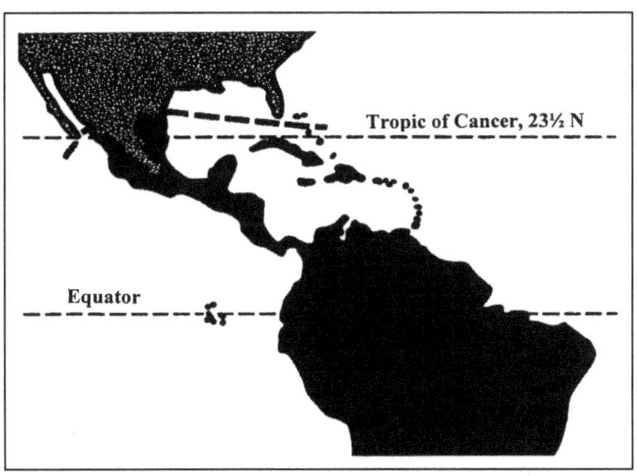

Sketch of the northern boundary of the Neotropical realm

The biological region Neotropica, referred to above, includes South America, the West Indies, and Central America; it extends northward on the North American continent to a little beyond the Tropic of Cancer. It includes coastal Peru, the Andes, and virtually all of Mesoamerica (except the southern part of the central Mexican plateau)--in short, the most technologically-advanced areas of pre-Columbian America.

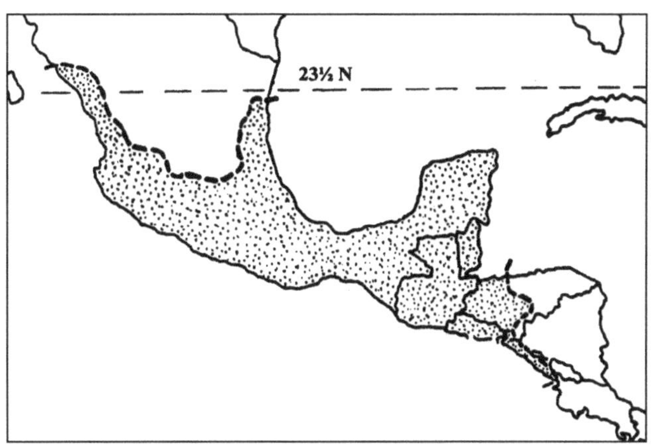

Sketch of Mesoamerica

Mesoamerica is an archaeologist's term, applied to the area of relatively complex pre-Columbian cultures extending from northwestern Mexico to western Honduras and Nicaragua.

Figure 1. Neotropica and Mesoamerica

Scientific chemistry, especially *inorganic* chemistry, began, at the earliest, with publication of Robert Boyle's *The Sceptical Chemist* in 1661, more than a century-and-a-half after the European discovery of America. (Therefore, none of the conquistadors or early chroniclers cited here could have had an inkling of such changes.) In contrast, *organic* chemistry (that is, the chemistry of carbon compounds) dates from well over a century later.

Modern organic chemistry is a highly specialized field, with a large and distinctive vocabulary. This discussion, however, is meant for those with only an elementary knowledge of the subject, and relatively few technical terms are used. In a number of figures, I have tried to sketch such information as is necessary to interpret the text.

Chemical science has greatly improved the understanding of folk arts and technologies. While many of the chemical reactions referred to here are relatively simple from the standpoint of modern chemistry, some of the processes used by Native Americans are surprisingly complex--for instance, those involved in the making of rubber, the dyeing of fabrics, the smelting of metals, and the electrochemical gilding of copper objects.

With regard to historical sources, the Spanish of such early authors as Fernández de Oviedo y Valdés, Francisco Hernández, and Cieza de León, I have translated into English, except where adequate translations are already available. In the interests of accuracy, most sources are quoted directly (and at some length) rather than being paraphrased. For purposes of placing these and other sources in a chronological framework, it is important to note that many of the publication dates listed in the References Cited are merely those of the editions consulted, not necessarily those of first editions.

Unless stated otherwise, italicization and bracketed comments in quoted passages are mine.

CHAPTER 1

FIRE

The human species is the product of long evolution in the Old World, where it remained until late in its history. In 1516, Peter Martyr applied the name 'New World' to the western hemisphere. (His work is titled *De Rebus Oceanicis et Novo Orbe.*) Clearly, this description would have been more accurate when the earliest immigrants arrived (probably largely from Asia across the Bering Straits): indeed, these immigrants found a hemisphere as completely without human influence as some distant planet. Likely, this first colonization of the Americas took place between 10,000 and 20,000 years ago; at present it is thought to have begun some 12,500 years ago, but that date is tentative. Since no artifacts *definitely* made by these first arrivals have been found, the range of their arts and technologies is, of course, somewhat uncertain.

Fire-making is the most ancient and basic of the chemical arts. The human species was already making fire when it first appeared in the fossil record some 100,000 years ago. Since that time, no chemical process has had greater practical significance, nor has remained so entwined with religious ritual. The control of fire made possible the spread of the human species from its ancestral home in the Old World's low latitudes to the cold Bering Straits. Only by possessing this skill was the species able to settle one contrasting biome after another in America, and press southward to Tierra del Fuego. In much of the Americas (especially in the technologically-advanced societies of the Inca, Maya, and Aztec) fire-making and fire-keeping were frequently heavily ritualized procedures. Men were often the fire-makers, but it was women who were responsible for keeping sacred fires burning.

Fire-Making Methods and Tinder

Burning is due to a vigorous chemical reaction between a combustible material and free oxygen in the air; this is a well-known example of the oxidation process. More generally, the term oxidation is applied to a chemical reaction in which either oxygen is added to an organic molecule, or hydrogen is removed. (Fire is simply the name given to the abundant heat and light evolved during that process.) Perhaps a less familiar term is reduction, the reverse of oxidation; in this reaction, either oxygen is removed from a molecule or hydrogen is added. In reduction, heat is absorbed rather than being released.

To make fire, it is necessary to raise the temperature of the combustible material to its ignition point. Raising the temperature of wood to such a point by rotary friction was almost the universal way of making fire in pre-Columbian America. The device used was a simple hand drill: a shaft of hard wood (with one end set in a depression made in a piece of softer wood) is twirled between the palms. When the heat of friction makes the depression glow, bits of tinder material are placed in contact with it, then blown upon to start a flame. Although laborious, it is said that fire can be made by this method within a few minutes.[1]

Another way of kindling fire (probably older than use of the fire drill) is to strike a rock containing iron pyrites against the edge of a piece of flint or quartz so that sparks fly into a pile of tinder. (Pyrite is an iron sulfide, FeS_2; because its crystal-faces are shiny and brass-yellow, it is often called 'fool's gold.') This method was rare in the Americas, though it was used by the Yahgans of Tierra del Fuego (a people of very limited technology) and by the Greenland Eskimo. This technique was probably discovered by the earliest humans—or even by their hominid ancestors—long before discovery of the Americas: chipping stones is one of the oldest human tool-making activities, and certain stones produce sparks when struck together. A similar method of making fire—striking flint with a piece of steel until sparks fly into a tinder box—was used in Europe until the mid-nineteenth century when phosphorus matches were invented there. (The match, like the hand drill, lights fire by friction.)

Everywhere, suitable tinder materials were sought out. In agricultural areas, cotton—almost pure cellulose—was sometimes used as tinder.

Fuels

Where firewood was scarce—for instance, in deserts and above the tree line on mountain slopes—alternate materials had to be found. For example, the Pueblo Indians of Arizona and New Mexico, among whom maize was a staple crop, often burned corn cobs. In arid Mexico, the dry leaves of maguey (agave) were burned, 'and in most places this is the poor man's firewood' (Motolinía 1950, 274).

In the treeless high Andes, fire-makers depended for fuel on *ichu* grass (*Stipa ichu*), such compact shrubs as yareta (*Azorella* sp.), and—since before Europeans brought

[1] In their markets the Aztec sold 'concave mirrors' (Sahagún 1956, 3:150). It has been suggested, albeit on slight evidence, that similar concave mirrors made by the Maya were 'burning mirrors' (Robicsek 1978, 106)—that is to say, mirrors used for generating fire by focussing the sun's light on tinder. Such mirrors are also claimed to have been used at Machu Picchu in the Andes (Bingham 1948, 29).

As mentioned in the Introduction, 'organic chemistry' is the chemistry of carbon compounds. In addition to the element carbon itself, these compounds consist principally of hydrogen, oxygen and nitrogen. Like other elements, these four elements have a definite combining capacity; that is, 'atoms of the elements in organic compounds can form a fixed number of bonds. The measure of this ability is called valence' (Solomons 1996, 4). The carbon atom itself has a valence of four; the hydrogen atom, one; oxygen, two; nitrogen has a valence of either three or five (three, in all of the compounds to be discussed here).

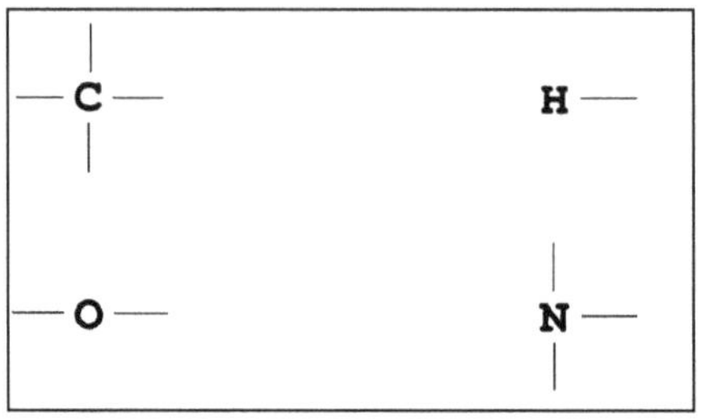

In the above figures, a single valence (a single bond) is represented by a dash.

Figure 2. Carbon Compounds

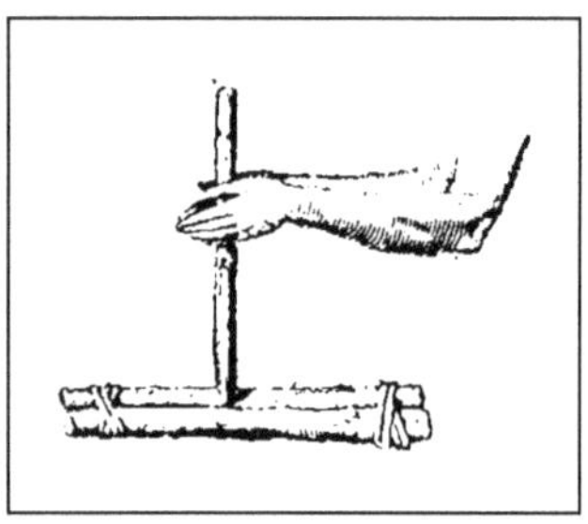

Left: A simple hand drill (Oviedo 1959a, vol. 5, lamina II, 2). Right: A fire-making Maya priest, from Codex Dresden (after Thompson 1972, plate 6)

Figure 3. Fire-Making with Hand Drills

sheep, goats, and bovine cattle to the New World, only the Andean area had domesticated ruminants, the llama and alpaca—especially on llama dung (*taquia*). 'There was no firewood, just sheep dung [i.e., llama or alpaca dung][2] and some roots that they pulled from below the ground' (Cieza de León 1998, 434). In the grasslands of the North American Great Plains, natural woody growth is alsoscarce (except along stream courses). The bison, which once grazed there in great numbers, digested huge amounts of grass, and in places its dried dung ('buffalo chips') was the most important fuel. Dry and slightly-weathered bison dung is easily lighted and produces a good deal of heat. In most of their range, the Eskimo had a hard time finding firewood. Near the coast, until quite recent times, they used the fat of seals and walruses, and cooked their food over blubber-burning lamps of stone or clay.

Despite the abundance of tree growth in the humid tropics, finding ready fuel in forests can be difficult for the uninformed.[3] Native people living there have learned to identify trees whose wood, because it contains readily flammable essential (volatile) oils, will burn while still 'green'—that is, when freshly cut.

[2] The llama and alpaca 'the Spaniards [at first] called sheep because they saw that they had wool and were so gentle and domestic' (Cieza de León 1998, 118).

[3] In many tropical rural areas, where the original forest has been cleared, three valuable firewood trees are easily found—namely, guava (*Psidium guajava*), *capulin* (*Muntingia calabura*) and *guácimo* (*Guazuma ulmifolia*). The three are widespread because seeds from their edible fruit are dispersed by both humans and wildlife.
As one would expect, tropical firewoods differ greatly in their burning qualities; for instance, *capulin* ignites easily and burns intensely (leaving a thick layer of ashes); guava, while giving off much heat, burns gradually. Of course, such differences apply also to firewoods in temperate zones, a fact well-known to those who still use wood stoves. For example, oak wood burns slowly, whereas the flames of poplar and willow wood last but a short time.
In pre-Columbian times, firewood was not cut in even lengths (there being no metal saws) nor were the pieces placed, cord-like, in uniform stacks; even today in Central and South America, such accumulations of firewood are rarely seen around Indian households. The typical Indian hearth is different, too; for instance, in Ecuador among the Quechua-speakers of Otavalo, 'Fires are built in the characteristic Indian way at the tips of the sticks' (Parsons 1945, 15). In another example, hearths among the Chocó are typically made up of three logs, equally-spaced, with their butt-ends drawn together, between which three stones are placed to support the round-bottomed pottery used for cooking.

Many chemical compounds contain only two elements, hydrogen and carbon, and are therefore called hydrocarbons. On combustion, hydrocarbons yield two oxides as end products—namely, CO_2 and H_2O.

The simplest hydrocarbon is methane, CH_4, commonly known as marsh gas; it is also called fire-damp in mines, where it sometimes causes destructive explosions. This simple hydrocarbon is the main constituent of the flammable 'natural gas' widely used for cooking in industrial societies. Methane is called a saturated hydrocarbon, because in its molecule one atom of carbon is combined with four hydrogen atoms, the maximum possible. The structural formula of methane is shown below.

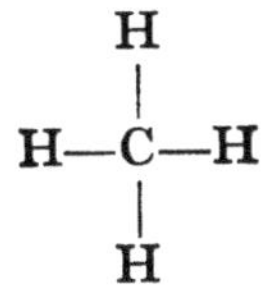

Methane

Although methane gas, itself, does not enter into the chemical arts that are considered here, the methyl group, CH_3— (which is left on removal of one atom of hydrogen from the methane molecule) frequently appears in the compounds to be discussed below. The essential (volatile) oils commonly give wood, leaves, and flowers their odor. The most important constituents of essential oils are the terpenes (hydrocarbons) and the oxygen-containing terpenoids. The basic structural unit of terpenes and terpenoids is isoprene, a hydrocarbon consisting of five carbon atoms attached to eight hydrogen atoms, C_5H_8. (Isoprene will be discussed more fully later.)

Such compounds as pinene, $C_{10}H_{16}$ (the commonest of all terpenes and the main constituent of turpentine)are found, for example, in the needles and wood of many conifers. That such compounds store copious amounts of energy can be seen by the intense fires that occasionally sweep over pine plantations.

The amount of heat liberated during combustion depends on the composition of the fuel, and, as one might expect, hydrocarbons give off more heat than already partly-oxidized compounds such as the carbohydrate cellulose. Although both carbon and hydrogen liberate heat in combining with oxygen, hydrogen liberates more than four times as much heat, weight for weight, as carbon. No other element exceeds hydrogen in the amount of heat that it yields upon oxidation.

Figure 4. Hydrocarbons

A few of these trees are easily identified. For instance, almácigo, *Bursera simaruba*, a member of the torchwood family, is readily recognized by its reddish-brown bark which peels-off in thin, flaky sheets; the wounded trunk exudes a turpentine-like fluid and a red-tinted resin. A 16th century manuscript on the Maya area notes that 'from this tree the [Maya] Indians make fire' (Roys 1931, 227).[4]

Nowadays, the Panamanian Guaymí know more than ten such trees, including species of the genera *Protium*, *Dipteryx*, *Inga*, *Parkia*, *Brosimum* and *Manilkara*. Take, for example, a giant *almendro* tree (*Dipteryx panamensis*):

> one need only make a bruise in the bark along a buttress, lay burning coals against it and, given two rainless days, the tree will burn through and topple. Without rain, the fire will burn on, consuming even the branches of the fallen tree. In rainy weather a log will sometimes smolder for weeks. (Gordon, 1982, 59-59)

Although wood is still a commonly used rural fuel, charcoal is the choice fuel in many places. In Latin America, charcoal is generally made in flat, shallow pits or in conical piles on the earth surface. Wood, preferably a hard dense wood, is stacked, covered with earth, and then fired. Heated thus in the absence of air, it is subjected to what is sometimes called 'smothered combustion.' Most of the wood's water, wood tar, and volatile gasses are driven off in the process; charcoal is the unburned carbonaceous residue, though it retains those mineral elements originally present, such as potassium and calcium.

Charcoal has great advantages over firewood as a fuel; the amount of heat liberated when charcoal is burned (heat of combustion) is considerably higher than when wood is burned. Charcoal has a lower ignition point than most woods; for example, 'it was early observed that charred wood ignites more easily than uncharred; thus, charcoal ignites at 580°F [304°C] and pine wood at 800°F [426°C]' (Hough 1926, 4). When ignited, charcoal does not burn with a smoky flame, but simply with a red glow; burning without flames, charcoal is also less of a threat to combustible, closely-packed habitations. Charcoal is much lighter than the wood from which it was made; the freshly-cut woods, commonly used in charcoal making, lose well over three-quarters of their weight, most of it as water. Thus charcoal is much easier than wood to transport from forests to urban centers.[5]

---

[4] Leaves from *Bursera* trees, 'frequently spray a mist of [flammable] volatile oils when broken [like fresh orange peels]' (Case et al. 2003, 195).

[5] In Mexico the well-known mesquite tree, *Prosopis juliflora*, which ranges from the American Southwest through the drier parts of Central America into South America, is a favored charcoal tree. Farther south in Peru and neighboring areas, several species of *Prosopis* are much used as well, principally *P. chilensis*.
In the humid tropics two other leguminous trees are much used. Both are plentiful in rural areas where, nowadays, they are often associated with the Spanish-introduced cattle industry:
(1) *madre de cacao* or *mata raton*, *Gliricidia sepium*, which is also used as a shade tree in cacao plantations and, nowadays, as a living-fence, and
(2) *guaje*, *Leucaena leucocephala*, probably native to southern Mexico but now widely scattered along roadsides and in abandoned pastures. (Cattle feed on its foliage.) The tree thrives in both the dry and humid parts of the tropics.

There are few accounts of indigenous peoples'charcoal-making written by the first Europeans to arrive in the Americas; since the practice was well known to them, they probably did not find it noteworthy. And so, though charcoal was certainly manufactured in many parts of pre-Columbian America—and had a variety of uses—it is unclear whether, except for Aztec Mexico, it was produced for use as a household fuel. Hernán Cortés reported that quantities of 'firewood and *charcoal*' were sold in Aztec markets (Oviedo 1959a, 4:45). Later in the 16th century, there were still numerous markets along public roads in Mexico where 'firewood, charcoal' were sold (Hernández 1986, 103).

What is sometimes called charcoal is often made up mainly of partly-burned carbonized wood and ash, but contains little real charcoal. (The crude 'charcoal' used for black color in ancient cave paintings is an example.) In prehistoric times, relatively pure charcoal was purposely made in both the Old World and the New—and in the western hemisphere, both in Mesoamerica and the Andean area. Its manufacture represents a great advance in chemical technology.

Among the oldest and most important uses of fire is, of course, the preparation of food.

Two other native trees are among the most used in making charcoal today; whether they were so used in pre-Columbian times is, of course, another matter:

(1) On the Pacific slopes of Colombia, *Hirtella carbonaria* (Chrysobalanaceae), takes its specific name from the fact that is used there for charcoal.

(2) Another important charcoal tree in the wet tropics is the aforementioned *guácimo*, *Guazuma ulmifolia*, a member of the linden family (Tiliaceae). It is commonly found in pastures where it provides shade for cattle; the animals also eat its fruit and leaves. During the Spaniards' first years in the West Indies, guácimo was the major source of charcoal for making gunpowder (Oviedo, 1959a, 1:254).

CHAPTER 2

PREPARING, COOKING, AND PRESERVING FOODSTUFFS

Like fire-making, cooking is one of the oldest chemical arts. Throughout the ages, roughly half the human population has busied itself with the processing and cooking of foodstuffs. Hence, the early foundations of chemistry were built in large part by women.

Cooking is so old that the uncooked diet of mankind's early ancestors can only be a matter of conjecture. The chronological sequence of the various cooking methods is also a matter of speculation; archaeological evidence on the subject is fragmentary, especially for the pre-ceramic era.

By the 15th century, many of the cooking methods in the Old World and the Americas were quite similar—and this despite the fact that when the first immigrants arrived in the New World, no pottery cooking utensils were made in either area.

Processing Raw-Materials before Cooking

In those areas where no agriculture was practiced, wild foods such as acorns, agaves, mesquite pods, and grass seeds were harvested for food. Maize (corn) and manioc (cassava) were the chief agricultural food plants of the pre-Columbian Americas. These two (along with the common potato, *Solanum tuberosum*, then confined to the Andean area) are the principal American contributions to the modern world's food supply. Maize is a large annual grass, and manioc a perennial shrub with some 5 to 10 starchy, swollen roots at its base. In the 15th century, maize had the widest range of any domesticated plant in the Americas: all the way from Quebec to central Chile. In much of the tropics, however, the manioc plant was, and is, the more important source of food. (In Spanish-speaking America, manioc is often known as *yuca*.) Numerous varieties of all three plants had been bred in pre-Columbian times—and, to some extent, plant domestication and the development of chemical arts have gone hand in hand. For instance, cookery and the utensils used became more elaborate as new varieties of maize were bred.

Milling and Other Methods of Food Fragmentation

It is often difficult to separate physical from chemical processes sharply. For example, the grinding of dry maize seeds (largely a physical process) may have strong chemical effects. If stored in a dry place, grass seeds, protected as they are by a cellulose cover (the pericarp), may remain unspoiled for years. Immediately after milling, however, their chemical condition is quite different. Compared to most other cereals, maize seeds contain an unusually high percentage of oil, and most of it is in the germ cell (i.e., the embryo). During milling the seed-cells are torn apart, the germ cell is crushed and its contents exposed to the air. As a result, oils in the meal oxidize and turn rancid within a few weeks. After milling, the meal loses much of its aroma within a few hours—evidence of an even quicker chemical change. Milling also affects the amount of niacin (nicotinic acid) present: 'Although many cereals have apparently high niacin contents..., *milling can have drastic effects* as a very large proportion of *the vitamin* is located in the germ' (Coultate 2002, 220).

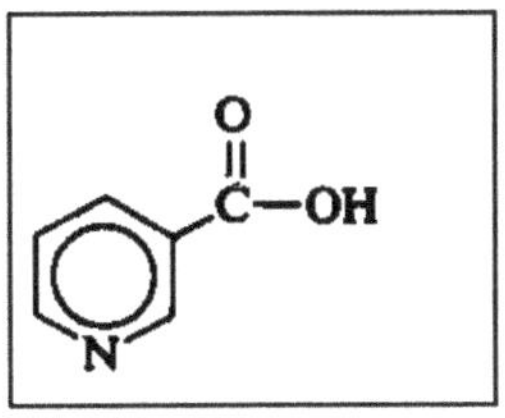

Niacin

Niacin is also called nicotinic acid. (Organic acids are identified by the presence of an acid or carboxyl group, 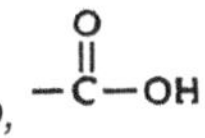 *, sometimes abbreviated as —COOH.)*

It should be noted here, with regard to the structural formula of niacin, that the concept of a benzene ring is fundamental to organic chemistry. The ring is represented in several ways. Of the three shown below, only the last two are used in this discussion:

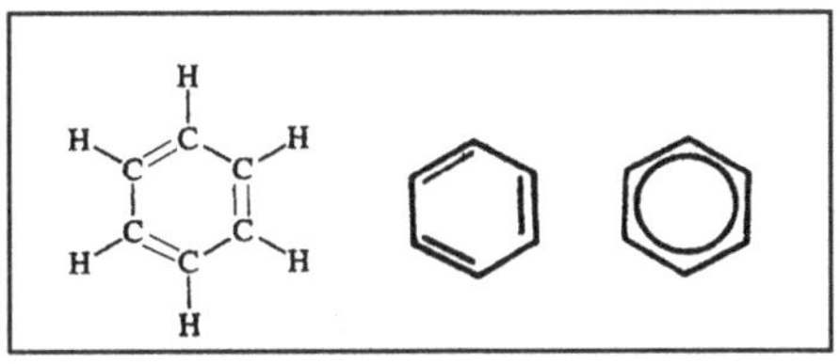

[In contrast with the hexagonal ring of benzene shown in the these three figures, there are many compounds having rings with a different number of sides. Nevertheless, the great majority of these ringed compounds contain either six or five sides. If, as in benzene, carbon is the principal element at each vertex, the ring is called homocyclic. If another element than carbon is present in the ring (for example, nitrogen, as in the case of niacin), the ring is called heterocyclic. Both heterocyclic and five-membered rings appear, for example, in nicotine—to be discussed later.]

Figure 5. Niacin and the Benzene Ring

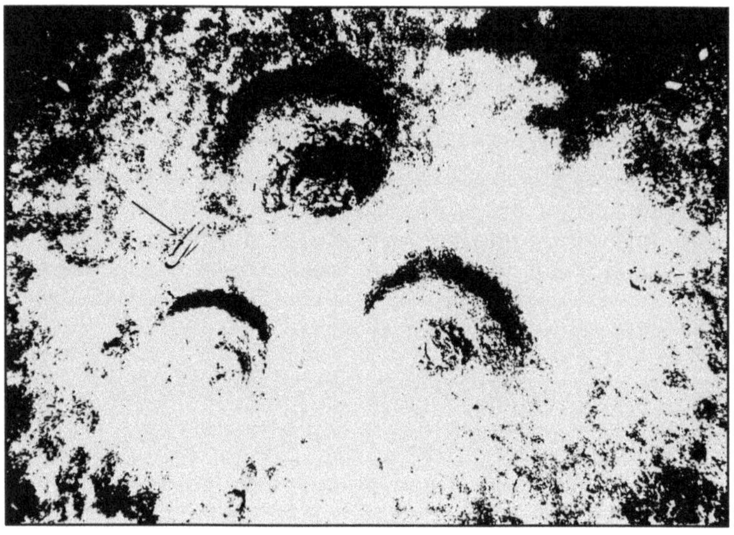

Left: Bedrock mortars made by the Costanoan (Ohlone) people in northern California for grinding acorns. (Note pocketknife for size.)
Right: Pestles and bowl-shaped mortars made by the same people.

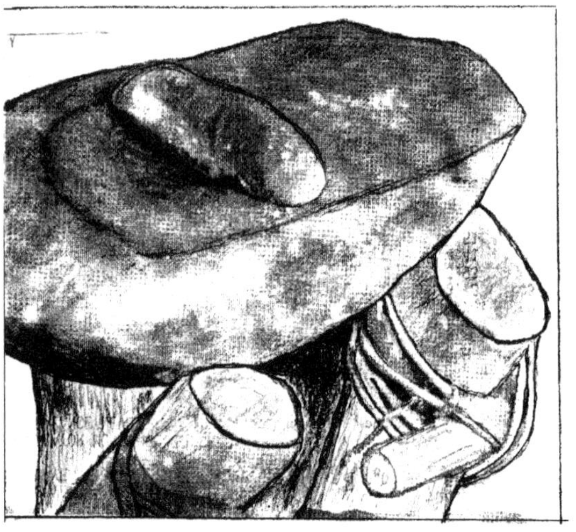

A mortar used nowadays by Zenú descendants in northern Colombia.

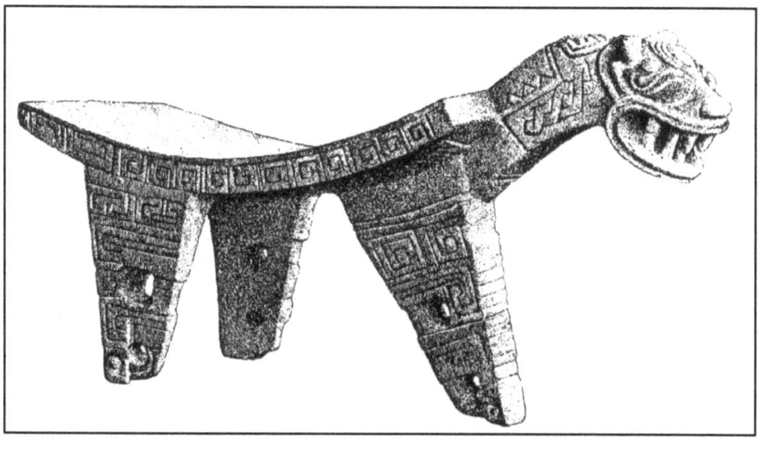

An ornate, archaeological, jaguar-shaped metate from Costa Rica (Holmes 1908, 129).

Figure 6. Milling Equipment

In pre-Columbian America, milling equipment was limited to pestles and mortars. The millwheel, itself, was lacking. In much of Latin America, and often in anthropological parlance elsewhere, pestles are known as *manos*; mortars, as *metates*. A variety of such tools was made: bowl-shaped and flat-surfaced metates; one-handed and two-handed manos; the disk-shaped rocker-pestle (also known as a rocker-mano or rocker-grinder), and so on. (Of all such milling equipment, these rocker-pestles are probably closest to the Old World millwheels.) The purpose of such tools is mainly either (1) to pulverize dry hard food or (2) to crush and fragment relatively-soft, moist food. In the case of maize, for example, the dry and hard kernels may be ground to a meal, or the kernels, freshly cut from the cob or softened by soaking, may be mashed into a dough-like paste.[1] The bowl-shaped metate was used for both functions; flat metates are usually used to fragment moist food. The rocker-pestle is generally used for pulverizing hard, dry materials.

Milling tools are usually made of stone, though occasionally of wood. Interestingly, their use in processing food is virtually confined to women. Metates are sometimes elevated on logs or wooden stands. However, some ancient mortars stand on three or four stone legs, and are often sculpted to resemble jaguars. With only stone tools, these metates were carved out of solid blocks of dark-gray vesicular basalt, a rock common in old lava flows. These elaborate mortars, found from southern Nicaragua to northern Panama, are unique to the New World (Nordenskiöld 1931a, 491).

Removal of Unpalatable and Toxic Substances

Some raw vegetable materials that are either unpalatable, scarcely digestible, or toxic become satisfactory foods after being cooked. Often, after such materials are cooked, more calories are accessible to the consumer.

Leaching and Rinsing

Starches, fats and proteins are the chief sources of energy in the mammalian diet. Starch, mainly in the seeds and roots of plants, is (except for cellulose which is indigestible by humans) the most abundant compound in organic nature, and the most plentiful of all human foods. Natural selection in the wild, however, has protected this great store of food with a variety of distasteful or toxic substances. It was, therefore, advantageous during the long domestication process, to select for those genetic strains of food plants which lack such distasteful or poisonous substances. Hence, among agricultural peoples, a preliminary detoxification of plant materials is generally less necessary than among those dependent on wild plants alone. The domesticated plant, bitter manioc (discussed below), is an outstanding exception. There are, however, other exceptions. Even the seed coats of the long-domesticated quinoa, *Chenopodium quinoa*, contain a bitter saponin which has to be removed before cooking. For example, in Ecuador, quinoa seed 'has to be washed well to get out its bitter taste. The wash water is not wasted: it is mixed with *penco blanco* (extracted from maguey) to wash cotton or woolen clothes' (Parsons 1944, 18 and 21). In any case, gatherers of wild plants must expend much time and energy in making plant materials accessible as food. On the other hand, because few mammals have flesh that is toxic to man, comparatively little time need be spent by hunters on processing—an advantage to those who, like the Eskimo, subsist mainly on products of the hunt.

Soaking in water is often the first step in preparing foods for cooking. Raw plant materials, especially seeds, require a shorter cooking time when subjected to such preliminary soaking. Soaking alone may also remove some unpalatable substances. Leaching, however, is a more efficient way of removing unwanted substances. In leaching the material is first usually ground, pounded, shredded, or grated. When water is flushed over this vegetable mass, it percolates downward carrying soluble compounds with it. The purpose of the initial fragmentation is both to rupture cells in the vegetable material and to expose more of its surface to the leaching water.

Although numerous wild plants contain toxic or distasteful substances, many do not; among the latter are various wild palms. The extraction of starch from palms is practiced throughout much of the world's tropics, and probably antedates agriculture. The procedure is best developed in southeastern Asia and islands of the western Pacific where sago starch is extracted from several palms, but principally from *Metroxylon sagu*. Starch extraction is common in South America as well. For example, starch from the 'moriche' palm (*Mauritia flexuosa*) is a staple for the Warrua (Warao) on the Orinoco Delta (Humboldt and Bonpland 1972, 3:278-279).[2] The spongy pith in the palm's trunk is pounded and then rinsed to separate starch from miscellaneous cellulose fragments. Starch extraction from the palm, *Arecastrum romanzoffianum*, is also described among the Tupi-Guarani tribes of Paraguay and Southern Brazil (Métraux 1963a, 83).

[1] In Mexico, it is often claimed that maize dough which has been worked on *basalt metates* gives the best flavor to tortillas. No doubt, numerous minute particles of the metate end up in the tortillas. Whether some chemical interaction between the basalt and the maize dough accounts for the improved flavor is hard to say.

[2] By the late 1790s, when Alexander von Humboldt and Aimé Bonpland traveled in the Americas, European chemists had entered the scientific age. For his time, Humboldt had a good knowledge of chemistry; in fact, he collaborated with the famous French chemist Joseph Gay-Lussac in determining that 'two volumes of hydrogen combine with one of oxygen to form water' (Findlay 1934, 109). Moreover, Bonpland often used Linnaean binomials, established by Karl von Linné, the Swedish botanist, the foundation for the modern taxonomic system. Thus their accounts are particularly noteworthy.

Vegetable tannins belong to a large group of organic compounds known as polyphenols. Indeed, according to some authorities, the terms 'vegetable tannins' and 'polyphenols' can be used synonymously (Haslam 1988, 6 and 8). The polyphenols are polymers of a simple compound called phenol, C_6H_5OH. All phenolic compounds have a hydroxyl group (that is, an –OH group) attached directly to a benzene ring in place of hydrogen. The structural formula of phenol is shown below:

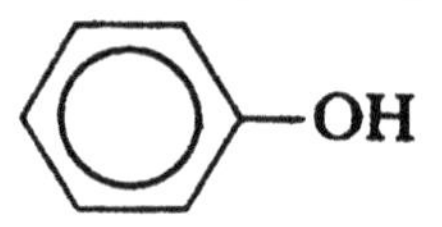

Phenol (also known as hydroxybenzene)

Tannins are widely distributed in the plant world and, although they make many plant products unpalatable, they contribute much to the flavor of others—for example, to such important beverages as chocolate, red wines and tea. Phenolic compounds 'occur in most fruits and most of them contribute to color and taste' (Belitz and Grosch 1987, 592); they are also present in several spices.

As the name itself indicates, the traditional use of tannins is in tanning, but they also may act as mordants in dyeing. Nowadays, tannins are also used medicinally for healing burns, as well as for their styptic action.

Figure 7. Tannins and Phenols

Extraction of starches by one means or another was a major development in the history of food technology:

> In the absence of cereals this simple expedient might well be deemed an epoch-making discovery, since it rendered possible the accumulation of a *permanent, readily transportable, food supply*, and thus protected man from the vicissitudes of the season and the chase...*Separated from the sugars and other readily soluble substances which retain or absorb moisture*, the starch of the taro [a Polynesian root], *cassava*, arrowroot, canna, and other root crops [as well as palms] *can be quickly and thoroughly dried and will then keep indefinitely.* (Cook 1903, 489)

Removing Tannic Acid by Leaching and Cooking

California Indians, dependent as they were on wild plants, were especially skilled in leaching. For many California tribes, oaks were the most important food plants—and in central California, acorns were truly the staple, as an early 19th century explorer said (Beechey 1931, 74). More than a dozen species of oak are native to the area. The acorn preferred by many central and northern Californian tribes was that of the tanbark oak (*Lithocarpus densiflora*).

Although acorns contain such valuable foodstuffs as starch and oil, they also contain tannic acid, a bitter and astringent substance. (Bitterness and astringency are related sensations, but bitterness is sensed only by the tongue's taste buds, whereas the astringent sensation affects the whole surface of the mouth.) To make acorns palatable, most of this substance must be removed: if consumed in large quantities, untreated acorns may cause acute indigestion. Fortunately, tannic acid is soluble in water. (Oaks contain unusually large amounts of tannic acid, not only in their fruits, but in their bark, leaves, and galls.)

Although the leaching and cooking procedures by which the California Indians removed the bitter taste from acorns are well known, the associated chemical reactions have received little attention, especially where they concern the tannins. This is a complex subject, and so some of the following remarks are tentative. (Because of their acid reaction, tannins are often called 'tannic acids.')

In central California, the acorns were first shelled and dried. They were then pulverized with a mortar and pestle, until reduced to a coarse meal. (Since dried acorns are too moist and oily to be properly ground, 'crushing' or 'fragmenting' seem better terms than 'grinding.') Easily-gripped and usually roughly cylindrical rocks served as pestles. Most mortars consisted of holes worn in stones of a movable size, but some were hollows in outcroppings of bedrock which women had gradually worked into holes. After the acorns were fragmented in these 'bedrock mortars,' the meal was scooped out by hand, and the depressions swept clean with a brush, often made of soap root fibers. (Bulbs of the soaproot plant, *Chlorogalum pomeridianum*, are surrounded by coarse fibers which were used for making brushes. The bulbs were also used as food—but, like acorns, they contain an unpalatable substance, a bitter saponin which is destroyed by cooking.)

Most leaching was done in what have been called 'sand filters.' Sand, near a stream, was often heaped in a small hillock; then a depression, lined with leaves, was made at the top and filled with a mass of pulverized acorns. Then, water was poured through the mass.
Among Californian tribes, the

> *commonest method was to pour hot water over the meal as it lay spread out in a basin of clean sand...This is the usual Californian method. Cold water apparently also*

removes the bitterness if given time enough. (Kroeber 1976, 88)[3]

After being leached, the meal was cooked in water-tight baskets. (Before European contact, the water containers used by most California Indians were closely woven baskets; pottery was made only in the southernmost parts of the State.) 'Stone-boiling' is the name applied to this process, although the food was often cooked at less than boiling temperature. ('Simmering,' that is, cooking at a little below the boiling point, may be a more appropriate term.) Although the resulting acorn mush was the staple food of most California tribes, many also made a kind of bread by cooking the acorn meal on hot stones or in pit-ovens. Such 'ovens' were little more than stone-lined pits in which a fire was built; when the stone 'lining was thoroughly heated, the fire was raked out and the food was placed in the hot pit, covered with leaves, and then banked over with earth...' (Balls 1972, 6). Pit-ovens, much like those made by California Indians, were used in parts of the southwestern USA and northwestern Mexico for cooking *mescal*—that is, the swollen heart buds of wild agave plants, a principal food item in those areas.

Substances Used in Detoxifying Plant Materials

Some Californian tribes used simpler methods of preparing acorns, in which there was no leaching. Among the Shasta tribe, along the border between California and Oregon, acorns of the maul oak (*Quercus chrysolepis*), 'were buried in the shell in mud until they turned black [or purplish]...The Shasta ate these darkened kernels cooked whole or *roasted them in ashes*' (Kroeber 1976, 293 and 88).[4]

Similarly, among the Pomo, a little north of San Francisco Bay, acorns (e.g., those of the tanbark oak) were shelled and put in the mud. When they 'turned bluish,' the acorns were 'pulverized and used for food' (Gifford and Kroeber 1937, 177).

Another method was used by some California tribes to remove bitterness from acorns: mixing the acorn meal with red earths or clay. (Iron compounds are chiefly responsible for the red color in soils.) When tannins and iron salts are mixed, they produce bluish-black colors. A related example follows: after iron tools had been introduced into California and were common there, Indians of the Sacramento Valley discovered that a black dye for their baskets could be 'made by placing rusty iron in water in which the bark of *Quercus lobata* [Valley Oak] had been soaked' (Balls 1972, 14). Likely such changes in color are due to ferrous salts in the red earth or mud reacting with tannic acids to produce bluish-black products. (Iron has a valence of either +2 or +3. The ending -ous is applied to those compounds in which the metal has a valency of two; the ending –ic is applied to compounds in which the metal has a valency of three.)

Among some California tribes the acorn meal was combined with red earth. For instance, the Pomo and the neighboring Miwok 'mixed red (presumably ferruginous) earth with the [acorn] meal' (Gifford 1936, 90). The following account is from (much-abbreviated) field notes on the northern Pomo:

> *Red* earth [was] sifted and [the] fine dust collected. Said to slake [i.e., to absorb moisture] *somewhat on exposure to the air*, which rendered it more suitable. Mixed with all species of acorns...Earth made into a paste and mixed with acorn meal. (Gifford and Kroeber 1937, 177)

Acorns were sometimes treated with another common substance, ashes. When plants are burned, most of their inorganic constituents remain in the ashes. In their recipes, tribes of both North and South America often specify ashes made from woods of particular tree species, or ashes made from the *barks* of those trees. Woods vary greatly in the amount of ash produced; for example, in the Paraguayan Chaco, an air-dried sample of wood from *Maytenus vitis-idaea* (an evergreen tree of the staff-tree family, Celastraceae) produced 18.4% ash, whereas the wood from a palm tree in the same area produced only 8% ash (Schmeda-Hirschmann 1994, 161). Of this ash residue, the principal constituents are often soda (sodium carbonate, Na_2CO_3) and potash (potassium carbonate, K_2CO_3). The relative proportions of the two compounds differ considerably depending on the species of the plant burned. Carbonates of sodium, potassium and calcium, when dissolved in water produce hydroxide ions (OH-) and are called *alkalis*: the carbonates of sodium and potassium are known as *caustic*, because they produce hydroxide ions in high enough concentrations to destroy living tissue. Calcium carbonate, which yields a lower concentration of hydroxide ions, produces a milder alkali.

Lye—a highly concentrated aqueous solution of potassium hydroxide, KOH (i.e., caustic potash) or sodium hydroxide, NaOH (i.e., caustic soda)—can be obtained by leaching ashes with water. The ashes of hardwoods are generally better for making lye than of conifers. On the other hand, in 16th century Mexico,

[3] Heat accelerates the leaching process. However, before Europeans had contact with California, most water used in leaching was probably unheated. Heating leaching water by stone-boiling adds a time-consuming step to the already lengthy process of preparing acorns. Moreover, using unheated water for leaching has the advantage that it leaves most starch and fat components behind in the acorn meal.
In the early 20th century, metal containers were already available throughout the state; thus, water was easily heated.

[4] A further description of the Shasta tribe's custom of treating various acorns follows:

> A step in the acorn preparation process that has often been overlooked is the removal of the membrane [especially rich in tannins] covering the kernel. The Shasta rubbed this off by hand, *which was probably the procedure elsewhere also.* (Kroeber 1976, 293)

maguey, a large perennial herb, was also burned because its 'ashes are very good for making lye' (Motolinía 1950, 274).

Hydrolysable tannins are called *gallotannins*, so named because they are especially common in gallnuts. Hydrolysis reactions are often catalyzed by aqueous solutions of alkalis such as ashes and lye, especially when the latter are heated. These alkalis participate in the hydrolysis of gallotannins, an action which breaks them down into simpler compounds. (The word hydrolysis—literally, 'water' plus 'decomposition'—refers to a mutual decomposition-and-exchange process, in which water reacts with a second compound. In this reaction, water molecules separate into H^+ and OH^- ions which are then added to the similarly fragmented molecules of the participating compound.)

In northern California, various tribes (e.g., the Yurok, Pomo, and Yuki) had the custom of burying acorns 'in a sandy place with grass, charcoal and ashes, and then soaking them in water...until they became sweet' (Gifford 1936, 87). (California Indians did not, like those of Mesoamerica and Peru, deliberately manufacture charcoal; thus, the 'charcoal' would have been only bits of charcoal, intermixed with fragments of charred wood and ashes.) Also, the Miwok 'sometimes mixed *ashes* of *Quercus douglasii* [Blue Oak] bark with the [acorn] dough' (Gifford 1936, 90).

There are other examples of the use of ashes in treating acorns. For instance, among the Shasta, acorns of the maul oak (and other oaks) were:

> sometimes roasted in ashes and eaten without any preliminary burying or boiling. However, burial whole in mud for several weeks was the customary treatment for these acorns. (Gifford 1936, 87)

In the northeastern part of the United States, whole acorns 'were leached by boiling in water to which lye [made from ashes] had been added' (Driver 1953, 57).

Like ashes, charcoal and clays are used to counteract bitterness in plant materials, and especially to detoxify certain alkaloid-containing plants. Many wild species of potato (that is, tuberous members of the genus *Solanum*) grow from the American Southwest to Chile. Unlike the common potato, domesticated in the Andes, most of these contain toxic quantities of bitter alkaloids. Untreated, their tubers are scarcely edible. (The alkaloids are discussed below.)

Charcoal and clays have unusual sorptive abilities, and it has long been known that, like charcoal, clays selectively adsorb certain alkaloids and other toxic substances. (Charcoal is still used in modern medicine as an antidote for the toxic effects of alkaloids.) Hence, food preparers

> *deliberately use the adsorptive properties of clay to bind toxins* in food...Detoxification basically involves adding clay directly to food plants during processing or at the time of ingestion, or soaking the plant product *in wet mud.* (Johns and Kubo 1988, 119)

Wild potato tubers, raw or cooked, are commonly detoxified by the addition of clay. For example, 'in some localities (among the Pueblos) clay was eaten...with wild potatoes *to mitigate the griping effect* of this *acrid tuber*' (Laufer 1930, 173). Among the Hopi in northern Arizona, the small tubers of *Solanum jamesii* are 'boiled and eaten with a talc of greasy taste' (Robbins et al. 1916, 73; the Hopi term for the substance means 'potato clay'). In New Mexico the Zuñi eat 'the tuber of *Solanum fendleri* (the so-called native potato) raw, and after every mouthful a bite of white clay is taken to *counteract the unpleasant astringent effect of the potato* in the mouth' (Laufer 1930, 171). The same effect is apparently produced by other minerals with clay-like sorptive properties. For instance, wild potato tubers are eaten by the Navajo—by whom, in order to 'obviate the *unpleasant, sour taste, or to prevent poisoning*, a bit of rhyolite tuff mixed with water is added in cooking...' (Bailey 1940, 286).

Besides the common potato, several species of 'bitter potato' are cultivated in Peru, and the custom of eating clay minerals with their tubers was noted shortly after the Spaniards arrived:

> The Indians of Peru call *pasa* [an Aymara word] a certain type of *clay* which is white with various brown spots like soap...and *they use it as a preferred sauce*—with which, dissolved and with salt, *they eat potatoes and other roots*, dipping them in this mud as if it were mustard; and *for this purpose it is sold in plazas of all the towns*... (Cobo 1956, 1:115)

Similarly, among Indians of highland Bolivia today, boiled potatoes are 'sometimes served with a white *kaolin clay*' (Sillar 2000, 111). Although the 'heat-stable, water-insoluble glycoalkaloids of the potato' are not destroyed by cooking (Johns 1986, 635), when eaten with such clays as *pasa*, their toxic effects may be negligible.[5]

Extraction of HCN (and Starches) from Manioc

Apparently all food maniocs belong to a single, highly variable species, but for practical reasons two main varieties (races) are recognized, 'bitter' and 'sweet' manioc. Both the roots of sweet manioc and bitter manioc contain a large proportion of edible starches, but those of bitter manioc cannot be eaten without special treatment. Present in its roots are dangerous amounts of a

---

[5] Glycoalkaloids have a combination of glycoside and alkaloid properties; they are frequently found in wild potatoes. *Pasa* is composed of clay minerals, predominantly of 'smectites, mixed-layered with illite' (Johns 1986, 640).

glucoside, which upon hydrolysis breaks down, releasing glucose sugar and hydrogen cyanide, HCN.

(Hydrogen cyanide is a poisonous gas. In an aqueous solution, it is called hydrocyanic acid or prussic acid, a highly toxic liquid.) Hydrolysis begins when the roots are macerated in water, in preparation for food. In processing bitter manioc's roots, hydrogen cyanide (a poison) and starch (a food) *are removed together*—that is, in the same operation.

Soaproot (or amote), chlorogalum pomeridianum, employed for use as a glycoside.

Glucosides, when hydrolyzed, yield glucose sugar—hence their name. However, a number of plant constituents yield other sugars than glucose; accordingly, a more general term, glycoside, is used for any of the class of compounds that yield a sugar upon hydrolysis.

Numerous glycosides are involved in the chemical arts: the dye compounds alizarin and indican; the anthocyanins (coloring materials in plants); the cardiac frog poisons; the saponins (i.e., soap-substitutes and foaming agents); and so on.

Saponins are widely distributed in plants. They are oily glycosides, bitter-tasting and irritating to the eyes and nose. When shaken with water they foam. Besides being cleansing agents, they are used as fish poisons and (nowadays) as insecticides. Despite their bitter taste, many saponins are not toxic when ingested.

Figure 8. Glycosides and Saponins

Notwithstanding its poison, bitter manioc was preferred in the 16th century by most native peoples to the east of the Sinú River (just west of Cartagena on the north coast of Colombia):

> There are also the roots *from which bread* is made in the islands of Cuba, Jamaica, and Española, but here they are of another quality, for those of the islands are dangerous and if one eats one of them one dies as one would by eating realgar [arsenic], but *those of this land of Sinú and the land beyond* to the west can be eaten raw or cooked... (Sauer 1966, 241, citing and translating Oviedo)

The roots of the sweet manioc are commonly eaten boiled—and only occasionally, raw. 'Only the sweet races were found in western parts of the Caribbean lands, in the intra-Andean lowlands, and along the Pacific. In none of these was cassava bread made from sweet yuca' (Sauer 1966, 241). (Maize, however, was the staple in those areas.)

The most effective method of removing the HCN poison is by peeling, grating—or crushing—the manioc roots, followed by leaching and cooking. Cooking volatilizes the poison. (Even the vapor that escapes during boiling may be injurious.) Peeling the manioc root is helpful because most HCN is in its outer parts. Grating or crushing the root ruptures its cells, thus making leaching more effective. Leaching is usually accomplished by a preliminary soaking, followed by either rinsing or pressing the fragmented pulp. Fermentation of manioc tubers removes much HCN—though fermentation, alone, apparently does not destroy all of it.[6]

When the Spanish arrived in the Antilles, they found that roots of the bitter manioc were processed in special manioc wringers—that is, long tubular basketry presses:

> *In order to make bread of it*, which is called cassava, *the Indians grate it and then press it in a strainer, which is a sort of sack about ten palms or more in length and as big as a man's leg*. The Indians make this bag from palms which are woven together as if they were rushes. By twisting the strainer...the juice is extracted from the *yuca*. This juice is a powerful and deadly poison, and one swallow of it will produce sudden death... (Oviedo 1959b, 16)

The pressing device is also well-known among a number of Amazonian tribes. It is operated in various ways, and is sometimes called a 'sleeve press' or *tipiti*; it is

> constructed by diagonal weaving so that it can be stretched lengthwise at the same time compressing the contents...As the cylinder is stretched its diameter diminishes and the juice is squeezed out though the basketwork to be collected in a container below (Lancaster et al. 1982, 26).

---

[6] There are other cases of rinsing and fermentation being employed to detoxify food. In the 19th century, the Seminoles of Florida extracted a starchy material from the poisonous roots of *Zamia pumila* (Cycadaceae), the so-called Florida arrowroot. Not only is the chemical composition of food material altered by fermentation, but the composition of toxic substances in the food is changed as well. In the case of arrowroot and the Seminoles, the toxin is crystals of calcium oxalate. After the roots had been washed, chopped in small pieces, and pounded in wooden mortars,

> the whole mass was taken to the creek near by and thoroughly saturated with water *in a vessel made of bark*. The pulp was then washed in a straining cloth, the starch of the Koonti [i.e., the *Zamia* plant] draining into a deer hide suspended below. When the starch had been thoroughly washed from the mass the latter was thrown away, and the starchy sediment in the water in the deer skin *left to ferment*. After some days the sediment was taken from the water and spread upon palmetto leaves to dry. When dried, it was a yellowish white flour, ready for use. (MacCauley 1887, 514-515)

Since calcium oxylate is soluble in water, any toxin not leached away by rinsing would likely be decomposed during fermentation.
The needle-like crystals of calcium oxalate when ingested cause difficulty in swallowing and extreme irritation to the mouth and gut. (Calcium oxylate is also found in many aroids—for example, yautia, *Xanthosoma sagittifolium*, the New World analogue of taro.) 'Two reasons for this irritation have been suggested: mechanical irritation by the crystal itself or chemical irritation by a curare-like drug associated with the crystal' (Sakai et al. 1972, 178:314-315).

Bitter manioc is processed for three main products: a coarse flour (*farinha*); *a relatively pure starch*; and a *watery liquid*. 'Contrary to the practice of some other Amazonian groups, all three products...are used as food by the Kuikuru' (Dole 1978, 224). (The Kuikuru are a Carib-speaking people who live in central Brazil in the Upper Xingú area.) Description of these products follows:

(1) *Coarse flour*. This account from the middle Amazon describes the making of *farinha* from pulp left in the sleeve press:

> The moist pulp comes out in long, crumbly cylinders. These are broken up and roasted [griddled] on a *large, round slab with high edges*, over a low fire, while the smoking mass is continually stirred and tossed into the air with a wooden spoon. When the operation is over the cassava *looks like fine yellow gravel. In this state it can be kept for over a year*, if properly prepared. The natives usually keep it in cone-shaped baskets covered with large waterproof leaves... (Tastevin 1943, 132) (Father Constant Tastevin, a Catholic missionary, worked in Amazonia between 1906 and 1926.)

One of the earliest reports from eastern Brazil describes such a 'large, round slab.' The slab or griddle 'in which they bake the [manioc] meal is of burnt clay, shaped like a large dish' (Hans Stade 1874, 131). Such griddles are still used in the northwest Amazon: 'Large flat plates to bake the cassava cakes on are made of earthen ware...' (Whiffen 1915, 96). Some of these griddles are indeed large, 'up to one and one-half meters' across, with elevated rims to keep the flour from spilling off when stirred (Dole 1978, 238).

After cooking, which removes most of the remaining HCN, the material may be made finer by pulverizing it with a wooden mortar and pestle.

(2) *Relatively pure starch*. The juice squeezed out of the sleeve press is collected in a large container. When this liquid is allowed to stand, starch settles as a sediment, and when the liquid is decanted, relatively pure starch is obtained.

'Since pure starch flour lacks [cellulose] fibers it provides more calories per unit of volume [than *farinha* does], and lacking any sugar, it is even more durable than *farinha*' (Dole 1978, 239-240). In the humid tropics, durability is an especially valuable property: other high-energy foods such as plant oils quickly turn rancid, and the proteins of animal flesh putrefy.

> In addition to being extremely durable, the starch flour is the most convenient of native foods...A particularly useful property is that when heated lightly the grains 'explode'...[and become edible] without further cooking. (Dole 1978, 240)

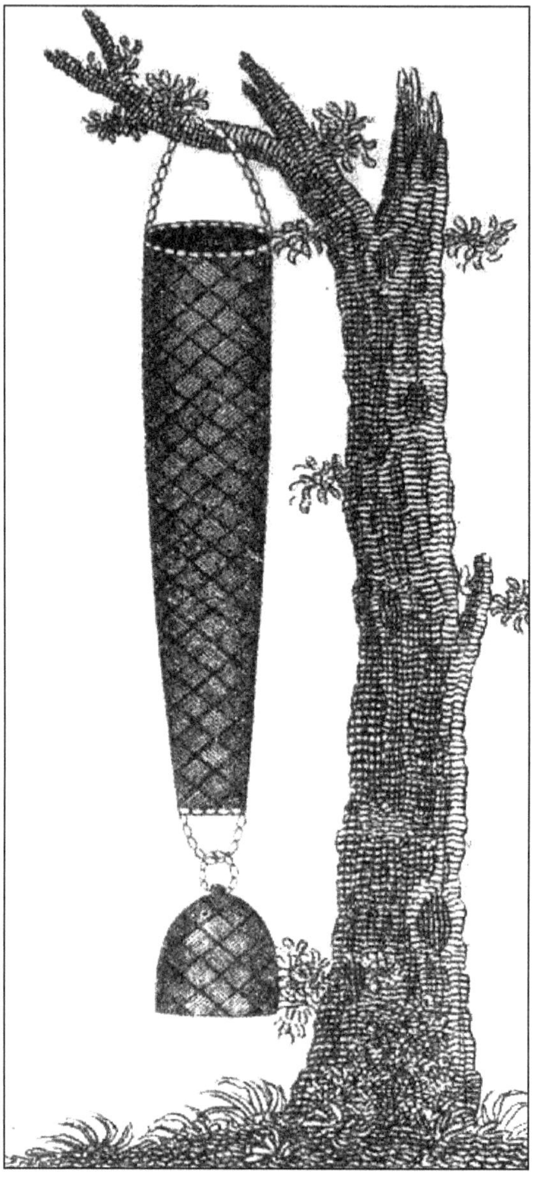

Among the Caribs, these twisted manioc wringers had weights hung on them to continue their squeezing action (after Rouse 1963, plate 94).

Figure 9. Basketry Press used for Straining Bitter Manioc

Starch, a white powder made up of minute granules, has a further advantage. If accidentally wetted, as often happens when weather is rainy, it undergoes little chemical change. This is true even if the water is warmed by a tropical climate. Though starch is almost insoluble in cold or slightly warmed water, when it

> ...is suspended in water and gradually heated (up to about 55°C), the granules swell and imbibe 50% or more of their weight of water. *This swelling is reversible, and after cooling and drying, the starch seems essentially unaltered.* However, if heating is continued in the range 60 to 80° most starches undergo an irreversible swelling or gelatinization... (French 1973, 1055)

In common practice, if still-wet starch is heated on a griddle, its granules swell—some bursting to form translucent lumps or globules. If this whole agglomerate is placed in boiling water, the manioc starch undergoes radical molecular transformation: from being a semicrystalline, granular solid, it swells considerably and becomes an amorphous jelly. This jelly-like mass is now widely known as tapioca—a word derived from the Tupi-Guarani language.

(3) *Watery liquid.* Left after decantation, the liquid is also valuable—and this despite its high HCN content. After being subjected to prolonged boiling in order to evaporate the remaining toxin, the liquid is consumed as a beverage or as a sauce (to be discussed later). Small amounts of sugar can be found in the liquid. (As noted above, hydrolysis of the glucoside in the roots of bitter manioc frees sugar and HCN. There is also left a little residual starch, which hastens the liquid's thickening while boiling.)

Although the sleeve press saves labor in leaching, it is not universally used in processing bitter manioc. For example, the Kuikuru remove the hydrogen cyanide by first peeling a thick layer of skin and cortex from the manioc roots, which are then washed and grated. The pulp is thoroughly rinsed and strained through a special, fine-meshed basketry mat spread over a large cauldron. After being thoroughly rinsed, the pulp is cooked directly, or pounded into flour for making bread. The rinsing fluid, when the starch sediment has settled, is decanted.

> Rinsing the pulp maximizes the yield of starch flour. In the rinsing process sugar leaves the pulp along with the starch, but *since sugar is soluble in water it is removed when the liquid is decanted, leaving a product that can be thoroughly dried and stored indefinitely without spoiling because in the raw state it does not absorb moisture and therefore does not ferment.* (Dole 1978, 239-240)

How it happens that bitter manioc is preferred over sweet manioc (even though, in a few places, both varieties were grown) is quite uncertain, especially since sweet manioc can be used for many of the same purposes. According to Oviedo, growing bitter manioc has an advantage over cultivating the sweet variety in that the former is avoided by the root-eating animals that raid fields (Oviedo 1959b, 16). Thus, the bitter manioc has a higher yield. The fact that bitter manioc contains more starch has been suggested as another explanation (Dole 1978, 239-240); however, it has not been shown that bitter races consistently have a higher starch content than sweet (Lancaster et al. 1982, 14).

An assortment of accessory tools is made for the treatment of bitter manioc. These include the toothed grater boards used for shredding, an invention that has been improved upon 'from simple spiny roots up to well-made boards artistically encrusted with quartz splinters' (Nordenskiöld 1931a, 501); finely-plaited shallow baskets and colanders for sifting flour; and special pottery, some of which is decorated to indicate its purpose (Dole 1978, 222).

Treatment of Maize with Alkalis

Two alkaline substances were, and are, often used in preparing maize for human consumption: (1) limewater, a clear and colorless solution of slaked lime in water, which provides calcium hydroxide, $Ca(OH)_2$; (2) either a mixture of ashes and water, or lye itself (the liquid obtained by leaching wood ashes), which provides sodium hydroxide and potassium hydroxide.

Calcining Limestone and Slaking Lime

The principal source of lime is limestone. This is sedimentary rock, predominantly of the mineral calcite ($CaCO_3$), which outcrops on many parts of the earth's surface.

Depending on its content of impurities, when limestone is heated above 750°C,[7] carbon dioxide is driven from the carbonate, leaving lime (calcium oxide, CaO) with various amounts of contaminants: $CaCO_3 \rightarrow CaO + CO_2$

Lime is also called 'quicklime,' 'burnt lime,' and 'caustic lime.' When lime is combined with water, slaked lime or calcium hydroxide (a fine white powder) is formed—a reaction in which large amounts of heat are liberated: $CaO + H_2O \rightarrow Ca(OH)_2$

The term *calcine* is somewhat confusing because it is not limited to the processing of limestone. If heated first, many other rocks or minerals are either reduced to powder or made easier to crush. Such heating or calcination should be to a temperature below the mineral's melting point, but high enough to cause loss of molecular water in hydrated minerals or to cause oxidation or reduction of the compound.

Unslaked lime and slaked lime were manufactured on a large scale in Mesoamerica, and both were sold in 16th century Mexican markets (Sahagún 1956, 3:143: '*vende la cal viva, y otras veces muerta*').

If the limestone is a relatively-pure calcite and completely calcined (at heat of 898°C), there is considerable shrinkage in *volume*, as well as 'a theoretical loss of *weight* of 44% as the CO_2 is evolved' (Boynton 1980, 160 and 167). Thus, especially under primitive conditions of transport, the best place to burn

[7] The temperature of calcining a rock varies widely. Limestone 'requires that temperatures in the range 750-850°C be sustained for several hours or more during its calcination' (Gourdin and Kingery 1975, 134).

limestone is where it is mined. Moreover, transporting burnt lime, rather than slaked lime, to town markets has an advantage for the countryman, since slaked lime may be almost one third heavier than quicklime (Eckel 1928, 122). For example, in the State of Chiapas, among rural Mixtec people, 'occasionally the cooked rock itself is carried and slaked near the market site to avoid carrying the extra weight of the water which is absorbed during the slaking process' (Pike 1980, 7-8). Great quantities of lime are still made for treating maize, as well as for making mortar for buildings.

Nixtamal, Masa, and Tortillas

In Mesoamerica, ripened maize cobs are often stored in well-ventilated corncribs. When taken from storage, the cobs are stripped of their dry, hard kernels. The kernels are then put in a pot of water containing either lime, lye, or ashes—that is to say, in a weak alkaline solution. There, they are heated about an hour at a temperature usually somewhat below the boiling point. Due to their high starch content, the kernels swell to several times their original size. As a result of this swelling and the corrosive effect of the alkali liquor, the pericarp ('skin') of the kernels breaks off and floats to the top. These skins are skimmed off and, being mainly indigestible cellulose, discarded along with the liquor. The swollen maize kernels are called *nixtamal* in Mexico, and hominy in parts of Anglo-America. (The word *nixtamal* is from Nahuatl; hominy is of Algonquian origin.)

Although ashes and lye were also used in processing maize in Mesoamerica, lime was the main source of alkalis there. (Probably tribes living to the north of Mesoamerica, like the Tarahumar in northern Mexico, learned lime-burning from the south within historical times.)[8] On the other hand, in studying the Maya languages, linguists have found that the 'earliest word for *slaked lime*...was closely related to the word for *ashes*, suggesting that ashes may have preceded slaked lime in this function' (Sharer 1994, 588).

The *nixtamal*, after being thoroughly rinsed with fresh water, is mashed on a metate into a dough called masa. Egg-size pieces of this dough, taken between wetted hands, are flattened and patted into disks no more than a few millimeters thick and usually some 10 to 15 cm in diameter. These thin disks, widely known as *tortillas*, are griddled a minute or so on a round, flat ceramic-plate called a comale. (Comales are sometimes more than two feet in diameter and, since they are thin, are easily broken.) While cooking, the tortillas develop a flexible skin on both sides, allowing them to serve (when twisted or folded) as spoons or plates for bearing sauces or other foods. This unleavened bread has long been the staple food of Indian Mexico. On the other hand, the rinsed *nixtamal* (hominy) may be dried and stored, or boiled for immediate consumption. If boiled, the dry kernels soften and swell yet further.

The reason sometimes given for such a time-consuming process as the alkali treatment of maize is that it dislodges 'skins' from the seeds. Probably, however, the underlying and more important reason for the practice is that it makes maize a more healthful food. Without such treatment, maize has serious disadvantages as a food staple: where people eat few other foods, the incidence of pellagra (a deficiency disease caused by lack of niacin) is high. A scientific explanation of the chemistry involved was provided some time ago (Bressani et al. 1958).

Compared with most other cereals (and especially with legume seeds), maize seeds are deficient in niacin. Much of the niacin in maize seeds is tightly bound to hemicelluloses and cannot be liberated by mammalian digestive processes. Under alkaline conditions, however, the bound niacin is liberated by heating. The amounts of some essential amino acids, likely to be absent in starch-rich diets, are also increased.

When lime rather than ash is used in the alkali treatment, there is an additional nutritional advantage: the lime is a good source of dietary calcium (Bressani et al. 1958, 772). Processing maize by the alkali treatment also changes the flavor and odor of some cooked products (tortillas, for example), a detail that native American cooks have probably been aware of for centuries.

Treatment of Maize with Saltpeter

The Aztecs also treated maize with saltpeter—that is, with potassium nitrate, KNO_3 (in Spanish, *salitre* or *nitro*). Saltpeter, which has a distinctive pungent taste, was described by the Aztecs as 'very sour...it puts one's teeth on edge...It is white' (Sahagún 1950-82, 12:243).

In one of their ceremonies, the Aztecs atoned to the maize plant for what it endured in being turned into a human food. The ceremony tells an additional bit about the chemistry of Aztec maize preparation:

[8] Elsewhere (as among the tribes of the Great Plains and eastern United States) ashes and lye were the substances used for that purpose (Katz et al. 1974, 768-769). For instance, to make hominy the Mohawk Iriquois placed corn seed in a wooden mortar, '*added a small amount of hot water and some ashes,*' and then ground the mixture to a coarse meal (Onion 1964, 65).
The alkali treatment of maize was probably discovered in Mexico. In its simplified form (that is, using only ashes or lye, not lime) the process spread from Mexico with the cultivation of maize (Bennett and Zingg 1976, 33).
In their list of tribes using lime, Katz and his co-authors include the Tewa and the Zuñi of the American Southwest, but I could not find evidence for this conclusion (at least in prehispanic times) in the references they cite: even in 1916, among the Tewa 'Ashes of corncobs are boiled with white corn in order to make it swell' (Robbins et al. 1916, p. 29). True, a 'lime-yeast' was used by the Zuñi in cooking some maize dishes (at least in historic times), but Zuñi cooks seem to have used ashes or lye for making their hominy. The Tarahumara are also listed as using lime as their source of alkali, but this is only a recent feature, a 'Mexican intrusion' (Bennett and Zingg 1976, 33).

> ...maize was made to rest in the *eighth* year, because it was said that we tormented it greatly in order to eat it, when we used *chili* on it, when we *salted it*, when we *treated it with saltpeter*, when it was *treated with lime*. It was as if we had killed it; therefore we revived it. It was said that thus the maize was made young (again). (Sahagún 1997, 69)

A similar personalization of maize is found among the Tewa of New Mexico (Robbins et al. 1916, 83).[9] In the passage above, the 'eighth year' may be significant because maize seeds ordinarily remain viable for roughly that time.

The chemical purpose of treating maize, or other plants which were treated similarly, with saltpeter is uncertain. (Although saltpeter is used as a food preservative today, especially for meats, it seems unlikely that it functioned as such in this case.) In some cases, the saltpeter may have been added to remove an unpleasant taste. For example, the foliage of an unidentified herb called *quiltonilli*, with broad leaves and black seeds, was 'boiled with saltpeter (*tequixquitl*); the solution is expressed before eating' (Sahagún, 1956, 3:295). Saltpeter is only slightly soluble in cold water, but in hot water it is quite soluble. Even in recent times at Tepoztlan, Morelos, a largely Indian town some fifty miles from Mexico City, 'green corn [is] boiled with saltpetre,' as well as a 'grass' (probably a forb) 'which is cooked in saltpetre' (Redfield 1964, 178 and 186).

Since the Aztec's markets made and sold saltpeter in several different colors and forms (Sahagún 1956, 3:157), apparently the compound was of some importance in their economy—but, aside from cooking and as a mordant for making dye (Hernández 1888, 212), saltpeter's uses in Aztec chemical arts are not well specified and somewhat of a mystery.[10]

Cooking Equipment and Methods

If, as previously suggested, fire-making and cookery are two ancestral chemical arts, pottery-making, which appears at a much later date, is a third. The craft of making pottery seems to have been developed mainly by women as they have gone about such traditional tasks as preparing and cooking food.

While cooking and pottery-making are probably ancestral chemical arts, there seems to be no publication on the origin and early historical development of the chemical laboratory. Even the modern laboratory is vaguely reminiscent of a kitchen—and the farther back in history one looks, the more obvious the similarity becomes. Plainly, the early laboratory's utensils and processes came from the kitchen. In short, the chemical laboratory seems to have developed as a mostly male variant of the women's kitchen. Moreover, the cook's recipes and the chemist's notes often have a similar purpose: to reproduce a previous event.

A number of cooking processes have been discussed already: stone-boiling acorns, roasting cornbread, boiling maize and manioc (cassava) preparations, and griddling tortillas and manioc cakes. (Tortillas and manioc pancakes are often said to be 'baked'; a more accurate word would be 'griddled.') The following comments enlarge on this subject, defining terms and giving additional details on cooking sites and equipment.

A confusing medley of cooking terms is used in English. From the standpoint of the physical processes involved and their chemical results, these terms are often either ambiguous or overlapping. The words 'roasting' and 'baking' are cases in point: use of these terms may even depend on the kind of material being cooked. For example, an oven might be used to 'roast a turkey' or to 'bake a cake,' but not vice versa.

Cooking is defined as the preparation of food by the application of heat. Yet the temperature at which cooking begins is an open question, because the process also involves the time during which food is exposed to heat. Many proteins are changed (*denatured*; i.e., their molecular structures are reordered) when heated to only around 40°C. Such changes are accelerated as temperatures rise. In liquid foods, pasteurization (heating to temperatures of between 55° and 70°C) destroys most fermenting bacteria, but with little significant change in the foods' chemical composition or flavor. (At 55°C, pasteurization may take half an hour, but at 70°C, only half a minute.) What is usually called 'cooking' begins only a few degrees higher than pasteurization temperature. Depending on the time it is exposed to heat, meat cooks at about 60°C, or 140°F (Lundberg and Kotschevar 1965, 5). The proteins in an egg 'start to react together to 'cook' the egg once the temperature is above about 75°C' (Barham 2001, 21). Then the egg white changes irreversibly from a liquid and translucent solution of proteins and becomes a solid white mass. (Although chickens were introduced by the Spanish, the

---

[9] The town 'of San Ildefonso is swept before the [harvested] corn is brought home, '*because corn is just the same as people* and must have the plaza clean, *so that the corn will be glad when we bring it in*'' (Robbins et al. 1916, 83).

[10] Sahagún does not say where the Aztecs obtained saltpeter; he says only that 'it is *heaped up in places* where there is plenty of it' (Sahagún 1956: 3:157). In the Old World, saltpeter

> is formed by the bacterial decomposition in the soil of waste nitrogenous matter (excreta, urine). In warm countries...this process takes place with great readiness, and in the neighborhood of the dwellings...the ammoniacal compounds formed by the decomposition of waste animal matter are converted by the nitrifying bacteria to nitric acid [HNO_3], which forms potassium nitrate [saltpeter, KNO_3] or calcium nitrate with the potash salts or lime presnt in the soil. By the addition of wood ashes (potassium carbonate), the calcium nitrate is converted to potassium nitrate and calcium carbonate. The nitrates are extracted from the soil by water. (Findlay 1934, 371)

Likely the Aztecs got their supply from similar sources. Europeans of the 16th century used saltpeter extracted in this way for various purposes, but chiefly for the manufacture of gunpowder (Findlay 1934, 373).

eggs of domesticated turkeys and ducks were raised in the Americas long before the Conquest.)

Both in Mesoamerica and Peru, small ceramic stoves were made; an example of the simplest type was found among the Zapotec Indians of Oaxaca. A large pot (*olla*) 'is sunk in the ground until the rim is just even with the surface: a good fire of coals is made in it'; the *olla* is then covered with a *comal* for baking tortillas (Starr 1900-1902, 2:61). A more complete stove is made by the Zoque Indians in Chiapas:

> In the cookhouse, on the floor near the rear, a small oven [i.e., a stove] is constructed; it is no more than fifty centimeters long and its height and breadth are less; it is oval or elliptical, and looks like a huge limpet or an armadillo shell attached to the floor. A pottery plate covers its opening. (Starr 1900-1902, 2:62)

In both coastal and highland Peru firewood is often scarce, and a baked-clay stove, well-designed for fuel economy, was used; the Inca 'were surprised at the way the Spaniards wasted fuel' (Garcilaso de la Vega, 1966, 216).

The following description of a stove comes from Peru, shortly after the Spanish arrived:

> In all of the houses, no matter how small they may be, there is a stove behind the door; it is shaped like a very small clay furnace, no more than one span high; it is entirely covered, except for a small hole for stoking the fire, and on top there are two or three round holes where the pots are set. This stove uses very little wood, and they never put anything in it but two small sticks placed straight in; as the sticks burn they stir them up. And just one of our ovens consumes more firewood than twenty of the Indians' houses. (Cobo 1990, 195)

Similar ceramic stoves are still common among the Peruvian Aymara in the Andean highlands:

> The common rural hearth at ground level is an oblong half dome, *molded from clay by hand*. The fire is fed through an opening at the floor, and on top there are two or three holes in which the cooking pots are placed. (Johnsson 1986, 47; also Tschopik 1950, 208)

Roasting, Toasting, and Parching

These terms are often used interchangeably. The term 'roast' has several different meanings; the one used here is to cook mainly by *radiant* heat. In traditional cookery, heat is transferred to food by three means: radiation, conduction and convection. Of these, radiation is the fastest: heat energy emitted by an open fire or live coals is transmitted through the air (with slight effect on its temperature) and absorbed by the cooking food.

An example of roasting is provided by the Fuegian Indians who pressed bent sticks into the ground near their campfires to hold skewered strips of meat over the flames (Coon 1971, 183-84). The fact that such roasting is commonly used by peoples with a simple technology suggests that this is among the oldest methods of cooking. Barbecued meat is another an example of roasted food.

In all such methods of cooking, fats are melted and removed. However, with the passage of time, fats are likely to turn rancid. Often the purpose of barbecuing is not only to cook food, but to smoke and dry, and thus to preserve, it. For instance, in the 16th century, the Indians of eastern Brazil

> ...when they want to cook any food, flesh or fish, which *is to last some time*, they put it four spans high above the fire-place, upon rafters, and make a moderate fire underneath, leaving it in such manner to *roast and smoke, until it becomes quite dry*. When they afterwards would eat thereof, they *boil* it up again... (Hans Stade 1874, 133)

On the other hand, fats usually fall into the fire, and thus constitute a loss of valuable food. (The English word *barbecue* is derived from *barbacoa*, a Taino-Arawak name for an elevated frame of sticks for supporting meat or fish over a fire; however, various other American tribes also roasted fish on wooden grills.)

Although the term roast is usually applied to meat, vegetal food (for example, a cob of fresh maize) may also be roasted. (A synonym sometimes used for roasting is broiling, except the latter term is restricted to meat.)

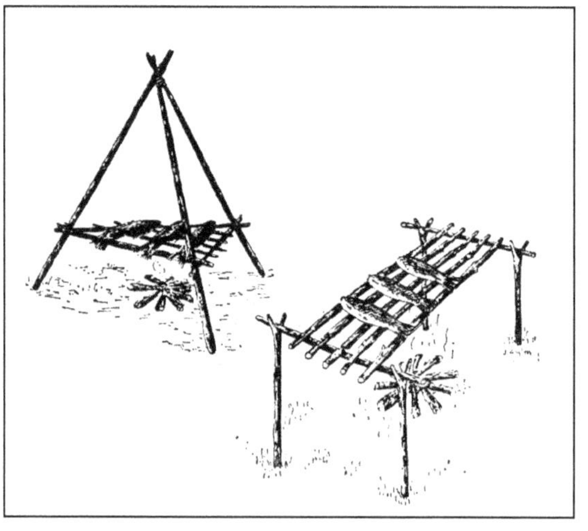

Figure 10. Barbecues (Steward and Faron 1959, 314)

As terms are used here, the difference between roasting and toasting is slight. Among non-agricultural peoples, wild grass seeds were commonly toasted, and this suggests that it also is one of the older methods of cooking. For instance, in California and parts of the Southwest wild grass seeds were toasted by tossing them with live coals in basketry trays.

The tray was dexterously kept in constant motion so that neither seeds nor basket would be burned (Driver and Massey 1957, 245). Toasting or roasting is also a way of softening hard seeds, such as dry maize kernels[11] and beans, so *that they can be more easily ground.* (Toasting also denatures enzymes in the seeds, so that the ground meal does not spoil so readily.)

Although the term 'parch' is often used interchangeably with toast or roast, here the term is applied to the process of partially cooking food by placing it on a much-heated surface. Even though parching or griddling may have been done occasionally on stone surfaces in the past, over time the procedure became heavily dependent on pottery utensils. Whereas in roasting the transfer of heat is mainly by radiation, in parching and griddling it is mainly by conduction. This is a fuel-efficient way of cooking: heat from the fire's flames is concentrated on the heated surface and conducted quickly to the food material, so that cooking time is shortened. The food is close to the source of heat, and no water is needed.

An old and widely used Indian beverage is made from maize seeds. They are ground and parched, then stirred in water. (In Mexico, the drink is called *pinole.*) Among the sixteenth-century Maya,

> they also parch the maize and grind it, and mix it with water, thus making a very refreshing drink, throwing in it a little Indian pepper [*Capsicum*] or cacao. (Landa 1941, 90)

In Peru, the Inca also made parched cornmeal. Maize seeds were 'toasted in *clay casseroles pierced with holes*...It is the most usual ration of food that they take with them on journeys' (Cobo 1990, 198). As with many other tribes, among the Tewa, a people of Pueblo culture, parched cornmeal is a 'conventional food of travelers' (Robbins et al. 1916, 92).

Popping corn is a parching process; it is distinguished from making pinole in that only whole-grained maize seeds are parched. When exposed to heat, a number of grass seeds puff-up a bit due to expansion of the water they contain. Nevertheless, popcorn is the only grass seed (indeed the only variety of maize seed) that was extensively popped for use as a food in early times. Curiously, this trait of domesticated grasses seems have been selected-for only in the Americas. No doubt, toasting wild grass seeds antedates the popping of maize seed, and popcorn seems to have been one of the earliest maize varieties that was domesticated.

Popcorn is a specially-bred variety of flint corn with relatively small, hard, and vitreous kernels. When popcorn seed is exposed to heat, its water-content vaporizes and expands. At about 215°C, the tough pericarp ruptures, the seed bursts open with a popping sound and puffs-up to many times its original volume. The starch in the seed is sufficiently heated so that it needs no further cooking. The characteristic odor of newly-cooked popcorn suggests that considerable chemical change has taken place.

Since maize itself was unknown in the Old World before discovery of the Americas, popcorn was, of course, new to Europeans—as one can see from the following Spanish note on a procession honoring an Aztec god:

> [The people scattered before the god's statue,] a toasted maize called *mimóchitl* which is a kind of corn *that bursts when parched and exposes its pith and looks like a very white flower.* (Sahagún, 1956, 1:64)

In Peru, the Inca also made much use of popcorn: 'They toast maize in *perforated pans of earthenware*...They toast *a certain type of maize until it bursts and opens*...' (Cobo 1956, 2:244; Cobo does not say whether the perforated pans had handles). The same utensils may have been used for both toasting and popping.

Popping maize seeds requires no more than the application of heat, but a container is necessary to keep the exploding seeds together. Long before the Inca empire existed, handled ceramic containers with holes (thought to have been used for popping corn) appear in the archaeology of Peru (Sawyer 1961, 273 and 497).[12]

Baking

Occasionally, especially in recent times, 'the terms roasting or baking refer to cooking foods uncovered in an oven with no liquid added' (Lundberg and Kotschevar 1965, 89). This method of cooking was quite uncommon in prehispanic America, since genuine ovens—stoves with

[11] A rocker-pestle is often used for this purpose, especially in South America where the alkali treatment of maize is absent. The rocker-pestle consists of a heavy, roughly disk-shaped rock, usually picked from a river bed where it has been smoothed and rounded by stream action. The rock is stood on edge on a broad flat stone or occasionally on a large slab of hard wood. Commonly operated by one hand (though occasionally two), the mano is rocked from side to side over the material to be fragmented. The Inca used such rocker-grinders to pulverize dry maize seeds (which are hard to crush), as well as some mineral substances (Cobo 1990, 195).
Like the millwheel, the rocker grinder is a labor-saving device. Grinding with an ordinary mortar

> is done by the force of blows, but the moon-shaped stone grinds whatever comes under it by its own weight and the Indian women can easily handle it because of its shape, rocking it to and fro and occasionally heaping the grounds in the middle of the flat stone with one hand so as to grind them over again, while the other hand is left free to hold the grindstone... (Garcilaso de la Vega 1966, 498)

With this tool it is easy to grind dry maize seeds *without toasting them first.* Rocker grinders are still commonly used—for instance, among the Aymara in Bolivia (Johnsson 1986, 47); in northeastern Peru (Tessman 1972, 496); and as far north as among the Térraba of western Panama (Gordon 1982; see photograph on p. 69).

[12] Such 'corn poppers' appear in the Peruvian Paracas-Nazca culture, several centuries before Christ: "Corn Poppers" are *fairly common* at Juan Pablo [in the upper Ica Valley of southern Peru]' (Sawyer 1961, 497).

enclosed chambers where food was cooked by the dry heat of circulating air—were either rare or non-existent.

On the other hand, among the Pueblo peoples (who were accomplished potters), something like baking was sometimes done in adobe- or stone-lined pits. For instance, the following statement describes the Tewa way of cooking:

> [The pit is rectangular,]...18 to 24 inches deep, cut in the rock outside the house and lined with slabs of stone. In this pit a fire is made; when it is hot, the embers are reduced to fragments and the [ceramic cooking] *vessel* is set among them; *the opening is closed* with a slab of stone, sealed with clay, and a fire built on top. Next morning the vessel is taken out... (Robbins et al. 1916, 91)

This seems to accord with current use of the word 'baking,' since the food (a bread) is cooked in an enclosed chamber.

Boiling and Steaming

When food is cooked in boiling water, heat is transferred to it mainly by the convectional circulation of heated water. Steaming, on the other hand, is usually defined as a process in which the food is cooked over (not in) boiling water. That is, to steam food is to treat it with water in the form of hot invisible gas or of hot vapor.

Boiling and the Containers Used

Boiling is the basis of many chemical arts, and was almost certainly first applied to food materials. The procedure demands a water-tight container. Although this method of cooking was commonly practiced by peoples who had ceramic containers that could withstand exposure to direct flames, various other containers were used: baskets, bent-bark boxes, depressions in stone slabs, and hollow, hard-shelled fruits. For example, the Eskimo used soapstone (steatite) as a container, with a depression carved in its surface (Coon 1971, 184).[13]

In fifteenth-century America, pottery was the principal container used for cooking in agricultural areas. Nevertheless, in several other areas, food was cooked in non-ceramic containers. In forested areas of the higher latitudes in North America, hunting tribes made water-tight containers of tree bark. In cooking, such containers 'must be held near enough the fire to boil the water, and far enough from flames to keep it from burning...[This method was] also used in northeastern Asia' (Coon 1971, 185).

In South America, the Botocudo of southeastern Brazil used sections of green bamboo (probably *Guadua* sp.) as cooking vessels:

> The bamboo vessels, filled with water into which were put beans, fish, or some other kind of food-stuff, were placed slantingly over the fire [to be warmed by heat radiation]. The length between two joints was used, one joint not being removed, but left to serve as the bottom of the vessel. The fire only blackened these vessels with soot, and *the water was readily brought to a boil*. When the food was done, the bamboo was split in two halves which were used as plates. (Linné 1925, 10)

Stone-Boiling

As mentioned earlier, the practice of stone-boiling is probably best known in California where few tribes made pottery. However, the earliest written record of stone-boiling in the Americas comes from Indians living not far from the lower Rio Grande:

> They have little to eat save *fésoles* [beans], squashes, and a little maize. They have no [baked clay] containers in which to cook, but they make a gruel (a kind of thin soup) in certain large calabashes in the following manner. They make a fire in which they heat a number of clean rocks or pebbles. They then pour water into a calabash and, after that, put in the hot stones which cause the water to boil. They then pour in the bean meal [*harina de los fésoles*], and throw more hot stones in with it, until the gruel is well cooked... (Oviedo 1959a, 4:310. This passage is translated from Oviedo's Spanish account of Cabeza de Vaca's experiences.)

(Because dry maize and bean seeds are hard, and thus difficult to grind, they were commonly toasted before being ground into meal; that is, they were already partly cooked before being stone-boiled.) Great Plains tribes boiled food by heating rocks in their campfires and used containers made of bison rawhide.

In California, large baskets, woven so closely that (when their fibers were wetted and swollen) they held water, were used for containers (Beechey 1831, 57 and 76). Stream-rounded, fist-size stones were heated and dropped in baskets partly filled with water; when cooled, they were replaced with freshly heated ones. Food materials (chiefly acorns and, near the coast, marine mollusks) were added and the heating process continued. To prevent stones from burning holes in sides of the basket, the food was stirred frequently and little-by-little brought to the boiling point or close to it. The cooking stones were transferred between fire and basket with large wooden calipers. (Fire-cracked heating rocks are plentiful in shell middens along the California coast.)

Stone-boiling in baskets, like boiling in general, is

> particularly advantageous for processing large quantities of mollusks. Shucking shellfish, especially smaller bivalves

[13] Steam-treatment of tree bark and other woody materials to make them more pliable is an ancient practice. Canoes made with steam-heated bark were widely used in the Americas. Woody materials (mainly cellulose) and animal horns (mainly a hard protein) can be softened and molded by first steaming them. In northeastern North America several hunting-and-gathering tribes used molded bark containers for stone-boiling food.

> such as clams [and mussels], is a laborious task, but shells open automatically when heat is applied and the tissue binding the two valves loses its elasticity. (Ikawa-Smith 1976, 515)

This was even more true of small-apertured univalves like the dogwinkle (*Thais* sp.) which is abundant in coastal California middens, and from which juices (broth) can best be extracted by boiling.

Stone-boiling was rarely practiced in areas where pottery was made: after the introduction of pottery, the practice was generally discontinued. In some places, however, the practice continued; for example, among the Mohawk Iriquois, baked 'clay vessels, some nearly 4 feet high...were undoubtedly used *along with bark vessels as cooking pots*' (Onion 1964, 65). Moreover, the lack of external burn marks on some pre-Columbian pottery made by the early Maya suggests that 'cooking was probably done by placing hot stones inside large vessels to heat liquids and solid foods (such as tamales)' (Morley and Brainerd 1983, 367). In South America stone-boiling is only known in the extreme south among the Yahgans and their neighbors (Nordenskiöld 1931b, 86).

In the deserts of Sonora, where water is scant, some Indians did no boiling of food. Boiling is also a special problem in the higher altitudes of Peru and Bolivia, because vegetation for fuel is scarce there and because of the much-reduced boiling point of water. Since boiling requires more fuel than various other ways of cooking, another difficulty with this method of cooking is the unusually high specific heat of water. (The specific heat is the number of calories required to raise the temperature of one gram of a substance 1°C; that of water is one of the highest known.)

Pre-conquest settlement in the Andes reached elevations of some 4545 meters (about 15,000 feet).[14] At that altitude, the boiling point of water is around 84.4°C (184°F). At an elevation of 3050m (close to 10,000 ft.), the boiling point of water is about 90°C (194°F). Yet much of the Altiplano (the high plateau) of Bolivia and Peru is over 3656 m (some 12,000 ft.) above the sea. Such numbers alone do not give an idea of the *increased time* that it takes to cook food in boiling water. It is estimated that among the Aymara, high altitude 'increases the cooking time by about 50 percent, compared to sea level' (Johnsson 1986, 46).

The Aymara of Bolivia make various cooking adjustments to the high mountainous environment in which they live. Before being boiled, a number of starchy foods undergo a preliminary cooking by being parched. For example, around 'the age of five or six urban children master such skills as toasting rice [an introduced crop], *always done before boiling* at this altitude' (Johnsson 1986, 48 and 49).

As noted above, starches when heated undergo gelatinization at between 60° and 80°C. This, however, is only the *initial gelatinization temperature*: 'Studies with x-ray diffraction show that complete conversion [from the insoluble semicrystalline condition] *to the amorphous state does not occur until temperatures around 100°C are reached*' (Coultate 2002, 33). Thus, on much of the Altiplano, where the boiling point of water is less than 90°C, starchy food is probably not thoroughly cooked by boiling alone: a preliminary toasting is necessary.

The emergence of pottery was an event of great importance in the history of the chemical arts, including the art of cookery. Obviously, a knowledge of ceramics, and the making of non-flammable containers and utensils, changed cooking practices considerably. For instance, the Pueblo peoples, with their knowledge of ceramics, were able to cook a much more diversified cuisine than the California Indians. Perhaps the greatest change, however, was in terms of ease and work-efficiency: in Mesoamerica and Peru, where simple ceramic stoves were common, the cook's work was much shortened.

Most antecedents of ceramic water-containers are flammable; probably many of these were used for stone-boiling. Two liquid-containers continue to be important, even though they are not presently used as containers for cooking—namely, the tree gourd (*Crescentia alata*, often called a calabash) and the vine gourd (*Lagenaria siceraria*, often called a bottle gourd). In tropical parts of the Americas, tree gourds (known there as *jícaras*, *totumos*, etc.) are still very commonly used: unlike metal utensils, they are cheap and readily available. The nearly-globular fruit has a hard, woody rind which may be as much as a foot in diameter. In rural areas, cut and shaped into bowls, ladles and spoons, the tree gourd is among the most useful of kitchen items. However, for carrying water to be drunk on journeys (or while working in the fields), the vine gourd is preferred because the liquid stays cooler: the shell of the vine gourd, being slightly porous, allows small amounts of water to seep through and evaporate, and so its contents are cooled.

The vine gourd must have played a more important part than the tree gourd in the early development of chemical arts. Though originally domesticated in the Old World, it is one of the oldest plants to be cultivated in the New. Since it is an annual, its fruits can be produced in a timely fashion, unlike the calabash whose trees takes several seasons to mature. Also, unlike the tree gourd, it can be grown in temperate climates. The many varieties of the

[14] Human occupation, based on shepherding, reaches its highest altitude in the Andes of southern Peru at 17,000 ft. (ca. 5151 m) above sea level, just below the snow line. Settlement at this altitude is the result of ladino appropriation of better land, lower on the slopes. At this elevation the Indians find only scanty forage for their animals. Fuel is mainly grass: 'It is too high an altitude for even the tola bush [*Baccharis* sp.]' (Brunhes and Deffontaines 1952, 468). Elsewhere, the highest human settlements are generally associated with mining.

vine gourd plant produce fruits with remarkably diverse shapes, including nearly globular, bottle-shaped, hourglass-shaped, and crook necked—often with knobby protuberances.

Griddling and Roasting

Mexican and South American cuisines differed considerably in the 15th century—as they do even today, especially in the preparation of maize dishes.

As discussed already, the tortilla is a thin disk, made from alkali-treated maize and griddled. The tamale, on the other hand, is a ball or small loaf of maize dough. In Mexico, it is made from the same *nixtamal* that is the starting point of tortillas. The tamale dough is wrapped in corn husks or leaves, made pliable with hot water, and boiled or steamed. Sometimes, bits of other food—such as beans or meat—are placed within. A third dish prepared from nixtamal is atole, a hot thick gruel made of maize. It was a typical item in the Mesoamerican cuisine, and is still the common breakfast of Indian households in much that area.

Both tamales and tortillas were made in pre-Columbian Mexico, but tortillas have long been the predominant food there. In contrast, Indian peoples to the south of the Maya area in Central America (for example, the Térraba, Guaymí and Cuna), as well as the Chocó yet farther south, still know nothing of making tortillas and comales, or of treating maize with alkali. Moreover, even the Maya had received such knowledge only in relatively 'recent' times (recent, at least from an archaeological standpoint).

Maize preparations eaten in ancient Kaminaljuyu: a stack of tamales on the left; on the right, tortillas. (After Kidder et al. 1977, 235—as identified by Taube 1989, 34)

Figure 11. Ancient Maize Preparations

Tortillas were not made by the Classic Maya, who lived between 250 and 900 A.D; at least comales are not found in their old habitation sites. (Kidder et al. 1977, 208; Taube 1989, 33; and Thompson 1938, 597). Tortillas, the manufacture of ceramic comales, and, more importantly, the lime treatment of maize are linked historically; all appear to have been spread together from Mexico. Over time, however, the tortilla partially supplanted the tamale among Maya peoples.

Differences between the cookery of Mesoamerica and South America are illustrated by the Quechua speakers of Andean Ecuador, to whom

> the Mexican maize tortilla is unknown, and the lack of this standard dish greatly differentiates Ecuadorian Indian cooking, food-serving, and housework from the cookery and domestic activity of Mexico. Fingers or spoons take the place of the tortilla dipper...The same condition prevailed in Inca Peru, where...*boiled or toasted maize* was generally eaten... (Parsons 1945, 22)

Moreover, in Peru a special maize flour was made for caciques and other 'people of quality' by grinding the dry kernels; the meal was sieved though a piece of cloth to remove the bran (Cobo 1956, 1:161). This procedure is quite different from the way maize was commonly prepared in Mesoamerica.

Something like a tortilla was made by the Tewa of New Mexico, but in pre-Columbian times, real tortillas were not made north of Mesoamerica. Despite their knowledge of ceramics, the Tewa made no comales; they used a large flat-surfaced stone instead. In making their thin, paper or wafer bread, *mowa* (the *piki* of the Hopi), they fashioned

> a rectangular slab of fine-grained stone, about 3 feet square, laboriously hewn and polished..., which rests on stones at the ends or at the four corners...; it is heated by a fire built beneath it. A soft liquid dough or batter is prepared in a mixing bowl, and when the stone has been thoroughly heated and wiped with a greasy rag, a small quantity of the batter is spread over the surface by a quick, sweeping motion of the hand, leaving a *thin, even layer*. In a *few seconds* this layer of dough is so far cooked that *it can be peeled off entire by one of its corners...*; *the semitransparent sheets* are folded in four, and sometimes the four-fold sheets are rolled into cylinders. (Robbins et al. 1916, 89)

Quantities of this bread were also made by the Zuñi: 'The piles of sheets laid one above another are often so high that they look like huge bales of fine wrapping-paper' (Cushing 1920, 338).

In Cueva country, along the southeast coast of the Panamanian Isthmus, Oviedo gives the first detailed description of making a tamale-like food. He notes that the Indian women grind the maize kernels with a stone mortar and pestle.

> As they grind, without stopping they *occasionally add a little water*. This produces a sort of paste or dough, portions of which are formed into a ball (bollo), a small loaf or roll a little less than a span in length and several fingers thick...Wrapped in a corn husk or the leaf of some other plant, the *bollo* is boiled. After being boiled it is taken from the pot or caldron and allowed to cool a little, but not

> completely. *If they are not wanted boiled, these bollos are placed in the hot embers and roasted until dry.* This turns them into something like white bread, with a crust on the outside, and the inside crumbly and soft. The leaf in which the *bollo* is boiled or roasted is stripped off, and the bollo is eaten while somewhat hot because if it cools it has neither so good a flavor nor is so easy to chew, and the colder it becomes the drier and rougher it turns. This bread, boiled or roasted, doesn't last more than two or three days because later it mildews and sours, and cannot be eaten (Oviedo 1959a, 1:228-229).

Bollos, balls of boiled maize dough covered with the shucks of corn cobs or with wild banana leaves,[15] are still a characteristic food of rural Colombia and Venezuela. Cornbread, as Oviedo says, is quite crumbly. Because maize dough has less gluten, it is less cohesive and kneadable (less capable of being pressed, folded and stretched) than the dough of the Old World wheat. The following early note on the Maya indicates as much:

> They have not yet been able to succeed in making a flour which they can knead like wheat flour, and if at times they make it like wheat bread, it is good for nothing. (Landa 1941, 91)

Steam Cooking

Water exists in three states: liquid, gas, and solid. When a change of state occurs, large amounts of heat are either released or absorbed. When water is evaporated, this heat passes into a hidden form, stored in the gas, which is called *the latent heat of vaporization.* This latent heat cannot be measured directly, but can be quantified as the number of calories required to change one gram of liquid into vapor at the same temperature. An unusually large amount of heat is required to evaporate water. The calories required, however, depend somewhat on the temperature at which the process occurs. For water boiling at 100°C, about 540 calories per gram are needed, the highest value known for any substance. In the reverse process, when steam is condensed to a liquid, an equal amount of heat is abruptly released, and may be transferred to any food present. This accounts mainly for the fast-cooking action of steam, as well as for the intense pain felt from a steam burn.

Steam-cooking was known among native peoples in South America, Mexico, and the American Southwest, at least in historical times. Only people with a knowledge of ceramics (or in later days, those with metal containers) could have practiced this kind of cooking. For example, earthenware utensils for steam-cooking have been observed among the Chiriguano in Bolivia. These contrivances, used 'in cooking a certain dish of maize,' consist of two containers. They are (except for being made of earthenware) very like those in general use in industrial societies today: a bowl with a perforated bottom containing the maize, set atop a larger earthenware pot in which water is boiled (Nordenskiöld 1920, 54-55).

In rural households in arid and fuel-scarce, high-altitude Bolivia, the Aymara (using a single container) steam potatoes, rather than boiling them. There

> cooking is done over open fire, [and] children learn exactly how *to stuff the cooking clay pot with hay*, before placing the potatoes on it. Informants claim that this is done to prevent the pot from bursting, *to save water by steamboiling* and to hinder the potatoes from burning, should the water evaporate completely by mistake. In addition, *a smaller amount of water saves fuel since it starts boiling quicker,* (Johnsson 1986, 48)

Not only is a smaller amount of water used in this procedure, but the latent heat released in steam-cooking helps to shorten cooking time.

In Mexico, *tamalitos* ('little tamales') are made by wrapping maize dough in cornhusks, and then steam-cooking them for about a half an hour to gelatinize their starch.

Steam cooking was also practiced by the Pueblos early in the 20th century.[16] By that time, however, despite the Pueblos' rather conservative culture, they had adopted a number of European traits. For example, the domed 'bee-hive oven' was present. This oven—still seen in the Southwest and in many parts of Latin America—was introduced by the Spanish; it is said to derive ultimately from ancient Mesopotamia.

Whereas there is a possibility that steam-cooking is an introduced trait among the Pueblo, their way of steaming is quite distinctive and unlikely to have been introduced from Europe. An example is the way the Zuñi steam-cooked maize dumplings:

> Considerable care was required in the manufacture of these. Fine flour was boiled in water until paste had been formed. Into this paste enough meal was mixed to make a stiff dough, and of this dough *little balls or pellets were rolled out and spread evenly over a yucca sieve or screen of sticks connected at the ends.* A large pot half-filled with water was set over the fire, inside of which a smaller vessel, partially filled with water and weighted with pebbles to keep it steady, was placed. Upon this smaller pot was laid the sieve or screen holding the balls of dough, the larger pot then being covered with a slab of stone and kept boiling until the dumplings were *thoroughly cooked by steaming.* (Cushing 1920, 299-300)

---

[15] Bananas and plantains, which belong to the genus *Musa*, are plants introduced from the Old World. The related genus *Heliconia*, sometimes called in English 'wild banana' (known widely in Latin America as *bijao*), has very similar leaves, but is native to the western hemisphere. The leaves of *Heliconia* are a common wrapping material in tropical America—especially those of *Heliconia bihai*. (Bijao is a familiar regrowth plant. The domestic banana and plantain, on the other hand, are only found where planted.)

[16] Two detailed reports on Pueblo cookery are available: one, on the Tewa, published in 1916 (Robbins et al.); the other, on the Zuñi, in 1920 (Cushing).

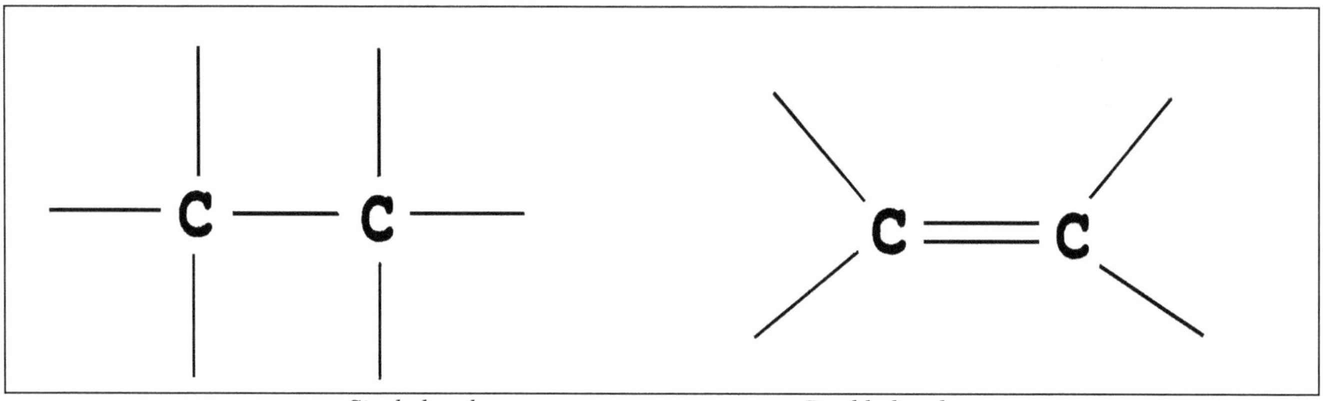

Single bond *Double bond*

In ordinary usage, the word "saturated" means "unable to hold or contain more." In organic chemistry, however, the term saturated means that a molecule contains the maximum proportion of hydrogen ever found in combination with carbon. (A good example of a saturated compound is the hydrocarbon methane; its structural formula is pictured in Figure 4.)

Fats and oils are classified as either saturated or unsaturated according to whether they have carbon-to-carbon double bonds in their molecules. Those that contain only single bond carbon linkages are called saturated. Those that contain double bonds are called unsaturated, because they possess fewer than the maximum number of hydrogen atoms—and, through the process of reduction, are capable of adding more. As noted earlier, oxidation liberates heat energy. Reduction, on the other hand, absorbs and stores energy, but yields it all back again upon oxidation.

Oils typically contain more double bonds than do fats: The number of double bonds, and the degree of unsaturation, increases progressively from the hard and soft fats (such as tallow and lard), to vegetable oils, and to fish oils. Unsaturation in a fat generally lowers its melting point and thus tends to make it a liquid at normal room temperature.

The double bond also represents chemical instability. Thus, saturated fats are more stable (that is, chemically less reactive) than unsaturated fats: the more double bonds present, the less stable the fat. For example, unsaturated oils turn rancid (oxidize) faster than the more completely saturated fats.

Figure 12. Fats and Saturated Compounds

Rendering

Although fats and oils are chemically very similar, they differ physically in that oils are liquid at room temperature, while fats are solid or semi-solid. Thus, an oil is simply a liquid fat. Oils are more abundant in plants (mainly in their seeds and fruits), whereas fats are probably more plentiful in animals. (Food oils and fats are called 'fixed oils' because they are non-volatile and cannot be boiled away. Thus, they should be distinguished from the essential oils.)

Fats and oils are lighter than, and insoluble in, water. Therefore, when fatty animal tissue or crushed plant seeds are immersed in boiling water, liquid globules of the freed fats and oils float to the surface where they can easily be skimmed off. This process is called rendering. It is more difficult to extract oil from plants than fat from tissues of animals. In plants the minute globules of oil are surrounded by rigid cell walls, often made up mainly of cellulose and lignin. To release the oil, these cell walls must be thoroughly crushed. In animals, on the other hand, 'only heating is needed to release fat from adipose tissue...The fat expands with heating, tearing the adipose tissue cell membrane and flowing freely' (Belitz and Grosch 1987, 472). Because they are more highly reduced (i.e., contain more hydrogen or less oxygen), fats and oils when digested liberate more than twice as much energy as the same quantity of carbohydrates, starches and sugars. Thus the 'molecules of fats and oils are marvelously effective devices for storing hydrogen, the ideal fuel, in a condensed form' (Read 1960, 114).

Because plant growth is scarce in snowy polar regions, herbivore mammals there hoard fats. (Such animals are capable of making saturated fats from plant products by the reduction of edible carbohydrates.) Carnivores, and sometimes omnivores, are forced to acquire fats from other animals. The Eskimo, living there, must use animal fats for food, fuel and lighting.

Oils are most plentiful in tropical vegetation. There are, however, numerous animal sources of oils and fats in the tropics. Among these are those derived from aquatic animals such as the oil from fish, the oil from turtle eggs, and the fat from manatees.

In the historical literature of the North American prairies, there are many references to rendering the fat ('grease') from the bones of several mammals. A Canadian archaeologist reports on slaughter sites of the American bison: 'I was impressed by the enormous quantities of comminuted bones...The bones had been broken up into very small pieces, and this had obviously been a regular practice rather than an occasional one' (Leechman 1951, 355). Fat from these bone fragments was rendered at simmering temperatures. This is because 'bubbles form but do not break as in boiling' (Lundberg and Kotschevar 1965, 33), and so roiling at the surface of the water is avoided.

According to an Old Crow informant in the Yukon, bone grease

> is made from caribou and moose bones. After the meat is cut off, the bones are left for one day, which allows them to dry a little. If the bones were left for two or three days, the bone grease made from them would taste too strong to be pleasant. A caribou skin from which the hair has been removed is laid on the ground and an anvil stone is placed in the middle of it. The bones to be broken are placed on the anvil stone and are smashed into little pieces, '*as big as finger nails*,' with the back of an axe. In the old days stone hammers were used for this. The broken bones are then put in a kettle with a little cold water and placed on the fire. As soon as the water comes to a boil, cold water is added (snow in the winter time) *so as to keep the water simmering rather than boiling violently. The purpose of this is to allow the oil and grease rendered out of the bone to float on the top, which it would not do if the water were boiling vigorously*. The grease is skimmed off and put in a separate vessel, usually the small inner part of a caribou's stomach. Here it will keep quite well *for as much as two or three years*. Some of it was used for making the best grades of pemmican... (Leechman 1951, 355)

In South America great quantities of oil were once rendered from turtle eggs. Humboldt comments on the oil produced in the late 1790s by Indians along the Orinoco River:

> This animal oil...keeps the better, it is said, in proportion as it has undergone a stronger ebullition [boiling]. When well prepared, it is limpid, inodorous, and scarcely yellow. The missionaries compare it to the best oil of olives. (Humboldt and Bonpland 1972, 4:487-88)

In southeastern parts of the USA,

> ...hickory nuts were rendered for oil, as were walnuts. Southeastern Indians *crushed hickory nuts, shell and all, into a meal* that was then submerged in water to extract oils. (Sassaman 1995, 227)

Similarly, in Mexico, *chia* oil was rendered from the lightly roasted seeds of a *Salvia* species—usually from *S. hispanica* (Cahill 2003).

Deep-Fat Frying

The term 'deep-fat frying' is a misnomer. The procedure actually consists of *boiling* food—but boiling it in oil, rather than in water. Thus it is a different process from ordinary frying. Frying depends largely on conduction for the transfer of heat from flame to food, whereas deep-fat frying depends more on convection: heated oil circulates upward around the entire surface of the cooking food.

In areas where water and fuel are scarce, and at high elevations, oil would seem to be a better medium for boiling food than water. Oil has a much lower specific heat (averaging 0.5) than water (1.0), and so gets hotter than water when the same amount of heat is applied. Moreover, water boils at a temperature which is dependent on its elevation above sea level, while in contrast, plant and animal oils boil typically at around 180°C whatever their elevation above sea level may be. Thus using oil reduces the cooking time. A third point in its favor is that oil, unlike water, does not boil away, and can be used repeatedly.

On the other hand, if an oil is overheated, undesirable tastes or odors arise, because some constituents begin to decompose through oxidation:

> Smoke is the sign of decomposition. The smoke point of a fat or oil is normally in the range of 200-230°C during prolonged frying, and it decreases in the presence of decomposition products. When it falls below 170°C, the fat is considered to be spoiled. (Belitz and Grosch 1987, 493)

Furthermore, at high temperatures, oils and fats readily catch fire, whereas water merely boils more intensely when the heat input is increased. Thus boiling food in water requires much less of the cook's attention.

Deep-fat frying is likely a late development in the history of cooking. It seems to have been an uncommon method of cooking in pre-Columbian Amerindian culture, probably because most vegetable oils (the fats most widely available in low and middle latitudes) turn rancid quickly. If food is to be immersed in heated oil, a deep and fire-resistant container is needed; thus, pottery is probably a prerequisite for this kind of cooking. (In pre-Columbian times, metal vessels were not used in cooking.)

Frying

In this method of cooking, the food material comes in contact with a heated surface, generally spread with a thin layer of oil, to keep the food from sticking. Like deep-fat frying, actual frying seems to have been a rather uncommon method of cooking among Native Americans. This is surprising, at least where fuel and water are scarce, because frying cooks mainly by direct heat—that is, by conduction. Thus, it makes efficient use of fuel. In frying no container is needed, only a flat, fire-resistant surface that will hold a thin layer of oil. Thus frying can be practiced even where no pottery is made. For instance, the Chumash living along the coast of California lacked pottery, but nonetheless 'made soapstone vessels and when they broke them they used the bottom pieces *to fry food on*' (Coon 1971, 184). Actually, since little food sticks to it, soapstone need not be covered with oil. In 1916, a maize preparation made by the Tewa was 'fried in mutton grease'; however, the sheep is an introduced animal.

In South America, among the Colorado who live in the Ecuadorian forests west of Quito, food 'is prepared by *steaming*, boiling, or broiling [roasting]. It is *never fried* in grease' (von Hagen 1939, 34).

Among the Aymara in the high Andes, where fuel is notoriously in short supply, 'boiling is preferred to frying, in part owing to the scarcity of cooking fuel and cooking fat' (Johnsson 1986, 48). The reason given for this preference is puzzling, because boiling is a less fuel-efficient method of cooking. The explanation may be that various starchy foods are parched (that is, already partly cooked) before being boiled. Of course, conditions were rather different in pre-Columbian times: a part of the native population was affluent and, for them, fats would have been easily available from domesticated camelids.

Food Additives

The food additives used by pre-Columbian Americans consisted, for example, of natural colors to make food more attractive, sweeteners to improve taste, and preservatives to prevent the growth of microorganisms.

Coloring with Vegetal Pigments

One of the most common colorants in the plant world is anthocyanin, a water-soluble glycoside that appears especially in the red, blue, and purple colors of flowers, fruits, and leaves.

In the Americas, amaranths (*Amaranthus* spp.)—which commonly contain red to purple anthocyanins in their leaves, roots, and seeds—were once a major source of coloring for water-based foods and drinks. The Aztecs cultivated amaranths for use in dyeing foods. In Mexico, even today, this practice continues. (The seeds are also an important food there.) The Hopi, too, prepare a red and yellow paper bread 'by mixing vegetal dyes in the dough'; the red dye they get 'from the seeds of *Amaranthus palmeri*' which they also cultivate (Robbins et al. 1916, 89, ftn.). The Zuñi also used other coloring materials: they made white cornmeal even 'whiter with a little kaolin; black, or rather a sable purple, with certain charred and powdered leaves' (Cushing 1920, 333-337).

Yet more widely used than amaranths for coloring food is achiote, a bush or small tree, *Bixa orellana*. (The word 'achiote' is derived from Nahuatl, the Aztec language, but the number of names by which the plant is known gives an idea of how widespread its use is. Among the more common names are annatto or arnatto, words derived from the Carib language; the French word roucu comes from urucu in Tupi-Guarani.) One of the many early descriptions comes from the Maya:

> There is a little tree which the [Maya] Indians are accustomed to raise in their houses, which bears prickly husks...They open at the proper season and have inside little seeds, which they use... (Landa 1941, 200)

The achiote's grape-like seeds are enclosed in a reddish, wax-like covering which contains its principal colorant, the carotene bixin. Bixin, like other carotenes, is insoluble in water but soluble in oil. (However, variable amounts of norbixin, a water-soluble form of bixin, are also produced from the seed's covering, the quantity depending on the extraction process.) Since anthocyanins are water-soluble and bixin is fat-soluble, the two coloring materials serve complementary functions in cookery. The Aztec, for example, used both.

The achiote seeds are usually soaked in hot water; then the water is poured off and evaporated. The resulting paste is dried to the consistency of putty. In native markets, it appears as small rolls folded in leaves or, after being dried still more, it may be cut into small cakes. Carotenes (named after the compound that imparts an orange color to the Old World carrot) give the color to achiote, which probably originates in South America. Carotenes (and a closely related compound, lycopene) also give a red color to tomatoes and red peppers, two plants native to Mesoamerica.

Achiote was used by the Maya 'for giving color to their stews...' (Landa 1941, 200). In tropical South America, too, achiote is commonly used as food coloring. For example,

> throughout the Ecuadorian highlands, achiote is put into food, *though it adds no flavor whatsoever*. No Quechua Indian will eat food unless it is so colored, and even a few mestizos demand its use in cooking. (von Hagen 1939, 25)

Achiote is, however, a good source of the fat-soluble vitamin A, which is often deficient in many Native American diets. Nowadays, achiote is used commercially for coloring salad dressings, butter substitutes such as margarine, colored butter on popcorn, and so on.

Coloring with Alkaline Ferments

Probably the greatest contribution to the Zuñi's cooking technology in a long time was the following:

> [The] most notable advance was perhaps the introduction of *ashes*, or of very *finely ground lime*, called *a'-lu-we*, mingled with salt into *fermented mush-yeast* to overcome its acidity. The *most prized leaven* of his [i.e., the Zuñi's] time, however, was *chewed sa'-ko-we* [a coarse maize meal] mixed with moderately fine meal and *warm water and placed in little narrow-necked pots over or near the hearth until fermentation took place, when lime flour* and a little salt were added. Thus a yeast, in nowise inferior to some of our own, was compounded. (Cushing 1920, 294)

Two illustrations of achiote (Bixa orellana). (Left: Schultes 1984, 25, after Piso 1658; Right: Whiffen 1915, plate XLII),

Figure 13. Achiote (or Annatto) Plant

As noted in Figure 4, terpenes consist of isoprene units. (The isoprene unit is one of nature's favored molecular building blocks, appearing in hundreds of natural compounds.)

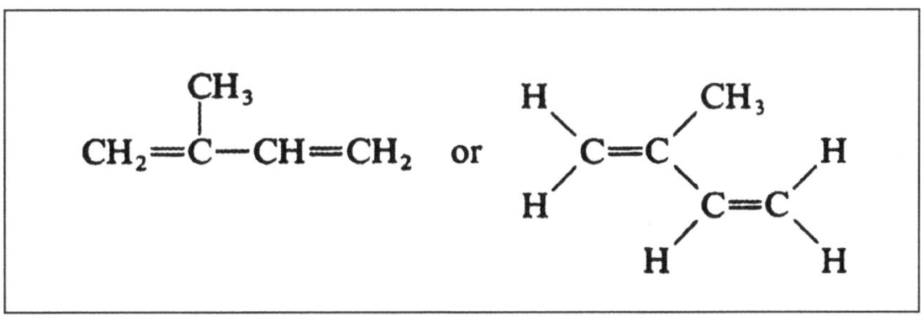

Above are two representations of the structural formula of isoprene. The branched end of the five carbon atoms—where the methyl group is attached—is called the 'head' and the other end, the 'tail' (Solomons 1996, 1103).

Terpenes are classified according to the number of isoprene units they contain. Monoterpenes are composed of two isoprene units and thus have 10 carbon atoms. Sesquiterpenes have 15 carbon atoms; diterpenes have 20 carbon atoms; triterpenes have 30 carbon atoms; and tetraterpenes 40 have carbon atoms. Pinene and limonene are monoterpenes; carotene and lycopene are tetraterpenes.

Figure 14. Carotenes, Terpenes and Isoprene

Though the mixture is here called a 'yeast,' the ferment (because of the cereal in it) is more like a sourdough—albeit there is no reference to the ferment's being stored until needed, as sourdough is. (Indeed it is doubtful, especially from reading the report of Robbins and his co-workers, that the mixture was always fermented.) In any case, one feature in the making of this mixture was definitely not an introduction from Iberian culture: there is no record of the Spaniards' masticating a cereal meal to make yeast. However, the pre-mastication of starchy food materials was widely practiced in pre-Columbian America.

Regarding the lime mentioned in the above quotation, the Pueblo, like the Tarahumar, may have learned how to fire limestone from the Spanish. Supporting this view is the following note: in early historical times, although limestone was readily available in the Pueblo area, 'to save trouble, the lime is more often bought from 'Mexican' peddlers' (Robbins et al. 1916, 93, ftn.). Probably ashes, not lime, was the older alkali constituent of the Pueblo's masticated maize: ashes were included in many, if not most, of their numerous corn recipes.

The Pueblo's 'sourdough' seems not to have been used as a fermentation-accelerator or starter, as it was in

Mesoamerica, perhaps because the masticated maize was not fermented before the Spanish arrived in 1540. As discussed below, starters have been long and widely used in the production of native American beverages. For instance, the Aztecs used sourdough as a starter in making one kind of atole:

> They also prepare another kind of atole that the Mexicans call *xoco atole*, or sour atole, which they make by mixing a pound of *sourdough* with two pounds of ground and cooked maize...The sourdough they prepare in the following manner: *they take black maize, make it into a dough, and keep it for four or five days until it is sufficiently acid* [i.e., sufficiently fermented?], and then mix it with the ground and cooked atole...later, in another container, it is served with salt and chile. (translated from Hernández 1888, 217-218)

There are several possible chemical reasons why the Pueblo might have added either this ferment or masticated corn to their maize preparations: (1) for coloring; (2) for sweetening; (3) for a stronger gelatinization of the starch content; and (4) for leavening bread.

(1) The Pueblo peoples seem to have been intensely interested in giving color to their food. Even as late as the early 20th century, the Zuñi commonly cooked with six different colors of maize seeds: 'No meal, however sumptuous, is ever eaten without one or another of them' (Cushing 1920, 336). Maize, the most widely-grown crop in pre-Columbian America, was generally of high religious significance there, and strains had been bred with multi-colored seeds; no other seed crop, and probably no other plant, has seeds that are colored so variously. The Pueblo even associated the color of maize seeds with cardinal directions: yellow seeds with north, blue with west, red with south and white with east (Robbins et al. 1916, 82).

Although food is tinted with vegetal colors in many parts of the world, the Pueblo practiced a more complex (and perhaps unique) chemical procedure for changing the color of maize preparations. For example, the Zuñi added the alkaline ferment to corn meal:

> In addition to its *leavening qualities*, this yeast had the remarkable property, when added to the meal of *blue corn or black, to change the color* during cookery to a beautiful green hue, or, mingled with *yellow-corn flour*, to render it *light blue.* (Cushing 1920, 294-295)

The same effect can be obtained, however, with either ashes[17] or lime alone (Robbins et al. 1916, 89 and 93). Probably ashes are the original additive.

Most colors in the seed coats of maize are due to the presence of anthocyanins. That these change color with variations in the pH is now well known. For instance, anthocyanins in weak 'alkaline solutions are violet, but turn to green in the presence of strong alkali' (Lee 1983, 445).

(2) Pueblo people did not keep the domesticated New World species of honeybee, as did tropical tribes farther south; consequently, sources of sweetening were scarce in their area.

The sweetening effect of pre-mastication is caused by the enzyme ptyalin in human saliva, which changes some of the starch content of maize seeds into sugars.

Tewa paper bread, described above, 'was formerly made of dough *sweetened by mixing with it chewed meal or stale mowa broken up fine and chewed*' (Robbins et al. 1916, 89). Also, the Tewa cooked 'dumplings made of corn-meal *mixed with ashes, sweetened, formerly with chewed meal*, now with sugar, and boiled' (Robbins et al. 1916, 90).

A Zuñi preparation was 'made of yellow cornmeal, a portion of the batter of which was *sweetened* either by *previous mastication and fermentation* or by the admixture of dried flowers' (Cushing 1920, 300). Because of their nectar content, even flowers were used as sweetening.

A further note on sweetening is given as part of a recipe for a kind of bread made with a sticky maize dough, and cooked in a pit oven:

> We now come to the greatest delicacy *in the way of bread* known either to the *older* or the recent Zuñis...Leaving salt out of this recipe and adding to it dried flowers, licorice-root [a species of *Glycorrhiza*], wild honey, or, *more frequently than any of these, masticated and fermented meal, this buried bread was made sweet...* (Cushing 1920, 303-304)

(3) The possibility that the ferment was added to provide extreme gelatinizion is suggested by one Zuñi preparation. They made a boiled, congealed, and thoroughly gelatinized maize dumpling, described as one of the more complex products 'of the ancient Zuñi kitchen:'

> These [dumplings] were made by hulling corn, then grinding it with water, precisely as colors are ground by artists with oil. This batter, fine and sticky, *well seasoned with lime-yeast*, was wrapped in broad shucks carefully folded over and tied at the ends, *then boiled*. The batter was *solidified* by the boiling and when done resembled to a great extent *well-cooked gristle or tough gelatin*. Great quantities of these *rubber-like dumplings* were made at a time, as the process of their manufacture was tedious and laborious. (Cushing 1920, 301)

Why the 'lime-yeast' had this effect is a puzzle. The following is perhaps an unlikely explanation: Some albumin—a complex of proteins—is present in many seeds. Though the albumin content in maize seeds is low, in wheat seeds it is more than 3½ times higher (Belitz and Grosch 1987, 500).

---

[17] The ashes of the shadscale (*Atriplex canescens*) 'are preferred' by the Tewa to change the color of their wafer bread (Robbins et al. 1916, 89). Plants of the goose foot family (Chenopodiaceae) are known for their high content of sodium carbonate; this characteristic of the family is best exemplified by the Old World saltwort (*Salsola kali*), formerly the major commercial source of the compound. Our word 'alkali' is derived from the Arabic name for that plant.

Albumin is hardened when mixed with slaked lime; thus, possibly these dumplings were made with wheat flour.

Despite their knowledge of the properties of lime-yeast, apparently the Zuñi made well-gelatinized dumplings without adding any ferment. For example, one of their dumplings was made

> by mixing fine meal or flour with an equal amount of coarse meal..., salting, and by the addition of cold water and the intervention of *excessive kneading* to form a stiff dough. This was divided into little pieces which were rolled into hard balls between the hands. A pot of water...was made to boil violently, the little balls were poured in. Instead of disintegrating, *they became harder and harder with the progress of the cooking.* (Cushing 1920, 298)

In short, the addition of the ferment seems to have been unnecessary—and, when added, its chemical function is uncertain.

(4) In the Old World leavens (concentrations of yeasts) have long been added to bread dough to make it rise. (Carbon dioxide gas, generated by the yeasts, enlarges air bubbles in the dough making the rising possible.) No leavened bread seems to have been made in the pre-Columbian New World, with one possible exception: Among the Zuñi,

> ...the rudest forms of *true bread* were made by placing in a bowl *fine flour* into which enough cold water was poured to make of it dough, and *sufficient lime-yeast to leaven it.* This was then *kneaded* and molded into thick cakes, which were *set away a short time to rise*, after which they were cooked on hot coals by frequent turning...or baked, buried deep under hot ashes. (Cushing 1920, 303)

Even before 1920, wheat—an Old World plant—was sometimes cooked by the Pueblo (Robbins et al. 1916, 91-93), and possibly this bread was made with wheat, rather than maize, flour. Perhaps another foreign influence may be seen in the fact that this bread was 'also *frequently baked*...in the large dome-shaped ovens'(Cushing 1920, 303). These are the introduced, 'bee-hive' ovens referred to previously.

Even supposing that the above-mentioned bread may have been made with wheat flour, the Zuñi made another bread, clearly specified as being made from maize. This was not one of their numerous boiled preparations, but was different from these in being *raised with yeast before baking*, therefore thicker and circular in shape...The dough of which these were formed was *invariably made of very finely ground white corn*' (Cushing 1920, 302).

Some statements made on leavening (in Cushing's otherwise excellent report) are questionable. For example, the Zuñi are claimed to have made dumplings

> in the form of little pellets of blue cornmeal, finely ground, mixed *with a considerable quantity of lime-batter or yeast*, kneaded into stiff dough, and wrapped tightly in corn-shucks. *During the boiling* these dumplings became not only extremely blue *but considerably swollen through the agency of the yeast.* (Cushing 1920, 299)

This comment on swelling during boiling 'through the agency of the yeast' is hard to accept: heating dough to temperatures much over 60°C will kill most yeasts. In contrast to Cushing, Robbins and his coauthors do not mention leavening, or even the fermentation of masticated maize.

The case for a New World leavening of bread before contact with Europeans is weak. The Pueblo probably learned the leavening of bread, and perhaps the process of making a yeast-ferment, from the Spanish. Of the four reasons given above, (1) changing the color of maize preparations and (2) sweetening are the most convincing.

Condiments

Early European explorations of the world were prompted not solely by a thirst for knowledge or a desire for gold, but also by a craving for spices and dyes. In the Americas, the Spanish found two of the world's best-known condiments, vanilla and chili pepper, as well as several valuable dyes.

Vanilla

In Mesoamerica, chocolate drinks were commonly flavored with the ground pods of the vanilla vine, a tropical orchid, *Vanilla planifolia.* As elsewhere in the area, vanilla pods were a trade-item among the Maya. When ripe, the pods are yellow at the tip. In a careful drying and curing process, the almost odorless pods turn chocolate-brown and become highly aromatic. During curing, the flavor precursors which are generally glucosides are broken down into vanillin (the principal flavoring component of the vanilla bean) and glucose.

Chili and Other Peppers

Hot sauces are made from *chili* pepper by grinding the pods and seeds of *Capsicum* species, which contain a pungent, biting spice. Capsaicin (the molecule is almost insoluble in water, but soluble in oils) is the active principle. *Capsicum frutescens*, which includes the Tabasco peppers, is probably the "hottest" species.

Sweeteners and Non-Alcoholic Beverages

Until the Muslim Arabs introduced sugarcane from Asia to Spain, Europe was largely dependent on honey for sweetening. Honey was a major item in commerce, as it was in the New World before the Spanish planted sugarcane there.

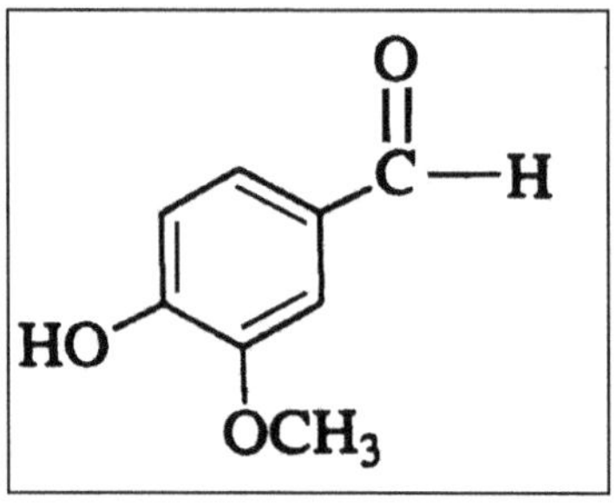

Vanillin

That vanillin is another phenolic compound is indicated by the hydroxyl group, —OH, attached directly to the benzene ring. Because of its attached aldehyde group, vanillin is also called 'vanillic aldehyde'.

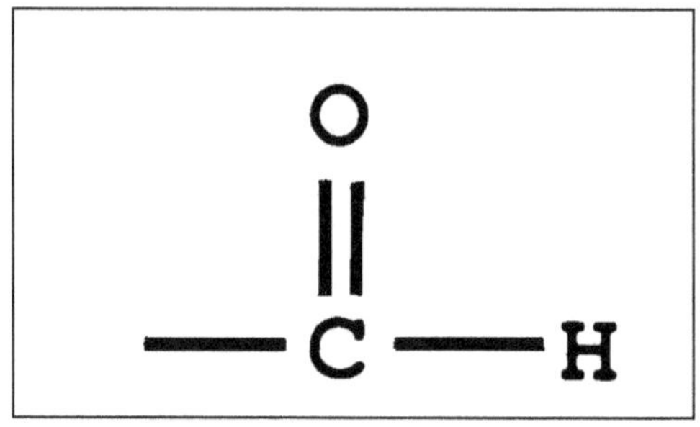

Aldehyde group

Complex aldehydes are often characterized by fragrant odors. The methoxyl group, —OCH_3, which also appears in the structural formula of vanillin, is frequently seen in the formulas of alkaloids (to be discussed later).

Figure 15. Vanillin and Aldehydes

Honey from Native Stingless Bees

In the Americas a number of bees of the genera Melipona and Trigona produce honey and wax. Unlike the common honeybee of Europe, *Apis mellifera*, these New World bees are stingless. A number of native bee-species are enticed into hives, and several, which have been kept for many generations, are now semi-domesticated (Schwarz 1948). The most important of these is probably *Melipona beecheii*, found from Mexico to Central America. Its cultivation was the basis of a commercial apiculture long before arrival of Europeans.

In the 16th century, Indians of Jalisco and the neighboring areas made beehives:

> which they keep in houses suspended in the air...for hives, they take a section of tree trunk and hollow it...and in one house there may be ten, and in another twenty or thirty suspended hives in which excellent honeycombs and honey are made...The bees are small and numerous, and no larger than flies; they don't sting and are harmless because they have no stingers...; since there are so many hives, there is a great quantity of excellent wax. (Oviedo 1959a, 4:270)

A similar description is given of a Maya apiary (Landa 1941, 193-94). The Aztecs at Tenochtitlán (now known as Mexico City) levied a tribute of honey on cities of the 'hot lands' farther south:

> Honeycombs were sent and even hives with bees inside; also great jars of white honey and others of yellow honey. Durán 1994, 206)

In Costa Rica there was a similar 'abundance of honey and wax' (Oviedo 1959a, 4:363). Suspended calabashes were used as hives by bee keepers in Venezuela.

Honey from stingless bees is still collected by native apiarists from Mexico to Argentina. Artificial hives are prepared by many rainforest tribes—for example, by the Guaymí in Panama. In the northwest Amazon, the Menimehe "make hives of hollow trees for bees to swarm in"; these hives are kept in their houses "so that a store of honey and wax is always at hand" (Whiffen 1915, 51 and 130).

Sweetening from Maize Stalks and the Maguey Plant

Although the deliberate manufacture of crystalline sugar, and its sale, is commonly associated with Old World plants (sugarcane and the sugarbeet), early Spanish authorities agree that sugar was also made from maize. Juice in the stalks of some varieties of maize has a high sugar content. In warmer climates, these varieties were quickly replaced by sugar cane which the Spanish introduced shortly after their arrival.

According to Hernán Cortés, the Aztecs made, and sold in their markets, "honey made from maize stalks" (Oviedo 1959a, 4:45). The sweet stalks of green maize were chewed, though it was warned one might "thereby have a toothache" (Sahagún 1950-82, 5-6:194). Maize with stalks having a high sugar content was also grown in Peru: "An excellent honey is made from the unripe cane, which is very sweet" (Garcilaso de la Vega 1966, 499; also Cieza de León 1998, 84). In the Cochabamba Valley of Bolivia, sweet young maize stalks "are sold today in the markets to be chewed like sugar cane" (Cutler and Cardenas 1947, 41).

The Aztecs boiled the sap drawn from maguey plants to make a syrup. The syrup maker 'cooks it, boils it in an *olla* [earthenware pot]. He fills large storage jars...cools it...it quakes, it quivers [i.e., it turns to jelly]' (Sahagún 1950-82, 4:74). With further boiling, the syrup became solid sugar: 'They also make small cakes of sugar from this maguey juice, but it is not so white nor so sweet as ours' (Motolinía 1950, 273).

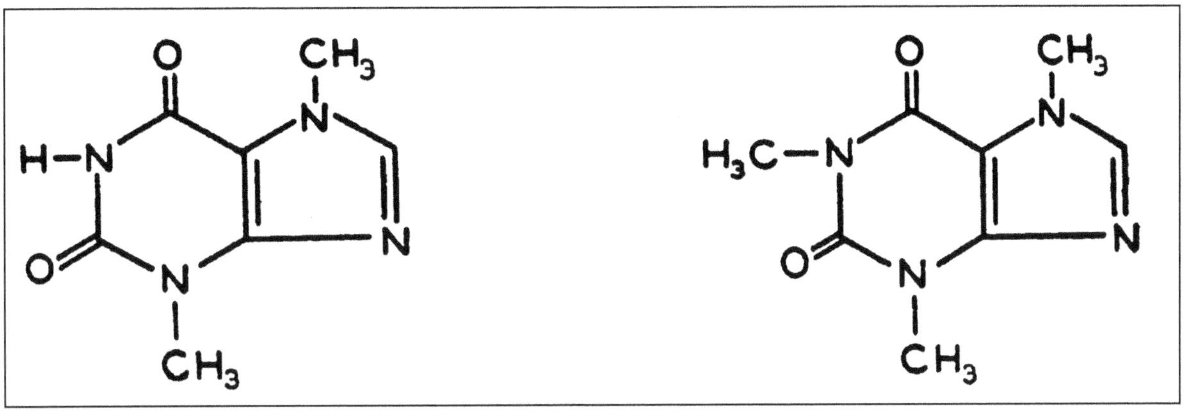

Theobromine *Caffeine or methyltheobromine*

A number of tropical American tribes had other plant sources for the stimulant caffeine than the introduced Coffea arabica *from Ethiopia via Arabia. For example, the guaraná seeds of* Paullinia Cupana, *a mid-Amazonian woody vine of the soapberry family, have 'the highest caffeine levels in the plant kingdom' (Muerer-Grimes et al. 1998, 293). Nowadays, these seeds are widely used in Brazil for preparing commercial soft drinks.*

Figure 16. Two Common Alkaloids

Sweetening from Manioc

On the island of Hispaniola, bread was made from 'bitter manioc,' after the starch had been extracted from it. The expressed juice (the decanted 'watery liquid' mentioned above) was alternately boiled and exposed to the sun in order that evaporation would concentrate its sugar content.

> The liquid which is extracted from the yuca is boiled several times and left in the open for several days. Then it *becomes sweet* and is used as honey or other syrup to mix with other foods. (Oviedo 1959b, 17)

Maple syrup

Indians of the eastern USA made incisions in maple trees, caught the sap, conveyed it to a receptacle made from reeds or curved bark, and then stone-boiled it. The sap was converted into sugar by letting it freeze, then removing the crystals of ice which are pure water. (This method of separating two liquid compounds is standard in the modern chemical laboratory. Whether the procedure was done in pre-Columbian times, I do not know.)

Chocolate Drinks

Chocolate, one of America's major contributions to the world's cuisine, is made from seeds of the cacao tree, *Theobroma cacao* (Sterculiaceae). The genus *Theobroma* has its origin in South America (Cuatrecasas 1964, 431), whence *Theobroma cacao* (apparently a hybrid) is thought to have been introduced by humans to Mesoamerica. The fruit of *T. cacao* is a ribbed pod—usually a little over a span long—with a thick and hard, fibrous shell. It contains a couple of dozen almond-shaped seeds ('beans')—enclosed in a sweet, mucilaginous pulp. The seeds of *Theobroma* species are 'rich in starch (15%), protein (15%), and oil (50%)' (Cuatrecasas 1964, 438). Those of *Theobroma cacao* also contain the mildly-stimulating alkaloid theobromine, as well as small amounts of the common and closely-related alkaloid, caffeine.

There are two main variety-groups of *Theobroma cacao*: *forastero* and *criollo*. Forastero seeds contain more polyphenols, including the glycoside anthocyanin; consequently, they tend to be purplish. Polyphenols 'make cacao beans taste astringent and bitter' (Gadsby 2002, 67). Criollo seeds are highly aromatic and more flavorful than those of forastero, but the trees have a lower yield and are more vulnerable to diseases. Since they contain little anthocyanin, the fruits and seeds are pallid. Probably in the lands subject to Aztec rule only criollo varieties were cultivated (Gadsby 2002, 67-68).

Although the processing of cacao seeds has been much studied in recent decades and is now highly mechanized, the basic procedures still follow the example set by 15th-century Mesoamericans. The fruit is harvested and cut open; the seeds and pulp are scooped out and sun-dried for a few days. Then, heaped and covered with leaves, they are fermented. During fermentation, a critical procedure in making a satisfactory product, the pulp surrounding the seeds is dissolved and drained away. Throughout the process, yeasts convert sugars in the pulp to alcohol; then bacteria convert the alcohol to organic acids, mainly acetic acid. 'Because of their low astringency, *criollo beans don't need a lot of fermentation to mellow them*' (Gadsby 2002, 67).

Next, the beans are toasted. This lessens their moisture content, the acetic acid is removed, and bitter phenolic compounds are further oxidized (Belitz and Gorsch 1987, 704). Fermentation and toasting change the color of the beans to a dark brown. Toasting also loosens the seed coats, so they are easier to remove.

After toasting, the seeds are ground into a paste which the Aztecs called chocolatl, the origin of our word chocolate. The word 'cacao' comes directly from *cacahuatl* (a Nahuatl word), although derived ultimately from the Maya language (Stuart 1988, 156).

A companion of Cortés described the making of an Aztec chocolate drink as follows:

> These seeds which are called [by the Spanish] almonds or cacao they pound and reduce to powder, and also grind other small seeds and put the powder *in certain jars with spouts*. Directly they add water and stir with a spoon, and *after it has been well beaten they pass it from one vase to another, which froths it, and this froth they collect in another jar kept for the purpose*. When they wish to drink it, they froth it with little spoons of gold or silver or wood, and it is then drunk, but *they have to open the mouth wide because it is froth*... (Narrative 1917, 41)

The foam or froth is formed by incorporating fine dispersions of gases in the beverage—for example, by beating the solution vigorously or by pouring it through the air from one container to another.

Although foaming-chocolate beverages were made from Mexico to Panama in the early 16th century, drinking them was apparently the exclusive right of caciques or other important people (Sahagún 1956, 3:156; Oviedo 1959a, 1:268). Recently, scrapings from the interior of Maya spouted vessels in Belize (dated between 600-400 B.C.) have been found to contain theobromine. Current research 'has indicated that cocoa is the only Mesoamerican plant containing *theobromine*' (Powis et al. 2002, 97; Hurst et al. 2002, 289). Accordingly, 'we have evidence that the Maya were consuming cacao as early as 600 B.C.' (Powis et al. 2002, 100). This is, at present, the oldest evidence of cacao use.

With regard to the chemical arts at the time of the Spanish conquest, three aspects of cacao preparation stand out: (1) the action of substances added to chocolate beverages; (2) the way that cacao beverages were made to foam; and (3) the way in which the cocoa butter was extracted and used.

(1) Although the chocolate beverage of the Aztecs was mainly a mixture of cacao paste with toasted maize flour and water, a variety of other plant materials were commonly added. These materials may have been added for a number of reasons: for flavoring, coloring, aroma, spice, stimulating effects, or to facilitate foaming.

The drink was sometimes flavored with vanilla, sweetened with honey (Sahagún 1956, 3:156), tinted red with achiote, and perfumed with aromatic substances—for instance, with sweet-smelling flowers (Roys 1931, 263). The beverage was often spiced with chili or mixed with petals of the sacred Aztec ear flower, *orejuela* (identified as *Cymbopetalum penduliflorum* of the custard-apple family, Annonaceae). 'When ground and added to *batido* they [the orejuela petals] impart a flavor strongly resembling that of *black pepper* with the addition of a *resinous bitterness*' (Popenoe 1919, 405). (*Batido*—meaning 'beaten'—is a chocolate drink made in Central America.)

In the State of Oaxaca, Mexico, the dried flowers of *Quararibea funebris* (*flor de cacao*) are used to spice chocolate drinks. They are also added as a scent, and to affect the drinks' consistency. Most parts of *Quararibea* trees have 'a peculiarly pungent, aromatic odor...The fragrance is strongest and most noticeable in the [pure white] flowers.' The odor 'is *extremely persistent,*' in preserved specimens, lasting for more than a century (Schultes 1957b, 250). Although the dried flowers are sold today in native markets, their use is 'not confined to the Indian population of Oaxaca' (Schultes 1957b, 258).

> As well as imparting to the beverage a peppery taste and aromatic odor, the flowers of *Quararibea funebris*, containing large quantities of *mucilage*, have a tendency to *thicken the water* in which the corn and cacao particles are suspended. (Schultes 1957b, 259)

In addition to spices and colorants, some chocolate drinks (especially those made by the Aztec and Maya) include plants that are physiologically active. A note on another plant added to cacao beverages comes from an account written in the early 16th century:

> There is a tree called *teonacaztli* [otherwise unidentified]; the flowers are very aromatic and beautiful; they have a strong fragrance and are of an *intense yellow color*; they are much used as a scent and *for drinking with ground cacao*—and if *drunk intemperately, it intoxicates.* (Sahagún 1956, 3:289)

(Obviously, the flowers were not those of *Quararibea*, since the latter are white.) The 'other small seeds,' mentioned by Cortés's companion, may have been added for the same purpose: intoxication. (Actually, large amounts of pure chocolate must be consumed to experience any stimulating effect.)

(2) Wild plants of the genus *Dioscorea* (the yams) were widely used in native America as medicines, and are now well-known for their saponin content.[18] The saponins,

[18] Species of *Dioscorea* were used in Indian Mexico as natural detergents, fish poisons, and contraceptives. Diosgenin, a saponin obtained from the wild yam, *cabeza de negro* (*Dioscorea mexicana*), is now used as starting material 'for a commercial synthesis of cortisone

foaming agents, form colloidal aqueous solutions that foam upon shaking. Rural people in Oaxaca sometimes add Dioscorea to their chocolate beverage:

> Occasionally, young shoots of a species of *Dioscorea* are added, and the beverage is beaten vigorously into a *thick albuminous froth*... (Schultes 1957b, 259)

Probably saponins were deliberately included when preparing chocolate drinks. A cacao paste of inferior quality (because mixed with too much maize flour and water) was sometimes sold in Aztec markets. In that adulterated condition, the paste could not generate foam without the use of some 'espumarajos' (Sahagún 1956, 3:156: 'y el cacao que no es bueno tiene mucha masa y mucha agua, y así no hace espuma sino *unos espumarajos*'). Such 'espumarajos' (froth-makers) were either chemical foaming agents or some sort of agitating implements—probably the former.

Apparently, there is some confusion about methods used in frothing chocolate beverages at the time of the Conquest. For instance, '*Spanish documents do not mention the method of pouring liquid chocolate from one vessel to another* to produce the coveted foam on the drink...' (Powis et al. 2002, 96). On the contrary, some of the earliest Spanish documents *do mention* pouring chocolate beverages from one vessel to another—for example, the companion of Cortés, quoted above, and Bernardino de Sahagún (1956, 3:156). As Oviedo states,

> they put the liquid chocolate in a bowl. Then, they put another container empty on the ground (or hold it in one hand and the bowl containing the chocolate in the other) and pour the stream of chocolate, from a height of more-or-less two palms, through the air from one container to the other; and so the foam rises to the top (Oviedo 1959a, 1:270).

As indicated above, trees of the genus *Quararibea* are closely associated with the making of chocolate beverages. Known in English-speaking areas as 'swizzle-stick trees,' their young shoots are used as 'frothing sticks or *molinillos* with which chocolate drinks are beaten' (Schultes 1957b, 263). In Panama, the new branches of *Quararibea turbinata* 'develop in whorls of five, regularly spaced. This characteristic is useful in making *molinillos*, the device made of slender twigs used in preparing chocolate' (Pittier 1908, 115). Similarly, there 'is a tree which they [the sixteenth-century Maya] *raise in their wells*...it is marvelous how it stretches its branches which grow on its trunk in a very regular way, for they grow in threes or more in clusters' (Landa 1941, 197); the tree has been identified as *Quararibea Fieldii* (Landa 1941, 197). Since the *Quararibea* was cultivated by the Maya, it was apparently needed on a regular basis.

It has been claimed that 'use of the *molinillo* was probably introduced by the Europeans' (Powis et al. 2002, 96). This, however, is by no means certain, since these sticks are also mentioned in the earliest Spanish sources. Sahagún states that at one of their banquets, Aztec merchants supplied much cacao, as well as large quantities of paddles or sticks for agitating cacao: 'se proveían de cacao, veinte cargas, o así; también se proveían de *las paletas o palos con que se revolvía el cacao*, hasta dos mil o cuatro mil...y de vasos para beber...' (Sahagún 1956, 3:45); some of the paddles were made of 'tortoise-shell' (Sahagún 1956, 3:35). Cortés's companion, cited above, concurs: 'they froth it with little spoons of gold or silver or *wood.*' On the other hand, knowledge of making *rotary molinillos*, some elaborately decorated, is a result of Spanish influence (Seler-Sachs 1909, 381; see Fig. 3 in Seler-Sachs).

A third method of making foam in chocolate beverages has been suggested: the use of spouted vessels. Though the details are not clearly understood, frothing may 'have been accomplished by introducing air through the spout' (Hurst et al. 2002, 289). As noted above, Cortés's companion said that the sixteenth-century Aztec used 'certain jars with spouts.'

The use of saponin-containing *Quararibea*-twigs for frothing sticks, as well as the wood of guayacan trees (*Guaiacum sanctum* or *G. officinale*) for cacao containers, is of chemical interest. The woods of both *Quararibea* and *Guaiacum* contain resins and essential oils whose odor is *exuded for years*, especially when the woods are heated. These compounds, which are insoluble in water, would readily dissolve in the warmed fat of chocolate Guaiacan trees, which also contain abundant saponins, and have a heavy 'light-yellow resinous wood.' The Maya made '*many medicines of it and even vessels and curious cups*'; a 'narrow jar *used in preparing chocolate* is carved from the wood' (Roys 1931, 311-312, quoting from an early manuscript). Later, fine shavings of guaiacan wood (lignum-vitae—i.e., wood-of-life) had a great reputation in European medicine. In short, the Maya and other Mesoamerican peoples may have used swizzle-stick trees for their chemical ability to promote frothing, and *Guaiacum* trees for putative medicinal qualities in their resins and gums.

(3) The Maya seem to have made their foaming beverage somewhat differently from the Aztec. Although the Maya also made—and still make—a chocolate drink which included cornmeal (Popenoe 1919, 408), they made

---

[perhaps the best known of the steroids] and sex hormones' (Solomons 1996, 1114).

The many hundreds of plants and animals used medicinally by the American Indians are dealt with here only incidentally, since their chemical preparation is much the same as with other substances in this discussion. They are the subject of a copious literature (for example, on North America written by V. J. Vogel, 1970, and on South America by E. H. Ackerknecht, 1963). Although commercial drug companies have found it profitable to canvass the folk remedies of native American tribes, unfortunately they (unlike the sixteenth-century physician, Francisco Hernández) have paid little attention to the non-medicinal practices of aboriginal chemistry.

another beverage of which the principal constituent was cocoa (cacao) butter: 'they get from the *cacao a grease which resembles butter*, and from this and maize they make another beverage which is very savory...' (Landa 1941, 90). Cocoa butter, the preparation of which required much time and patience, was extremely valuable, both as food and medicine.

Although cocoa (cacao) butter is solid at room temperatures, it melts at between 30 and 36°C, a little below the normal temperature of the human body. This light-yellow fat has unusual qualities; for instance, when it 'melts in the mouth, a pleasant, cooling sensation is experienced...' (Belitz and Grosch 1987, 478). Although most vegetable oils oxidize and turn rancid in a short time, cocoa butter belongs to a class of fats that show resistance 'to autoxidation and microbiological deterioration' (Belitz and Grosch 1987, 478).

To the south of the Maya, toasted maize seems not to have been mixed with cacao drinks. At least Oviedo's detailed account, describing custom in the 1520s between Nicaragua and the Chiriquí area in western Panama, does not mention cornmeal nor any other additive except a red colorant.

To prepare the cacao beans for drinking, the Chorotegans, in the Nicoya area of Costa Rica, first toasted the beans and removed their husks, then ground them intensively, sometimes adding a little achiote. Moreover, they ground the cacao beans 'on *a very clean stone*' (Oviedo 1959a, 1:270), no doubt to prevent the cacao fat from absorbing flavors of other foods.

The intensive grinding probably had the effect of reducing the cacao beans to extremely small (almost colloidal-size) particles which would remain suspended in the water that was added. In addition, grinding the mixture of cacao beans and achiote to a fine powder would cause the bixin in the achiote to dissolve in the beans' fat.

Apparently, south of the Maya area the butter and the rest of the cacao material were separated into two different drinks. The following is an early sixteenth-century account of the way in which the fat was extracted (a method of rendering more complicated than that described earlier when discussing the oils and fats used in frying). As translated from Oviedo's Spanish, 'fist-size lumps' were made of ground cacao paste. The lumps were then placed in a pot of water and cooked

> until the liquid is thick; then the pot is taken off the fire and cooled until it is lukewarm or a little warmer. Then, with a scallop shell or a spoon, a quantity (some five or six spoonsful) of the cooked mass is placed in a big pot holding some three liters of water; and over the paste or gruel of the big pot of water—*after the oil rises to the top—is placed a circular fabric of 'palm'* (something like the hotpads that in Flanders are placed under hot plates and silverware so they won't burn the table). *Then an Indian woman with well-washed hands, places the palm fabric against the risen oil, absorbing it, and wrings the fabric into a jar* or vase in which they want to keep this oil or precious liquid, which after five or six hours *cools and hardens*, and if colored with *bija* it is red, or *if not it is a golden color*...When the most important Indians drink this cooked cacao, it is little by little...The cacique takes three or four draughts which puts this grease all over his lips and chin...where it glistens like butter. (Oviedo 1959a, 1:270-271)[19]

This warm, viscous fluid was difficult to drink without some being spattered on the cheeks, chin, and nose—whence it was 'scraped off with a finger and eaten' (Oviedo 1959a, 1:270). That it glistened 'like butter,' and was of a 'golden color,' further indicates that the preparation was almost pure cocoa butter and contained no toasted cornmeal.A different method of extracting the fat was practiced to the south in Chiriquí, Panama, as described by Oviedo in the following passage:

> After the cacao beans are toasted and stripped of their thin husks, they are ground two or three times [into a liquid paste] *without adding a drop of water, using only their natural moisture*. And while they are being ground, a pot containing two liters, more or less, of clean water is put over a gentle and low fire. After the pot is brought to a slow boil, the cacao paste (being ground as described) is slowly poured in. Then, the mixture is *stirred with a little cane or a thin stick*, lifting the cane or stick several times to see that the material is well-cooked. It is finished cooking when nothing is fastened to the cane, and the latter comes out clear and clean—only covered with liquid, not with the cooked material. When this is the case, *the middle or bottom of the pot is tapped gently with the stick in order to get the oil to come to the surface. With a spoon, it is carefully skimmed off*, little by little, *being careful that no cacao is gathered with the oil* because the oil is the 'flower and the principal virtue', the remaining cacao being of much less value. Thus what is gathered with the spoon is placed apart. After taking as much oil as possible, it is put in a different pot away from the fire. The leftover [cacao] mixture is put in a separate pot with clean water. When stirred, the rest of the oil rises to the top...If one wants *to get out any oil that is left, a feather is passed carefully over the surface of the liquid and the oil it absorbs is shaken out*. However, the oil does not come out as clean as in the first method. The mixture of water and cacao that remains, depleted of its oil, is drunk for it is very healthful. (Oviedo 1959a, 1:272)[20]

---

[19] Since the quotation in the text may be too freely translated from Oviedo's old-fashioned Spanish, the passage is printed below in its original version:

> E después que está muy bien molido en una piedra de moler, pasado e remolido cuatro o cinco veces, echándole un poco de agua al moler, hácese una pasta espesa, e aquella masa guárdase fecha un bollo. E cuando lo quieren beber, ha de haber pasado, después que se molió, cuatro o cinco horas, a lo menos, para estar bueno, e mejor desde la mañana a la noche, e mejor está para otro día; e así se tiene cinco o seis días e más. (Oviedo 1959a, 1:269)

[20] Oviedo's original is quoted below:

> Así que, tornando al propósito, tostadas las almendras, móndanlas de aqella cáscara delgada, e muélenlas dos o tres veces sin gota de agua alguna, antes, de su propria humedad, está asaz líquida la pasta. E en tanto que se muele, ponen a un fuego dulce y lento una ollica que quepa una azumbre de agua, poco más o menos, e hinchen de buena agua limpia la olla hasta las dos

Food Preservation

How best to preserve foodstuffs is a universal subsistence problem. This is especially true of staple foods that are only abundant at certain seasons. The growth of food-spoiling microorganisms is inhibited by smoke, lack of water, low temperatures, and salt.

Smoking, Drying, and Pounding

Smoking, drying and pounding—separately, or in some combination—are probably the oldest methods of preserving animal flesh. Smoke, the composition of which depends much on the kind of wood burned, generally contains volatile phenolic compounds such as the cresols. Their source, by thermal degradation, is the polyphenol lignin, abundant in woody material. Such compounds have some bactericidal action. Thus, smoking not only dries meat, but changes its flavor, in what amounts to a 'phenol enrichment process' (Belitz and Grosch 1987, 278 and 279). In America, as in many parts of the world, fish were (and are) eaten with little other preparation than smoking. On the other hand, wood smoke 'has been shown to include over 200 different compounds...Of these the *most important antimicrobial* [and thus preservative] compound is almost certainly [the gas] *formaldehyde*' (Coultate 2002, 251), the simplest of all aldehydes.

Both meat and fruits are preserved by drying, often by application of heat. In the tropics of eastern Brazil, some four and a half centuries ago, the Tupinamba were described as preserving meat by drying and making meal of it:

> *they roast the flesh or fish above the fire in the smoke*, and they allow it *to become quite dry*; then they pull it to pieces, dry it once again on the fire in pots...Thereupon *they pound it small in a wooden mortar*, and they *pass it through a sieve, reducing* it in such manner to *powder*. This lasts a long time; for they have not the custom of salting fish and meal. Such meal they eat with the root-meal [manioc-meal]... (Stade 1874, 132)

partes; e después que ha hervido un poco, despacio, echan el cacao en ella (que está molido como es dicho), e con una caña delgada o un palito muy limpio, menéanlo alrededor, hasta tanto que, levantando el palillo o caña una e dos e más veces, se ve questá cocido después que ha hervido bien. E vése que está cocido en que en el palillo o caña no queda nada pegado del cacao, que sale limpio, e todo está líquido e cocido, e corre como agua. Fecho aquesto, dan con la caña, en medio de la masa u olla, para abajo, golpes pasico, como para que se abra; e por allí sale arriba luego el aceite, e con una cuchareta, sotilmente, cógese poco a poco, guardando que no coja el cacao con el aceite, porque el aceite es la flor e virtud principal, e lo que ya queda del cacao es acesorio e de menos valor. E así, aquello que se coge con la cuchara, se pone aparte. Después que desta forma que he dicho, se ha sacado lo más que ha seido posible, lanzan en una higüera, que está aparte, fuera del fuego, con agua limpia, el dicho cacao, después de sacado dél el aceite, la mitad o el tercio o cuarta parte del cacao, e en otra e otras higüeras, lo demás. E revuélvenlo, e luego se sube sobre el agua el aceite que quedó, que no se pudo sacar con la cuchara, e aquello debido, así fecho aquel caldo, es excelente e sanísimo. E si quieren sacar aquel aceite que, como dicho es, había quedado, toman una pluma sotilmente; e a de suso, cógenlo lo mejor que pueden; porque luego se pega a la pluma, andando sobre aguado, e sacuden la pluma donde lo quieren recoger, e se despide della el aceite, e vuelven por lo demás. Pero esto no sale tan limpio del agua e del cacao como lo que primero se dijo; e el agua e cacao que queda, sacado el aceite, bébese e es muy sanísimo. (Oviedo 1959a, 1:272)

Much the same process was known from the Orinoco to the Paraná Delta. In the early 19th century, the process of making fish flour was described along the upper Orinoco River. The Indians there

> ...*fry fish, dry them in the sun, and reduce them to powder without separating the bones*. I have seen masses of fifty or sixty pounds of this flour, which resembles that of cassava. When it is wanted for eating, it is mixed with water, and reduced to a paste. (Humboldt and Bonpland 1972, 5:547)

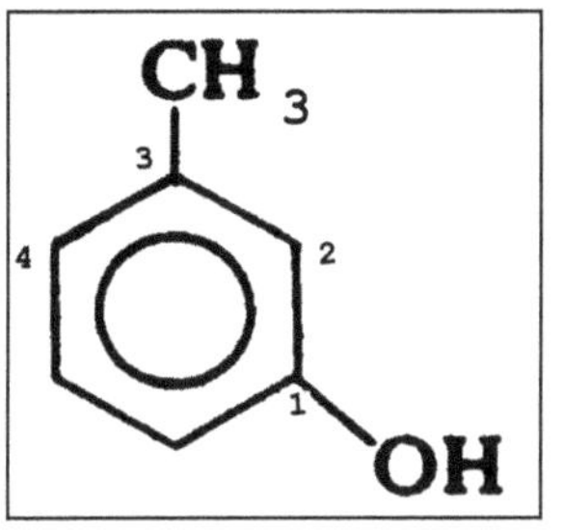

The common name for methylphenols is 'cresols,' a term taken from creosote. The small numbers shown on the benzene ring indicate where methyl groups may be attached.

Figure 17. One of the Cresols (3-methylphenol)

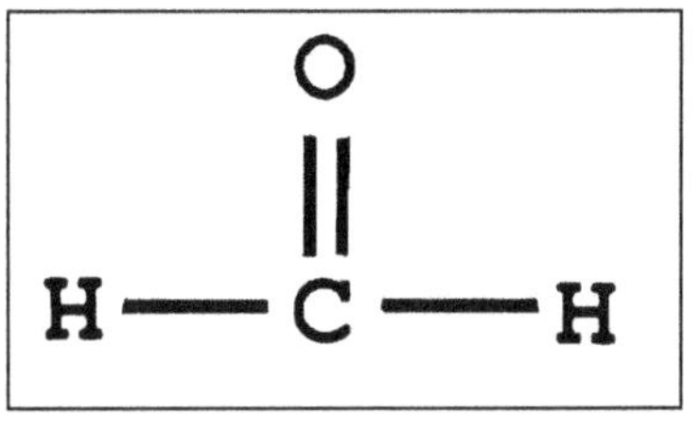

Figure 18. 'Formaldehyde, commonly used in aqueous solution. (Vanillin, figured earlier, is a more complex aldehyde.)

Other Indian tribes, living near the Paraná Delta, also ground fish into a long-keeping, edible powder (Lothrop 1963a, 182).

For many North American Indians, *pemmican* (a Cree word) was a characteristic food. After being dried by the sun or wind, lean meat was pounded into a paste. (Such pounding not only breaks down the fibrous material, but expresses some of the remaining water.) The paste was then mixed with hot fat or marrow grease and dried fruit, pressed into small cakes, and carried in rawhide containers.

Among the Plains Indians, after stone-boiling water and bits of bone, grease was rendered for making pemmican:

> The process began by smashing bones into small pieces with a mortar. Stones were heated to a high temperature and then transferred to a hide- or clay-lined pit containing

> water and the bone fragments. The grease was then skimmed from the surface and stored in skins or gourds. (Sassaman 1995, 227)

Saturated fats such as tallow (the almost tasteless solid fat of ruminant animals), marrow, and bone-grease are quite stable at room or colder temperatures. (In fact, as in pemmican, saturated fats may act as preservatives for other food.) In northwestern North America, tribes of the Columbia River Basin dried salmon, pulverized it, and then mixed it with melted tallow and berries. Great quantities of this mixture, transported in rawhide bags sealed with hard fat, were traded in the area. With care 'such pemmican would keep for years, although most of it was consumed within a single year' (Driver and Massey 1957, 245).

Salting

Common salt, NaCl, was likely the first antimicrobial chemical ever used. Salting (except for salt-pickling) is also a drying process, since salt draws water out of food material.

Along the Venezuelan coast opposite Margarita Island, the Indians caught and gutted immense quantities of fish at the beginning of the 16th century. With the supplies of salt

> ...already collected on the shore, they [the native inhabitants] salt the fish. Thus prepared, they are exposed on the sandy beach. The sun's rays are of such intensity that the fish are dried within one day...When dried, the fish are loaded on ships and covered again with salt...All of the surrounding regions are supplied with these fish... (Mártir 1964, 2:632)

The Maya, too, preserved fish by salting (Landa 1941, 190).

Pickling in salt solution was not so common as in the Old World, though practiced in a few areas. Granular salt, being strongly hygroscopic, has a drying effect and where the local climate has a rainless season, salting is often combined with drying. For example, an early sixteenth-century account of the Sinú (Zenú) country, near the north coast of Colombia, describes a trade in dried and salted animal products (Mártir 1964, 1:335).

Salt is added to foods to enhance their flavors, but many American tribes did not use much salt in their food: of 'all the people of the Globe the natives of America are those who consume the least salt' (Humboldt and Bonpland 1972, 2:248). (Although needed for the normal functioning of the human body, the minimum requirements are usually supplied by unsalted food. Aside from the need to preserve food, adding salt seems to be mainly a matter of culinary tradition.) The Indians of Peru

> consume very little salt in comparison with us [i.e., the Spanish]...; the dried flesh which they make and the fish which they dry—in order to preserve and carry them from one place to another—are without a grain of salt, and these they make in the following manner: in the hot, low country, if the meat or fish they have to preserve is for a short time, they use the barbacoa; but those of Peru, whether for a short or long time, *rinse the meat and fish with salt and put them to the sun to dry* on the coast, and to the *frost in the cold sierras.* (Cobo 1956, 1:113)

Alternate Freezing, Thawing, and Drying

Long before the 16th century a freeze-dried food, chuño (chuñu), was made in Andean South America. Chuño is made by freezing and drying potato tubers. At that time, the process was either unique or highly unusual.

In the high Andes, especially where the climate is too cold for maize, potatoes are the staple crop. 'Bitter potatoes' (several species of tuberous *Solanum* with high glycoalkaloid content) are raised there because of their resistance to frost (Werge 1979, 229-230; also Salaman 1949, 39). Frost-resistant species are grown up to altitudes of 15,000 feet (4572 m), some two thousand feet above where maize can mature.

In the preparation of chuño, the Indian cultivator manipulates the physical environment in order to produce chemical change in his food—even 'making use of his worst enemy, the frost' (Salaman 1949, 39). Potato tubers are put in the sun during the day and left to freeze at night. (Diurnal ranges in temperature are great at these altitudes.)

Father Cobo described the process as practiced by the Aymara and Quechua Indians of Peru about a century after the Spanish arrived:

> The tubers are spread out on the ground for a period of twelve to fifteen days during which they are exposed to the sun in the days, and to frost in the nights. At the end of this time, they are somewhat shriveled, but still quite watery. To expel the residual water [and to slough off the skins], they are trampled thoroughly, then exposed to the sun and frost again for another fifteen or twenty days after which they are as dry and light as a cork, very dense, hardened and so shrunk that of four or five fanegas of fresh potatoes there is left no more than one of *chuñu*...So durable is this *chuñu* that it will last for years without spoiling...[and] in the provinces of Collao the Indians eat no other kind of bread. For the caciques and other important people a more refined and highly esteemed *chuñu* is made in the following manner: after being dried in the sun and frost, the tubers are kept for two months in water, then exposed once more to the sun to dry, after which they remain chalky white throughout. This prized *chuñu* is called moray... (Translated from Cobo 1956, 1:168)

Reports written more recently (Salaman 1949; Werge 1979; Sillar 2000) agree substantially with this early account, although making *chuño* from bitter potatoes requires an additional step: detoxification. Tubers of most potato species cannot be consumed fresh, because of their glycoalkaloid content. (A glycoalkaloid is an alkaloid bonded to a sugar molecule. Such compounds are characteristic of the genus *Solanum*. Even the common potato contains glycoalkaloids, though not in harmful amounts.) After being trampled, the tubers 'are gathered up and taken to a stream to be soaked in cold, running water. The purpose of soaking is to reduce bitterness in taste by leaching out the glycoalkaloids' (Werge 1979, 231).

Chuño, being dehydrated, is light and thus easily transportable. Since its remains have been found in archaeological sites of lowland and coastal Peru, the material was apparently traded widely in pre-Columbian times. Now, however, 'chuño has little acceptance outside of the Andes,' where it is 'identified as a peasant and Indian food' (Werge 1979, 233-234). On the other hand, in montane Indian communities, freeze-dried potatoes 'form a major part of highland diet...it is a highly valued resource that fetches high prices, and is sometimes taken down to the valleys where it is exchanged for maize and other products' (Sillar 2000, 57). Chuño, which is mainly starch, 'can be stored for about two years prior to eating' (Sillar 2000, 57).

The word 'jerky' is derived from the Quechua word, *charqui*. Like *chuño*, traditional *charqui* is made by a freeze-drying process and a roughly similar procedure. Excess water is driven off llama and alpaca meat by freezing, drying, and by pounding (rather than by trampling):

> They ate more dried meat than fresh, and they prepared the dried meat *without salt* in the following way. They cut the meat in wide, thin slices. Then *they put these slices on ice to cure*. Once the slices were dried out, *they pounded them between two stones* to make them thinner. They call this dried meat *charqui*. (Cobo 1990, 198)

Jerky is still made 'among camelid herders of southern Peru' and constitutes there an important trade item (Miller and Burger 1995, 440).[21]

Adding Organic Preservatives

Putrefaction of animal and plant protein by bacteria produces a deadly ptomaine, a special problem in the preservation of foods. And so, in the humid tropics, another method of preserving food materials is to add organic preservatives to the cooking food and to boil it continuously. The active principle of peppers in the genus *Capsicum* (namely, capsaicin) seems to have some anti-bacterial action, and thus some preservative qualities. In any case, a mixture of vegetables and meat is kept from spoiling by being cooked in the well-known tropical American 'pepperpot,' to which generous amounts of strong chili peppers are added. Servings of food are taken from the pot from time to time, and the pot is rarely emptied completely: pepperpots 'are reportedly kept going for many years by adding additional meat and vegetables to the sauce as necessary and bringing it to the boil each day' (Lancaster et al. 1982, 24).

The Witoto pepperpot, in the northwest Amazon, is an example:

> Each family has a *big pot that simmers constantly* over the special fires. Into this go all things, and it is replenished daily from the proceeds of the kill...An Indian will help himself from the hot-pot at any time the fancy may seize him, or, for that matter, from any hot-pot at any time, so long as the owner thereof is present. (Whiffen 1915, 136)

An additional constituent of pepperpots is the liquid decanted when bitter manioc is pressed. For example, Nordenskiöld refers to 'the Indian invention of utilizing the juice squeezed out of the manioc, *for preserving meat against putrefaction in a tropical climate*' (Nordenskiöld 1931a, 492).

In 1854-55, the naturalist Henry W. Bates noted that on the Japurá and Tapajos Rivers various sauces were made, either to be eaten with fish or in which fish meat itself was preserved. One such sauce is

> in the form of a yellow paste...called Arubé, which is made of the poisonous juice of the mandioca root, boiled down before the starch or tapioca is precipitated, and seasoned with capsicum peppers...Tucupí, *another sauce made also from mandioca juice*,...is made by boiling or heating the pure liquid, after the tapioca has been separated,...and seasoning it with peppers and small fishes...It is generally made as a liquid...I have seen the Indians on the Tapajos, where fish is scarce, *season Tucupí with Saüba ants* [leafcutter ants]... (Bates 1962, 188)

Ants, mentioned in the foregoing passage, contain formic acid, the simplest of all organic acids. Formic acid from ants, oxalic acid from *Oxalis* (and perhaps other plant acids), and acetic acid from vinegar (abundant in those areas where alcoholic beverages were made) were among the few acids available to pre-Columbian Americans. (The mineral acids were probably unknown to them, nor were they known in Europe until the mid-seventeenth century, with the possible exception of sulfuric acid.)

[21] The Inca stored freeze-dried, and other, foods:

> Long-term storage alleviated the effects of drought and other natural and man-made calamities. Along the royal highways, the state built enormous warehouse complexes (2400 on one hillside near Cochabamba and over 1000 above Xauxa) which can still be studied. (Murra 1984, 122)

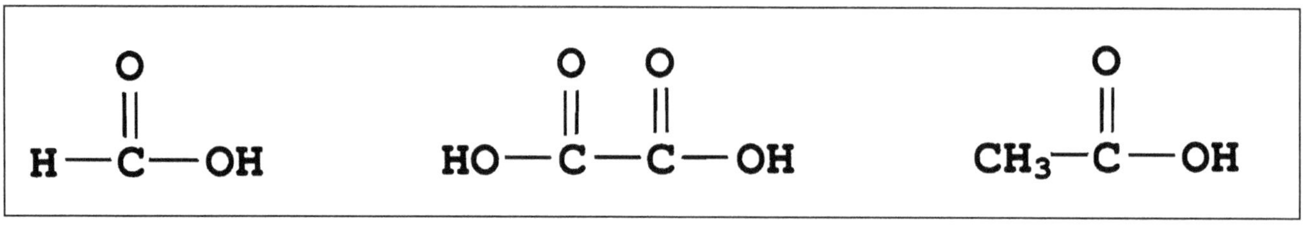

Formic acid (left), oxalic acid (center), and acetic acid (right).

Figure 19. Common Organic Acids

Hermetic Sealing

In western Arizona, archaeologists have found a complete example of hermetic sealing: an air-tight jar containing food (Euler and Jones 1956). The vessel is prehistoric, being dated at around 1300 A.D. Closed with a lid 'ground from the bottom of another pottery vessel' (89), it was 'tightly sealed with a resinous material' (87); the 'bond was tight enough to shut out air, and this would protect the food against bacteria, yeast, and molds' (90). The sealing material was identified as a 'lac' secreted by the wild scale insect, *Tachardiella larreae*, which grows on creosote bushes (89). The resinous lac, 'on heating becomes soft and viscous, but on cooling again becomes hard and quite tenacious' (89). This lac 'is closely akin to commercial sealing wax' (96) made from secretions of the anciently-domesticated scale insect of southeast Asia (the 'shellac' insect). 'Apparently the Southwestern Indians can be credited with an independent invention of sealing wax from lac' (89).

The jar contained numerous dark-brown, fibrous slabs of mescal, a cooked food prepared from agave plants, covered with dried syrup. Among tribes in the Southwest, there are various historic references to air-tight, sealed vessels containing perishable food (Euler and Jones 1956). In America, this technique may have been discovered about the same time as in Europe, or even earlier, but the underlying chemical 'theory was not understood until the work of Pasteur and others after 1850' (97).

Fermentation

The organic reactions involved in the chemical arts take place mainly in aqueous solutions, and extreme aridity tends to slow such reactions. Moreover, an elevation of temperature increases the speed and vigor of chemical change. (An increase in temperature by 10°C will double the rate of most chemical reactions.) In short, food preservation is easiest in cold and dry climates, and most difficult in the humid tropics:

> ...millions of people live in areas where the climate is *warm or tropical accompanied by high humidity...* [Moreover] some items of food may occur in abundance at only certain times of the year...The answer has often been to *ferment food* in such a manner that it may be preserved for months. (Hesseltine 1965, 182)

Fermentation has long been a means of preserving foodstuff and beverages in the American tropics, and even in some temperate areas. The following are among the chemical advantages gained by subjecting foods to fermentation: (1) In some foods, fungi synthesize enzymes (proteins which act as catalysts) which are able to hydrolyze protein and starch, thus increasing their digestibility. (2) The 'increase in niacin and riboflavin [vitamin B_2] content of *cereal grains* and legumes by fermentation has a great nutritional significance in those countries where deficiencies of *these vitamins* occur in the diet' (Wang and Hesseltine 1981, 82). (3) Fermentation usually adds markedly to the flavor of food. (4) A further advantage of fermentation is that some raw foods after they are fermented require shorter cooking times. Both food and beverages were often fermented in buried (or partly buried) earthenware containers.

To be sure, few legume seeds were fermented in pre-Columbian America, but one kind was used in making *chiga* bread. Some two hundred years ago this procedure was described among the Otomac and Maypures Indians living near the Orinoco River. Alexander von Humboldt speaks of their custom of burying the ground beans of a leguminous tree so that they would ferment, and then making bread of them (Humboldt and Bonpland 1972, 5:662):

> The seeds are taken out of the pods and *buried for some time in damp soil*; then after a certain degree of *fermentation* has set in they are taken out, washed, and pounded. The flour or starch is used...even today for bread-making... (Ernst 1889, 72)

The tree from which chiga bread is made is now classified as *Campsiandra comosa*.

As discussed earlier, in much of the Mesoamerican area Indian women make hominy (*nixtamal*) by soaking maize kernels in hot limewater. In places, however, rather than using the hominy to make tortillas, it is often ground into a coarse dough (*masa*), then shaped into balls of various shapes and sizes, wrapped in wild banana leaves, and allowed to ferment 'for 1 to 14 days or more depending upon the taste of the consumers and the prevailing circumstances' (Steinkraus 1983, 227; Miguel Ulloa and his co-workers are cited for most of Steinkraus's information on this subject). The purpose of wrapping the dough in leaves is to keep it moist during storage or transport, thus encouraging fermentation. Mixed with cold water, the moldy balls are consumed as a thick beverage. This preparation is known as pozol or posol. Although known widely by these variations of its original Nahua name, pozol is most commonly made well to the southeast of Mexico City. Pozol is a basic food not only among many native peoples of southeastern Mexico and parts of Central America, but also among the rural mestizo people of that area.[22]

The following is a comment on the pozol made by 16th century Maya:

> ...they give to workmen and travellers and sailors large balls and loads of the half-ground maize, and *this lasts for several months merely becoming sour.* And of that they take a lump which they mix in a vase made of the fruit, which grows on a tree [i.e., the tree gourd]...And they drink this nutriment... (Landa 1941, 90)

A note on the Maya from the beginning of the 20th century reads much the same:

> '*Posol* is carried by Indians traveling on foot;...at the brookside a jicara of water is dipped, and a handful of posol squeezed up in it...[It] *has a slightly sour taste, as if beginning to ferment*' (Starr 1900-1902, 2:72). *Pozole* is equally important to modern Maya. (Redfield and Villa 1962, 39)

Dependent, as it is, on air-borne micro-organisms (bacteria, yeasts and molds), the fermentation is spontaneous. The flavor and aroma of pozole seem to depend somewhat on which combination of these organisms happens to inoculate the dough, as well as on the fermentation time:

> During the first stages of fermentation, bacteria outnumber the yeasts and molds and are probably responsible for most of the acid production during the first 24 hours. (Steinkraus 1983, 228-229)

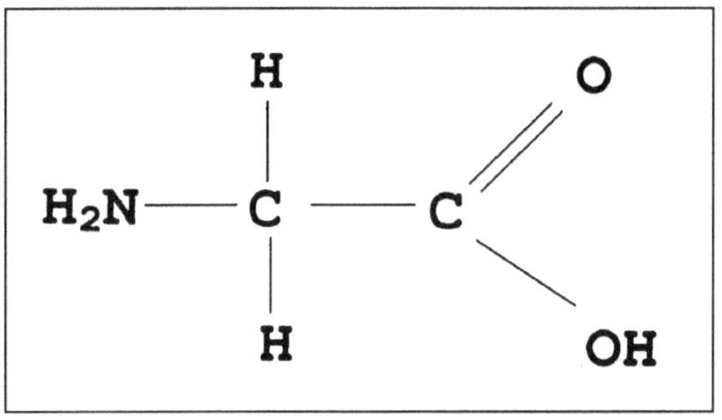

Glycine (aminoacetic acid)

*Amines are derivatives of ammonia (NH_3), and formed by the replacement of hydrogen with one or more organic groups. Like ammonia, they are weak bases. Amines are classified as primary, secondary, or tertiary according to the number of organic groups attached to the nitrogen atom. For example, **—NH_2** is a primary amine because only one organic group is attached to nitrogen; **—NH—** is a secondary amine; and **—N—** is a tertiary amine. Amines end in the letters 'ine'; for instance, all alkaloids (nicotine, caffeine, etc.) are amines. Amino acids are called amphoteric molecules, since they contain groups of opposed basic and acidic character, usually -NH_2 (an amino group) and –COOH (the carboxyl group). Amino acids are the building blocks of proteins, and lysine is one of the amino acids. The simplest amino acid is glycine. Like lysine, glycine is a component of food protein.*

Figure 20. Amines and Amino Acids

As a staple food product, pozol has certain advantages over tortillas: no further 'cooking [of the *nixtamal*] is necessary with pozol. In addition, the nutritional value of pozol due to the fermentation is higher than that of tortillas' (Steinkraus 1983, 233).

Several of the bacteria species found in pozol fix atmospheric nitrogen; these bacteria belong principally to the genera *Agrobacterium* and *Aerobacter* (Steinkraus 1983, 230). Thus, as a result of fermentation—as well as its original alkali treatment—pozol has a higher content of such nitrogen-containing nutrients as protein, niacin, riboflavin, and lysine than the maize seeds from which it was made (Steinkraus 1983, 232). (Lysine, an essential amino acid, is in short supply in maize protein.)

Nowadays, the Guaymí on the Atlantic slope of western Panama preserve the fruits of peach-palm by burying them in pit silos to ferment. A hole is dug about one to two meters in cross-section and a meter deep. The bottom and sides are carefully lined with wild banana leaves. The ripe, uncooked fruit is put in the hole, then covered with more leaves and dirt. When food is scarce, the fermented fruit is dug up, cooked and eaten—as much as a year later. Fermentation changes the taste of the fruit somewhat: it now has a sour-sweet flavor. Such stored

[22] It is said that various tribes in the American tropics wrapped pieces of meat from game animals in papaya leaves in order to tenderize it. The papaya plant contains papain and other proteolytic enzymes—that is, the enzymes which break down proteins into simpler compounds by hydrolyzing them, a process similar to that of digestion. Nowadays, papaya enzymes are used as commercial meat tenderizers, and today meat is wrapped in papaya leaves in Brazil (Pio Corrêia 1974, 5:60). It is sometimes said that this knowledge is derived from aboriginal tropical America, but the only documentation I find for this statement is in a secondary source: 'An early reference to the use of papaya in Brazil was that of the Dutch traveler Linschoten, who in 1598 described...*the use of leaves to wrap meat as a tenderizer*' (Prance 1984, 92, citing Pio Corrêia, vol. 4, 1974; however, I did not find this statement when glancing through Pio Corrêia's volumes). There is a further note on papaya fruit among the 16th century Maya: 'The Indians pick it green *and ripen it in ashes...*' (Landa 1941, 199), but I do not know the chemical benefit of such treatment.

fruit is often used, too, in making a fermented beverage. Pits, dug under houses, keep the fruit especially well (Gordon 1982, 115). A similar method of preserving the fruits of the peach-palm was used by Indians in Honduras and Nicaragua (Conzemius 1932, 91).[23]

In addition to preservation, fermentation sometimes serves to detoxify food. In the upper Amazon during the 16th and 17th centuries, during the season of floods, the Omagua preserved the tubers of bitter manioc (*yuca*), which spoil quickly after harvest, in much the same way:

> In order that there should be no lack of food during the season of the high-flood...they make a practice of harvesting the fruit of their new plantations in January and February. The maize that they have garnered they keep in their houses; the yuca and mandive [a kind of plantain?] they bury in pits well-provided with wide hatches, and so they keep them *underground below the water, not only for months but for two years or more*. From thence after the level of the River begins to fall they raw out all that is needed for their consumption, leaving the rest buried, and although this yuca and mandive *may rot, when pressed it becomes better* and of greater sustenance *than when fresh*, and from it they make their drinks, flour, and cassava-bread. (Fritz 1922, 50)

Well to the south, in the southern Mato Grosso, the Nambiquara treat bitter manioc similarly. Sun-dried balls of the pulp are 'buried at marked places. In times of scarcity, perhaps months later, the *half-rotten balls* are made into flat cakes, hastily cooked in hot ashes, and eaten' (Lévi-Strauss 1963a, 3:363). The following account describes manioc preservation as practiced by the Indians of eastern Brazil some four and a half centuries ago. In this case, the food was preserved by a combination of fermentation, smoking, drying, and pounding:

> They also take the roots fresh, and *put them in water, leaving them to rot* [ferment] *therein*; *then they remove them and place them above the fire in the smoke* allowing them to dry. The root thus *dried*...keeps a long time. When they want to use it, they *pound it in a mortar* made of wood, when it becomes white like corn-flour, therefrom they make cakes...They also take *rotten* [fermented] mandioka before they dry it, and mix it with the dry and the green [manioc meal]. From this they prepare and dry out a meal, which lasts fully a year, and which can be eaten as it is. (Stade 1874, 132-33)

[23] This custom of burying, fermenting, and storing vegetable foods has an interesting parallel in the methods used for preserving taro and breadfruit in Polynesia (Soderstrom 1937, 239; Cox 1980, 84-85).

Chapter 3

Alcoholic Beverages and Vinegar

The Spanish adopted the name 'chicha' from a native language (probably Arawak) and applied it to any native alcoholic drink, whether beer or wine:

> In New Spain [Mexico] the pulque is made of maguey; in Tucuman [Argentina] the chicha is made from algarrobas [the pods of a *Prosopis* tree]...In this kingdom [Peru], in addition to making chicha from maize, they also make it from quinua seeds and oca roots, from the berries of the molle tree...Also in other places they make wine from a certain liquid which flows from the heart of palm trees after they have been cut down...All of these chichas are intoxicating;...some are as strong as or even stronger than wine. (Cobo 1979, 28; also Cobo 1956, 1:162)

In South America, several species of the genus *Prosopis* are called 'algarroba'; the reference here is probably to *Prosopis pallida*, a close relative of the mesquite tree. Oca is an oxalis (*Oxalis tuberosa*), domesticated in the high Andes for its edible tubers. Molle is the Peruvian pepper tree, *Schinus molle*. (Nowadays, in Spanish-speaking America even various non-alcoholic beverages are called 'chicha.') In pre-Columbian times alcoholic beverages were almost unknown to the north of Mexico.

Maize was the most widespread raw material for making fermented drinks. Maize *chichas* were made in different colors: 'red, white, yellow, ash-colored and others' (Cobo 1956, 1:162). Today in Bolivia's Cochabamba Valley, the chicha made from 'a cherry-red to almost black maize which contains large amounts of *water-soluble anthocyanins*...is a rich Burgundy color.' Furthermore, 'young inflorescences of a purple amaranth are also said to be used to dye chicha' (Cutler and Cardenas 1947, 40).

Since distillation was unknown in pre-Columbian America, most fermented beverages had a low alcoholic content. Aside from their inebriating effect, the wide distribution and intensive use of these drinks has several explanations:

(1) They are often an important addition to the food supply; for example, among Quechua-speakers of eastern Ecuador: 'No household is without this precious food [i.e., manioc beer] for more than half a day, and it is rare that the household ever runs out' (Whitten 1976, 82).
(2) Food materials, that would otherwise have spoiled quickly, were converted by fermentation into alcohol, a rich source of calories—although, since fermentation is an exothermic process, some valuable food calories are lost.
(3) It may be, too, that—like beer in medieval Europe—their alcoholic content made drinking fermented beverages safer than drinking fresh water. Water sources were no doubt often polluted where human populations were dense. This may partly account for the aversion of Peruvian Indians to water at the time of the Spanish conquest:

> Water is their worst enemy; they never drink it pure unless they are unable to obtain their beverages, and there is no worse torment for them than being compelled to drink water (a punishment which the Spanish sometimes give them, and they resent it more than a whipping). (Cobo 1979, 27)

In parts of Mexico, the Indians drank much pulque, but little or 'no sweet water...neither from a river, spring, well, puddle, or lake...except when it rains...[When] it occasionally rains, from ditches in the ground they collect the rainwater in pools, and some Indians drink from these' (Oviedo 1959a, 2:26). Similarly, early accounts of the Maya refer to their drinking corn gruel all day, 'for they are not accustomed to drinking water alone' (Landa 1941, 90).

Those alcoholic beverages made by fermentation of substances rich in sugar—such as juices of fruits and berries, and the sap of maguey and some palm trees—are classed here as wines. In wine-making, sugars are converted by yeasts *directly into alcohol*. Though many wild fruits contain too little sugar to make wine, domesticated trees with sweet fruits were numerous. Beverage materials that require a preliminary conversion of starches to sugars (as with maize seeds, manioc roots, and the farinaceous fruits of palm trees) are here classed as beers. Admittedly, many alcoholic beverages do not fit well into this arrangement. For example, the term beer applies, strictly speaking, only to beverages made by fermenting malted cereals. (Moreover, one hardly thinks of honeymead as a wine.) Nevertheless, as defined by this rough classification, beers were likely the principal alcoholic beverage in native America, though wines were plentiful.

Beers

Most beer in the New World was made from maize and manioc. Of the two, maize-beer had a much larger geographical distribution. Beer-making involves two main stages: the conversion of starches into sugars, and the conversion of sugars into alcohol.

Malting and Brewing

The malting and brewing process involves a series of planned chemical changes. The first stage of traditional beer-making is the production of a malt, made by steeping cereal seeds (in the Americas, usually maize) in water until they begin to germinate. Then the germ (the embryo) of the seeds is killed by heat. (In the Old World, germinating barley seeds are traditionally used, but barley was unknown in the Americas.) In Peru, the development of malted drinks has gone hand-in-hand with the selection of maize varieties having special germinating characteristics (Nicholson, 1960). From place to place, a few other seeds were malted, including those of the quinoa plant.

During germination, cereal seeds produce an enzyme called diastase which acts as a catalyst in the breakdown (hydrolysis) of starch into a sugar—specifically, into maltose or malt sugar. The malt is mixed with warm water and allowed to ferment. During the fermentation or brewing process, two chemical changes occur: yeasts produce an enzyme called maltase which converts maltose into glucose. Then, through the action of another enzyme, zymase, glucose is converted to ethyl alcohol and carbon dioxide.

The following note by Oviedo describes how a malt beer was made from maize in Veragua in the early 16th century.

> They take a quantity of corn, according to the amount of *chicha* they desire to make, and *put it to soak. There it remains until it swells and begins to sprout.* Then there appear some small shoots on the part where the grain was attached to the ear on which it grew. As soon as the grain has sprouted, it is boiled in water, and when it has been boiled several times, the kettle or pot in which it is cooked is removed from the fire and the liquid is left to settle. It cannot be drunk on the first day...on the fourth day it is best. On the fifth day it begins to sour...[i.e., turn to vinegar] The Indians consider this beverage one of their chief staples... (Oviedo 1959b, 39)

(These Veraguan Indians, who lived along the Atlantic coast of western Panama, have disappeared, or at least much of their culture has.)

Far to the south, the Inca too had a drink made by soaking maize seeds in water until they sprouted, after which they were mashed and left to ferment in the same water. One of their stronger chichas, *sora*, was 'made from maize seeds which had first been buried several days until they sprouted' (Cobo 1956, 1:162). In other places, too, malted maize and water were put in a sealed container and buried. The probable explanation of such burying is that it allows better control of physical conditions: yeasts grow best in a uniform temperature and, unlike green plants, do not require sunlight. Recognition of this principle seems to have been extensive: burying fermenting drinks and other substances was widely practiced.

Malting and chicha-making are at their most complex in the Andes. For example, among modern Quechua speakers in the Cochabamba Valley of Bolivia maize is malted,

> by soaking the grains overnight in a pottery jar with enough water to cover them. The next day the grains are spread to a depth of about four inches on some leaves, and covered with several blankets. The temperature in one such bed was 34°C [93.2° F]. When the shoots are nearly as long as the grains they are placed in the sun to dry. In Tiquipaya, and also in Cuzco, Peru, it is a common sight to see these sprouted grains spread out on bright colored blankets in the dusty streets. When dry, the malted grains are ground. (Cutler and Cardenas 1947, 45)

In northern Mexico, malting is practiced by the Tarahumar Indians; the drink is usually called tesgüino. The following is a late 19th century account:

> Nothing is so close to the heart of the Tarahumare as this liquor, called in Mexican Spanish *tesvino*...To make it, the moist corn is allowed to sprout; then it is boiled and ground, and *the seed of a grass* resembling wheat [unidentified] *is added as a ferment.* The liquor is poured into *large earthen jars made solely for the purpose*, and it should now stand for at least twenty-four hours... (Lumholtz 1973, 1:253; Lumholtz's two volumes were published in 1902.)

Boiling, of course, would destroy the fermenting microorganisms. Likely, the pores of the special-purpose pottery would be a renewed source of such organisms.

Similarly, on the upper Guaporé river, near the border between southwestern Brazil and Bolivia, the Guaratagaja added the *leaves of a certain bush* 'as a ferment' to the boiled maize they kept in wooden fermenting troughs (Métraux 1942, 150).

Pre-mastication of Starches

In fermenting maize seed and other starchy materials, the malting procedure can be by-passed. As many peoples had learned, pre-chewing starchy source materials leads to a more successful fermentation. This is because, as noted already, mixing starches with human saliva changes the former to sugars.

It is surprising that a custom that seems so likely to spread contagious diseases should have been practiced so long and so widely. Its persistence may be explained partly by the presence in human saliva of a special enzyme called lysozyme which attacks bacteria. It may be, too, that many pathogenic bacteria and fungi are destroyed by alcohol during the fermentation or heating process. Moreover, only a minute amount of the enzyme is needed to bring about chemical changes within a very large quantity of fluid.

One of the earliest eyewitness-accounts of beer making from manioc in the Americas comes from eastern Brazil:

> The womankind make the drinks; they take the mandioca [manioc] root, and they boil great jars full. When it is boiled they take it out of the jars, pour it into other pots or vessels, and allow it to get cool. Then the young girls sit down to it, and chew it in their mouths, and the chewed stuff they put into a separate vessel.
>
> When the boiled roots are all chewed, they put the stuff again into pots, and once more fill them with water. This they stir up with the chewed roots, and then again they *heat* it.

> They have especial vessels, *which are half buried in the ground*, and they use them like we here do as casks for wine or beer. Into these they pour the chewed stuff, and close it up well; it ferments by itself, and becomes strong. They allow it in such manner to stand for two days, then they drink and become drunk therewith. It is thick, but of agreeable flavour. Each hut makes its own drink. (Stade 1874, 135)

Variations of this practice were, and in some cases still are, applied to other source materials: Carib women pre-chewed both manioc and sweet potatoes in making beer. In 16th century Peru, maize was treated in the same way. There, the most common beer was made from masticated maize seeds, thus speeding fermentation:

> The chicha most commonly drunk by the Indians of Peru is made from masticated maize...They do not masticate all of the maize from which they make chicha, but only a part of it, which mixed with the remainder *serves as a leaven* [that is, as a starter]. (Cobo 1956, 1:162-163)

Today, in parts of Peru and Bolivia, maize flour is treated in the same way; in some places, the 'salivated morsels are dried in the sun and sacked for storage and shipment' (Cutler and Cardenas 1947, 41). In Panama, the Guaymí pre-chew the farinaceous fruit of the peach palm, *pejibaye*, to make a fermented drink. Fermenting troughs, in which yeasts probably survive, are constantly re-used.[1]

As with many of the chemical arts discussed here, ritual programs once played a larger part in fermentation activities:

> It was not until the Europeans broke in upon the native civilizations, undermining and destroying their traditional customs, that ritual rules lost their force. (Bühler 1948, 2486)

The Jívaros and Canellos of eastern Ecuador make a fermented beverage and have ritualized the procedure: 'during fermentation in earthenware jugs, the women squatted around the vessels *singing magic chants to aid the process*' (La Barre 1938, 229). In Honduras, within the last century, Miskito Indians made fermented drinks by pre-chewing maize and manioc. The fermentation process was accelerated '*by drum beating*' (Conzemius 1932, 99-100). The Chaco Indians make a brew from algarroba pods. The fermentation process 'is '*hurried on*' *by various ceremonies, as the beating of drums, the shaking of rattles*, etc...' (Karsten 1926, 312-13). Such ritualization, accompanied by rhythmical music, is not uncommon and not limited to America alone.[2]

Fermentation-Accelerators

Fermentation-accelerators ('starters') are substances placed deliberately in foods or drinks to speed fermentation. Such substances hasten fermentation because they already contain concentrations of fermenting microorganisms, mainly yeasts. Their use extended from Mexico to the Amazon and Peru.

Wines

Even though grapes grow wild in North America, they were not used in making fermented beverages; other fruits were used instead. When sugars in fruit juices were deliberately fermented into ethyl alcohol, it 'was probably our first accomplishment in the field of organic synthesis' (Solomons 1996, 420). Wine-making was likely discovered by accident upon the spoilage of fruit collected in a container. In the Middle East wine-making from grapes began about 7000 years ago (Berkowitz 1996, 26). Little is known, however, about the early chronology of wine-making in the New World.

Fruit-Juice Wines

Almost any edible fruit containing sufficient sugar can be used to make wine or vinegar. In the American tropics many domesticated trees have sweet fruits—for instance, the sapodilla, which is especially sweet, and the hog plum. Such fruits spoil quickly. In wine (until it turns into vinegar) their calories remain available as alcohol; thus, wine-making can well be classified as a form of food preservation.

When the first Spanish arrived in Comagra, Panama, they were given 'wine, both white and colored, of different flavors' (Mártir 1964, 1:233).

The hog plum, *Spondias purpurea*, which is grown in both Central and South America, has a considerably higher sugar content than most European fruits (Kozioł and Macía 1998, 373). The tree's fruits are called *ciruelos* by the Spanish. According to Oviedo, who gathered much of his information in Panama shortly after the Spanish arrived,

[1] Some native American beer-makers are well aware of the effect of organisms on fermentation. For instance, the Canelos Quichua in the forests of eastern Ecuador make a strong beer from masticated manioc roots. The women go to the field 'to roast manioc, lug it to the house, and store it under damp leaves *to develop its fungus* for the making of fungus chicha...' (Whitten 1976, 171). The roots are

> ...taken to the household, stacked, loosely covered with balsa leaves, and allowed to sit...for about a week. *An orange powder-like fungus* (probably neurospora) is sprinkled in with the tubers. Orange, white, black, and yellow fungi develop in the tubers, and within ten days everything inside the skin has succumbed to the bright orange growth. (Whitten 1976, 88)

[2] For instance, in making their narcotic drink (*caapi*), the Barasana Indians, living in the Pira-Piraná area of Colombia, use bark of the *Banisteropsis* vine. The drink-maker throws the bark into a trough. Then, taking a club with a blunt end, he begins 'to macerate the stems with heavy rhythmical blows. While continuously stamping the mass he murmured and, once in a while, *broke into a chant*' (Reichel-Dolmatoff 1975, 158).
Similarly, in Europe 'music entered into alchemical conceptions and practice.'In the late 15th century, 'music was considered to exert a beneficial effect upon alchemical operations'(Read 1995, 71). 'The last and most complete record of an attempt to establish an alliance between alchemy and music is to be found in Count Michael Maier's *Atalanta Fugiens,*' an early seventeenth-century work (72). Alchemy, the 'divine Art,' was also strongly associated with a complicated mythology, mysterious symbols, and religion: in short, the ritual execution of chemical processes. In Europe, elements of this association lasted well into the 18th century (64).

the wine made by the Indians from ciruelos '...is of a fair *quality and keeps a year*' (Oviedo 1959a, 1:260). In western Panama, the Guaymí continue to make fermented drinks from the hog plum.

The Peruvian pepper tree, *Schinus Molle*, was widely planted in southern parts of the Inca Empire:

> In most of the settled areas one sees certain trees, large and small, which they call *molles*...From a small fruit the tree bears *they make wine*, or a very good beverage, *and vinegar*, and excellent *syrup* by crushing the amount they wish of this fruit in a vessel of water, and boiling it until it has partly evaporated, and it turns into wine or vinegar or syrup... (Cieza de León 1959, 115)

Palm Wines

Wine was also made from the sweet sap taken from the trunks of several species of tropical palm (e.g., *Acrocomia vinifera*). The juices of these palms, if allowed to collect and ferment for a few days in a hole cut in the crown-end of a felled trunk, produce a wine-like beverage. Spiced palm wine is mentioned in one of the earliest Spanish accounts from Veragua (Sauer 1966, 133, citing Ferdinand Columbus).

The following is von Humboldt's early-nineteenth century description of the process in the Sinú country of northern Colombia. Though palm wine was made in this case by people of only partly Indian ancestry, the method is substantially the same as that practiced by various tribes not far away:

> After having thrown down the trunk, which diminishes but little towards the top, they dig... an excavation...; and three days after this cavity is found filled with a yellowish-white juice, very limpid, with a sweet and vinous flavor...During eighteen to twenty days, this *wine of the palm tree* is daily collected; the last is less sweet, but more alcoholic and more esteemed...The natives affirm that the flowing is more abundant, when the petioles of the leaves which remain fixed to the trunk, are burnt. (Humboldt and Bonpland 1972, 7:442-443; italics in the original)

Today, large quantities of palm wine, 'toddy,' are also made in tropical southeast Asia. In Mesomerica, another wine—rather similar to palm wine—was made by fermenting the sweet juice of maize stalks.

Pulque and Other Drinks Made from Maguey

Certainly the best-known indigenous drink of Mexico is pulque, made from the fermented sap, *aguamiel*, that collects in cavities made in maguey plants, especially *Agave atrovirens* and *A. americana*. (The word *maguey*—now applied from Mexico to Chile to species of the *Agave* genus—was taken and disseminated by the Spanish from the Taino Arawak language spoken in the West Indies. Some half a dozen species agave are domesticated. One, *A. americana*, is known in English-speaking North America as the 'century plant,' because of the mistaken notion that it blooms only once in a hundred years.)

In the sixteenth-century, the Aztec held a feast to honor one of their gods:

> And in the courtyard of his pyramid they set a large open jar of wine [pulque], and those who were wine merchants filled it to overflowing, and all who wished went to drink. They had some canes through which they drank. (Sahagún 1950-1982, 3:36)

The procedure of making pulque requires some knowledge of plant physiology. The Otomí, who live in highland central Mexico, proceed as follows:

> ...scraping of the interior cavity wall [of the maguey] is repeated daily for two or three days, leaving the accumulating sap and the scraped material inside the cavity. On the third or fourth day the accumulation of sap and scrapings is cleaned out, tossed to one side, and the cavity scraped again. By this time the interior cavity wall should be turning a pale yellowish color. This color change signifies the onset of maximal flow of sweet aguamiel... (Parsons and Parsons 1990, 35)

To transfer the aguamiel from the plant cavity to a container, the Otomí and others make something like an enlarged pipette ('perforated at each end') out of a long, hollow, dried, vine gourd. This is 'used to suck the aquamiel from the central cavity of the plant' (Parsons and Parsons 1990, 42-43). (The device is called an acocote; for an illustration, see Steinkraus 1983, 331.)

To accelerate fermentation the Indians add 'a *little old and acid pulque*' (Humboldt 1966, 2:525) as a starter. Unlike pulque, tequila (another well-known Mexican alcoholic drink) is made by distilling fermented agave juice. Knowledge of distillation was almost certainly first introduced to the Americas by the Spanish; thus, its manufacture is post-Columbian. (With distillation one can extract alcohol from an aqueous mixture by first vaporizing it, then cooling the vapor to obtain alcohol in liquid form.) Though early 16th century Spanish records sometimes refer to *destilacion*, the word was used ambiguously: the verb *destilar* often meant simply to filter, evaporate, or decant.

In Mexico, other fermented drinks are made from maguey—for example, by the Tarahumara in Chihuahua:

> To prepare the liquor, the leaves are cut from the bulb-shaped stalk or heart, which looks like a hard white head of cabbage. These hearts contain a great deal of saccharine matter, and are baked between hot stones in earth mounds, being protected against contact with earth by layers of grass. When the Tarahumares want to make maguey wine, they leave the *baked stalks* in water in natural hollows or pockets in rocks, without any covering. The *root of a certain plant called frijolillo* [unidentified] is *added as a ferment*, and after two days the juice is wrung out with a blanket. (Lumholtz 1973, 1:256)

Figure 21. Agave plant on the left. To the right, the pot with dots above it is the pulque, or fermented beverage, symbol (Sahagún 1950-1982, vol. 2, fig. 42)

From maguey juice, the 16th century Aztec made not only pulque, but:

> honey, vinegar and sugar... The thickest parts of the leaves and the trunk, cooked in pit-ovens...are good to eat...By distillation [evaporation] they make the juice sweeter and thicker, until it condenses into sugar...From the sugar condensed from this juice, they *prepare vinegar* by dissolving it in water and then setting it in the sun for nine days. (Hernández, quoted selections by Lozoya in 1991, 92-96)[3]

Honeymead

The amount of alcohol produced by fermentation depends to a large extent on the sugar content of the materials used. Honey contains ample sugar, as Scandinavian honeymead-makers also had discovered. In the 16th century, the Maya made a fermented ceremonial drink called *balche*. It was made from 'honey and water and a certain root of a tree [*Lonchocarpus longistylus*], which they cultivated for this purpose, by which the wine was made strong and stinking' (Landa 1941, 92).

Among living Maya,

> Balche, the old ceremonial beer, fermented from the bark of the *Lonchocarpus tree, is used in many of the non-Catholic ceremonies*...Four pieces of bark, about a foot long, are *pounded with sticks* and placed in a jar with two jícaras of water and a *cup of honey*. It is left three days and then tasted. If it is not good, *more honey is added* and it is allowed to stand till it comes out yellow, good. (Redfield and Villa 1962, 38)

Among the Lacandones, descendants of the ancient Maya, who live in Chiapas at the headwaters of the Usamacinta River, balche continues to have ritual significance: the mixture is placed in a hollow log, and a long series of prayers are chanted 'during the fermentation of the ceremonial drink' (Tozzer 1907, 177 and 178-188).

[3] Xavier Lozoya, the editor, quotes large excerpts from Francisco Hernández's publications which are not available to me. Some of these excerpts, I have translated into English.

Field studies of the modern Maya demonstrate their control of the fermentation process by using plant additives. These studies verify to a considerable extent Landa's comments, quoted above. Maya fermentation procedures are:

> ...dependent on keeping contaminating microorganisms in relatively low numbers by the use of plant additives. These plant additives, typically in the form of bark strips of *Lonchocarpus*, *Hibiscus*, or *Heliocarpus*, and bundles of *Nicotiana* or *Datura* leaves, serve as a means for maintaining an inoculum of fermentative yeasts. The bark strips and leaf bundles are immersed directly into the fermenting beverages, where they become covered with a very dense, and relatively *pure inoculum of fermentative yeasts*. The strips and bundles are saved and used from batch to batch until their chemical potency is depleted...The chemicals in these plants retard the growth of many kinds of bacteria and fungi, but the yeasts apparently suffer no ill effects...these chemicals consist mostly of alkaloids, phenolics and glycosides... (Dahlin and Litzinger 1986, 731)

While some tribes depend on yeast spores that are simply present in the air or on leaves of plants, most seem to have repeatedly used special beer-making containers. Both practices are much the same as using starters. For example, large, rough-surfaced pots are kept and used repeatedly by the Colombian Chocó in their beer-making. The pots are used for nothing else, and water is never boiled in them. That so many tribes leave a mixture of water with maize, manioc, fruit, or honey to ferment in hollowed logs or old dugout canoes is an indication that such rough wooden surfaces are an ideal habitat for the fermenting yeasts.[4] Similarly, the air-borne spores offermenting organisms are likely to collect on *certain plants*. (Why this should be so is unknown to me.) In any case, parts of these plants (leaves, roots, seeds) are widely thought to speed alcoholic fermentation

Vinegar

The fermentation in which *yeasts* produce the ethyl alcohol in wines or beers is often (unless the beverage is consumed quickly) followed by a second fermentation in which *bacteria* (mainly of the genus *Acetobacter*) oxidize the ethyl alcohol when it is exposed to air, thereby converting it to acetic acid. (Vinegar usually contains from 3 to 6 percent of acetic acid—a colorless, pungent liquid.)

The earliest production of vinegar can hardly be thought of as an invention:

[4] Of course, they had no idea that fermentation was caused by microrganisms—nor did anyone else before the second half of the 17th century, when yeasts and bacteria were first seen by the Dutch microscopist Anton van Leeuwenhoek.

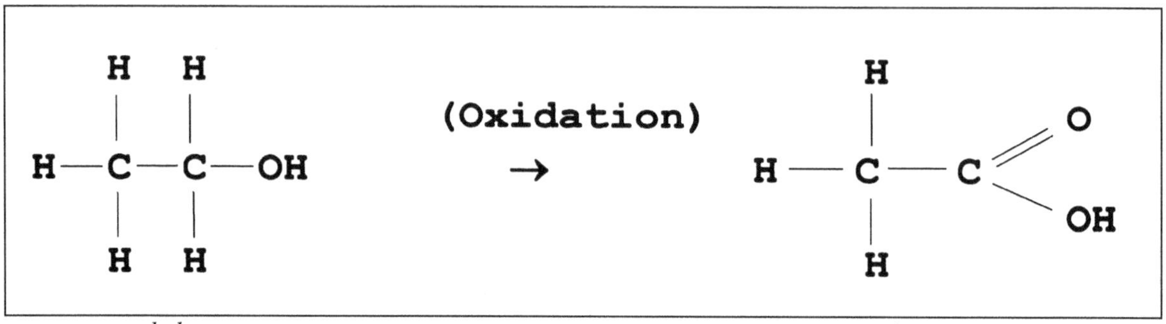

ethyl alcohol *acetic acid*

Figure 22. Making Vinegar

> Because of the readiness with which wine is converted into vinegar, it is evident that vinegar may have been produced when wine was first made. (Li 1948, 208)

Thus vinegar must have been one of the first acids known to mankind.[5] Vinegar had many uses in pre-Columbian America, but in early Spanish records it is usually mentioned only incidentally, no doubt because it was such a common material both in the Old World and the New. There are, however, a number of references to its manufacture. In Tierra Firme, 'a good vinegar is made from these ciruelas [hog plums]' (Oviedo 1959a, 1:261). In Mexico, 'Good vinegar, too, is made from this liquid [aguamiel from maguey]' (Motolinía 1950, 273). In Peru, vinegar was made from several plants. Maize flour was 'soured in the Indian fashion to make a very good vinegar' (Garcilaso de la Vega 1966, 499); in the southern Andes, fruit of the Peruvian pepper tree, *Schinus molle*, 'provides a splendid vinegar' (Garcilaso de la Vega 1966, 504; also Cieza de León 1959, 115); vinegar was also made there from maguey juice (Garcilaso de la Vega 1966, 506).

In the ancient Near East when fruit wines turned sour, the vinegar was made stronger by evaporation. Though no early sources specifically describe this method of concentrating vinegar's acetic acid content, there can be little doubt that it was used in the Americas too. As the following two citations suggest, it would have taken place on an almost daily basis and could hardly have remained unobserved.

(1) In the West Indies, vinegar was made from a syrup left over after processing poison manioc:

> Later this liquid [i.e., the unused syrup] is boiled and *placed outdoors*. It then turns sour and is used for vinegar without any danger whatsoever to the user. (Oviedo 1959b, 17)

(2) In Mexico (as noted earlier) vinegar was manufactured from the aguamiel of the maguey plant: 'To make vinegar, sugar produced by condensing the [maguey] juice is dissolved in water and *exposed to the sun for nine days*...' (Hernández 1888, 237)

In Peru, vinegar was used as a solvent for substances thought to have medicinal value; various minerals were mixed in vinegar and applied to inflammations in Peru (Cobo 1956 1:115 and 116). In Mexico, too, vinegar was mixed with ground cochineal and applied to wounds. Because of its acidity, vinegar is a weak bactericide. Thus, it is a good preservative and can be used to 'pickle' various perishable organic materials.

[5] In the ancient Near East and Egypt, vinegar was the only acid known. Moreover, the 'early Chinese, like the early Europeans, were acquainted with only one kind of acid, namely vinegar...' (Li 1948, 54).

CHAPTER 4

DRUGS AND POISONS

Numerous physiologically-active substances were discovered in the New World. The few mentioned below are divided roughly into stimulants and hallucinogens or toxins.

Stimulants and Hallucinogens

These universally interesting substances can hardly be discussed without reference to a group of compounds called alkaloids.

Alkaloids are complex compounds, found mainly in plants, that contain an amino group. Frequently the nitrogen atom is part of a heterocyclic ring. Most alkaloids are quite bitter, and many have intense physiological effects on mammals. As their name ('alkali-like') suggests, the alkaloids are mildly basic. Well-known members, whose effects were discovered in the New World, are nicotine, cocaine and curare (tubocurarine). As can be seen from its structural formula, nicotine has both a heterocyclic ring and a five-membered ring.

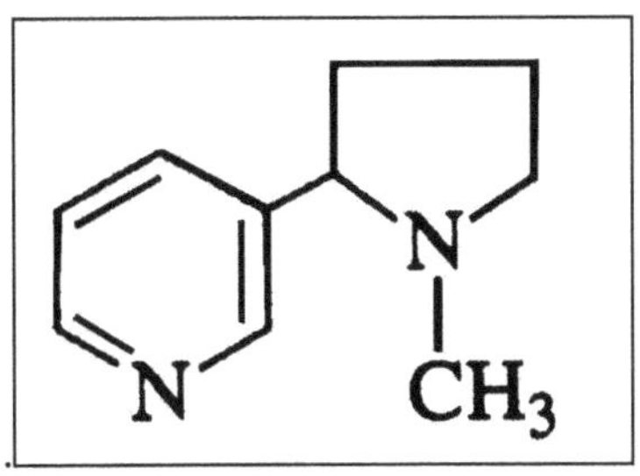

Nicotine

Indole is present in many alkaloids (Stowe 1959), especially those that are hallucinogens (Schultes and Hofmann 1973, 17-18). So-named because of its occurrence in indigo, indole has a benzene ring fused with a pyrrole ring:

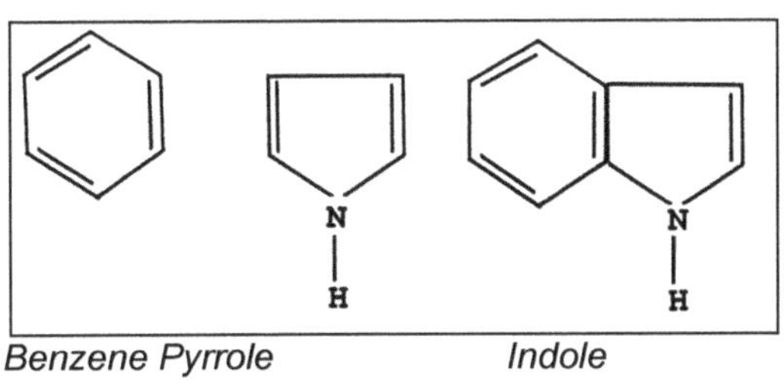

Benzene Pyrrole Indole

To be sure, indole also appears in non-alkaloid compounds, not only indigo but certain plant-growth hormones. Moreover, many alkaloids do not contain indole.

Figure 23. Common Components of Alkaloids

Though inhabitants of the Andes and the dry coasts of the Pacific may have been the source of most significant South American cultural innovations, it was in humid tropical forests east of the sierras that the majority of narcotics were discovered. While it is true that the flora there is much richer, cultural factors often underlie such differences. For example, biodiversity alone cannot explain the contrasting numbers of narcotic materials discovered in the eastern and western hemispheres:

> It is of interest that *the New World is very much richer in* [recognized] *narcotic plants than the Old* and that the New World boasts at least 40 species of hallucinogenic or phantastica narcotics as opposed to half a dozen species native to the Old World. (Schultes 1960, 147)

This is true despite the fact that the Old World includes a much larger land area, with an at least equally diverse flora. Moreover, the eastern hemisphere has been inhabited by the human species over a much longer time and with a larger number of individuals.

Before the 19th century, no other people had gone farther than the inhabitants of tropical America in identifying physiologically-active plant species, nor in investigating the effects of organic substances on human physiology. Beyond simple ingestion, the following procedures were practiced: (1) Smoking: This is an ingenious process for introducing physiologically-active substances into the blood stream. The active ingredient is volatilized, drawn with air into the lungs, and then copiously transferred by pulmonary capillaries to the blood stream. In the 16th century, this smoking procedure was almost unknown outside the Americas. (2) Snuffing: There may be good anatomical reasons for the effectiveness of snuffing. For example, certain veins in the nose 'communicate directly with the cranial cavity' (Wassén and Holmstedt 1963, 40). (3) Topical injection of a toad or frog hallucinogen into burn-wounds. (4) Administration of infusions made from tobacco leaves, or other narcotic plants, rectally with rubber-bulbed enema syringes (Métraux 1963c, 228).

> The Indians of the Amazon region have contributed two important inventions in the field of clysters: the *rubber-bulbed syringe*, brought to Europe by the Portuguese, and the use of narcotic clyster, *lately adopted in modern anaesthetics*. (Ackerknecht 1963, 629-630)

The Maya, too, used narcotic clysters (Furst 1976, 27-28).

The discovery of physiologically-active substances from plants may have been encouraged by native taxonomic methods: 'Our Indians [in Venezuela] named the trees, as usual, on *chewing on the wood*' (Humboldt and Bonpland 1972, 5:256).[1] A large number of plants were chewed and their juices presumably ingested:

> Everywhere that I have traveled in the Indies I have noticed that *the natives find great pleasure in keeping roots, twigs, or plants in their mouth*...[For example, in Quimbaya, Colombia] they cut slivers from a kind of small tree that is soft-wooded and always green, and keep them between their teeth all the time. (Cieza de León 1959, 259)

Tobacco Smoking

A number of tobacco plants was used in the Americas, but two domesticated species were by far the most important—common tobacco, *Nicotiana tobacum*, which has white-to-lavender colored blossoms, and a yellow-flowered species, *Nicotiana rustica*. The ancestral area of both is Peru and Bolivia (Goodspeed 1953, 397). *Nicotiana rustica*, which, in the 15th century, was cultivated from Quebec to Chile, has a much stronger effect; it is now mainly grown in industrial countries for the production of insecticides.

The leaves of these tobaccos, the best-known of American narcotic plants, were rolled into cigars and chopped for making pipe tobacco. The leaves were not only smoked, but chewed and powdered for snuffing as well. They were even scented and dyed. A copious paraphernalia was invented to make the plant's nicotine content effective: Cigar holders and pipes of various shapes, tubular and elbow-shaped, were constructed. Pulverized tobacco leaves were stored in special snuff boxes and introduced into the body with two-pronged snuff inhalers.

Chewed Tobacco and Chewed Narcotic Mixtures

As mentioned above, the leaves or seeds of a number of alkaloid-yielding plants were chewed with lime or ashes—in the Americas, especially the leaves of coca and tobacco. The custom is of chewing such plants with lime is widespread elsewhere.[2]

On the left is a clay smoking pipe excavated in the Huasteca area of Mexico; it 'is slipped with buff and decorated with red paint' (Ekholm, 1944, 475). In the center is a 16th century Aztec pipe (Sahagún 1950-1982, 2: Pl.36). On the right, a Maya smoking god (Cordan 1962, 179).

Figure 24. Mesoamerican Smoking Scenes

[1] The following account is another example of the common Indian taxonomic classification of plants by taste (a chemical test), rather than (as we usually do) by sight alone. When several Indians went searching for the right liana, one broke off a piece of one liana

> and chewed it for a moment; then he spit out the splinters and pulled at another vine...We then went to another tree nearby and then to a third...As far as I was able to ascertain, all three bundles of vines belonged to the same botanical species—*Banisteriopsis caapi*—but Biá [the leader of the group] insisted that they were three different kinds of yajé. (Reichel-Dolmatoff 1975, 158)

As already noted, the Barasana Indians of Colombia also gather *Banisteriopsis* bark to make their caapi (yajé).

[2] Narcotic plants were treated in much the same way across the Pacific: 'betel chewing' occurs among peoples living in southern Asia and on islands of the Indian and Pacific Ocean. There, too, limestone is calcined to make lime. Leaves of the pepper vine, *Piper betle*, are used as a wrapper for lime and a slice of a seed of the areca palm, *Areca catechu*, which contains the alkaloid, arecoline.
Such customs 'may be very ancient, possibly being survivals of a Paleolithic heritage [i.e., starting before human immigrants arrived in the New World]...The chewing of lime or ashes with some kind of a narcotic is such a trait' (Willey 1966-71, 22). It is improbable that lime was made so early.

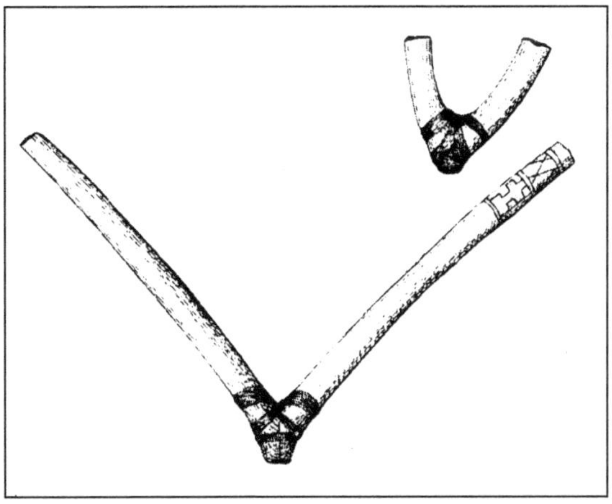

The lower snuffing tool is used with an assistant; the upper one is used unassisted (after Farabee 1922, 58). Although snuffing tools used at present are quite simple, in western Peru elaborate pre-Columbian snuffing equipment has been excavated; for example, gold snuffing spoons were made during the Chavín cultural period, 900-200 B.C. (For an illustration, see Quiller 1998, 1059.)

Figure 25. Snuffing Tools

The Aztecs described *Nicotiana rustica* and its use as follows:

> ...its blossoms are *yellow*. It is pounded with a stone, ground, *mixed with lime*...And it is *chewed*...It intoxicates one, makes one dizzy, possesses one, and destroys hunger and a desire to eat. (Sahagún 1950-82, 12:146)

Probably the first written description of the use of coca in the New World is by Amerigo Vespucci. He recorded the habits of islanders, likely off the coast of Venezuela:

> ...each had his cheeks bulging with a *certain green herb..., and each carried hanging from his neck* two dried gourds, one of which was full of the very herb he kept in his mouth; the other full of *a certain white flour like powdered chalk*. Frequently each put a certain small stick (which had been moistened and chewed in his mouth) into the *gourd* filled with flour. Each then drew it forth and put it in both sides of his cheeks, thus mixing the flour with the herb which their mouths contained. This they did very frequently a little at a time. (Cited and translated by J. E. Brooks 1937, 189)

Written almost five hundred years ago, this description of coca-taking applies with little alteration to today's Kogi Indians who live in Colombia's Sierra Nevada de Santa Marta—the same general area as that to which Vespucci referred (Wassén 1967, 234).

In much of the area between Colombia and Peru, quicklime was the most commonly used substance to chew with coca leaves. The first Spaniards to arrive in Peru noticed the custom (Cieza de León 1959, 259). In the early 16th century, the people living along the Spanish Main east of the Gulf of Urabá chewed coca leaves mixed with quicklime:

> From the age of ten or 12...they carry a wad of leaves the size of a nut on both sides of the mouth, the whole day without taking them out except to eat or drink. With this medicine they blacken their teeth to the color of charcoal...It is well to know how they preserve these leaves so that they do not spoil. Before grinding the dry leaves and reducing them to powder, they go into the mountainous forests, where...there grows an abundance of *mollusks* and *snails*, and taking an assortment of these they put them in a *furnace prepared for that purpose* with *a special kind of wood*, and subject them to the action of fire, obtaining in this way quicklime which they mix with the powder. And such is the strength of this lime that it burns and hardens the lips of whomever tastes it for the first time, as would happen to us if we rubbed our lips with quicklime; for the same reason calluses come to cover the hands of laborers because of constant use. It does not have the same effect on those who are habituated.' (Translated from Mártir 1964-65, 2:687-88)

Since there is some question as to whether a lime-burning furnace or kiln existed in South America in pre-Columbian times, the fact that it was used in Colombia or Venezuela to make quicklime is significant. (Mollusk shells tend to be a pure form of calcium carbonate, purer than any limestone, and a higher temperature may be required for their calcination—almost as much as calcite.)

A traveler in Peru in the 1840s notes (as did Mártir in Europe, much earlier) the problem of chewing quicklime, with ground coca leaves, painlessly:

> The application of the *unslaked lime* demands some precaution, for if it comes in direct contact with the lips and gums, it causes a very *painful burning*...owing to my awkward manner of using it, I cauterized my lips so severely that I did not venture on a second experiment. (Tschudi 1854, 314)

In most of the Amazon area (where limestone outcrops are rare and, except for snails and bone ashes, calcium carbonate is harder to come by), coca leaves are generally first toasted and pulverized; they are then mixed with 'finely sifted' ashes made by burning *Cecropia* or *Porouma* leaves (Schultes 1957a, 241). A modification of this method is found among the Tanimuka Indians in eastern Colombia, who have a chemically interesting way of aromatizing the coca-ash mixture with the resin of *Protium heptaphyllum*:

> Long and slender tubes or 'cigarettes' of rolled and partly dried leaves of *Ischnosiphon* [?] are *tamped half full with small lumps of the whitish resin.* The tip of that part of the 'cigarette' *containing the resin is lighted and brought to a glow by a gentle blowing through the tube.*
>
> In the meantime, several armfuls of dried leaves of *Cecropia* [an abundant regrowth plant] are set afire on the earth floor of the house and reduced to ashes. The ashes are then scraped together into a small, more or less conical pile. *Before the ashes are completely cooled, several Indians with resin-tubes insert the glowing ends of the tubes into sundry places in the ash-pile and blow vigorously.* The balsamic incense or smoke from the

> glowing resin *permeates the ashes*. This process, which fills the house with a pleasant myrrh-like aroma, continues for seven or eight minutes or until most of the resin in the tubes is spent.
>
> The ashes are then collected, sifted through a piece of fine, pounded bark-cloth and added to an equal amount of pulverized and sifted coca powder, the product is then ready for use. (Schultes 1957a, 244)

Tobacco and coca leaves are often ground and mixed with ashes in preparing them for chewing or snuffing. Certain plants are thought to be better than others for making the ash taken with the toasted and pounded coca and tobacco leaves. In Amazonia, the Jamamadis on the Rio Apitua 'are insistent that the snuff is *ineffective without the Theobroma* [a species of cacao] *bark ash* and said that they never take a pure tobacco snuff' (Prance 1972, 222). In another instance, tribes in the Vaupés of eastern Colombia collect 'dried leaves (of *Porouma cecropiaefolia*, the jungle-grape) to *provide ash additive*' (Hugh-Jones 1978, 54). The ashes of more than a half-dozen plants are sources of the alkalis mixed with coca leaves. Unfortunately, 'we still know very little about the chemical compositions of the ash residues of these plant materials...' (Plowman 1980, 253). To complicate matters further, the soil in which the trees are grown may also influence composition of the ash.

Although the chemical benefit of chewing alkaloid-containing leaves with lime or ash is unclear, it is unlikely that a custom so old and widespread (and so consistently followed) was practiced without some definite practical advantage. However, as yet, there has been only partial agreement on the question. The following is a sample of comments (varying somewhat in specificity) on the subject:

(1) 'The occasional *admixture of alkaline substances* such as ashes or calcined shells is probably of considerable importance in determining the *intensity* and general orientation of the hallucinatory experience.' (Reichel-Dolmatoff 1975, 21)

(2) 'The ash is added to the coca leaves to *activate the narcotic by providing an alkaline environment.*' (Prance 1972, 229)

(3) 'In the presence of strong alkalis like lime or potash, *alkaloids* occurring in combination in the coca leaves are *liberated*' (Cooper 1963, 552); '...the lime serving as a solvent of the alkaloid.' (Izikowitz 1930, 130)

(4) '...the people whom the Spaniards encountered on the coast of Peru *made use of caustic lime in order to stimulate the organs of taste*. The Goajiras at the mouth of the Rio la Hacha, and others, do the same at the present day. A deprivation of this irritant causes internal disorders of a general nature.' (Lewin 1998, 18)

(5) The following account suggests that the addition of ashes makes the alkaloids in the leaves taste less bitter: In the northwest Amazon, in order to prepare coca for use, its leaves

> are picked and fire-dried. They are then pounded with other ingredients in mortars...[The other ingredients are] procured by *reducing to ashes certain palm leaves*, [mixed with] *baked clay that is scraped from underneath the fire, and some powdered cassava flour...Whether these leaf-ashes are a form of calcium I do not know*. In the [Andean] Sierra powdered coca is mixed with pulverized *unslaked lime*, or with ashes of the *Chenopodium Quinoa*...The drug is carried in a bag, or a beaten-bark pouch, that is worn suspended round the neck. *The clay and palm-leaf ashes certainly neutralise the bitterness of the pure leaf...* (Whiffen 1915, 141)

(6) 'While nicotine—in contradistinction to intoxicants like coca (*Erythroxylum*) and betel (*Piper betle*)—does not require an alkalizing substance to liberate it, the presence of such an agent, nevertheless, accelerates and intensifies the action of the drug by increasing salivation.' (Wilbert 1987, 138, citing Hamilton 1957)

(7) Such 'alkaline source materials...[may] serve to 'sweeten' and potentiate the coca quid, apparently by *increasing the absorption of the alkaloids by the mucous membranes*.' (Plowman 1980, 253)

(8) 'The role of these alkaline substances during coca leaf chewing might be to provide *an alkaline medium* in the mouth to liberate cocaine from the plant material as the free base.' (Rivier 1981, 333)

Thus, the chemical effect of the numerous alkaline admixtures chewed with coca leaves remains highly uncertain because reports on the chemical composition of these admixtures are scarce and imprecise, and, especially, because

> it is difficult, indeed almost impossible to extend conclusions drawn from synthetic mixtures to in vivo systems. It should be kept in mind that Peru and Bolivia [with their large Quechua and Aymara populations] are the only countries where coca chewing is still a legal practice. Consequently, it is suggested that the work...should be repeated in Peru using modern analytical methods. (Rivier 1981, 332)

In the 16th century, the use of lime or ash mixed with coca extended from South America to at least as far north as western Panama. To the north and east of Panama, the use of this mixture continued, but with tobacco taking the place of coca. Ground tobacco leaves were sold mixed with lime in Aztec markets; in fact, in the 20th century this mixture was still used in parts of southern Mexico. For instance, the Mazatecs and Tzeltals 'on the road carry a little gourd or calabash...[containing *pisiete*]; this is powdered green tobacco, mixed with lime and *chili*; it is dipped with a little stick and chewed or sucked...' (Starr 1900-1902, 2:71; also 1:78).

Ingested Tobacco Extracts

Tobacco mixtures were also ingested. In South America, tobacco is sometimes used in the form of *ambíl*, which is made by boiling-down tobacco leaves in water, to form a thick dark liquid. Often this liquid is thickened further into a semi-solid paste by adding manioc starch (Wilbert 1987, 39-46). Ambíl is used by two widely separated groups of Indians: 'One group is composed of tribes...of the western Amazon basin. The other group...is located in the Sierra Nevada de Santa Marta, in northeasternmost Colombia' (Kamen-Kaye 1975, 57).

Ambíl is either drunk or licked. For example, the Jívaro 'imbibe [this] tobacco juice on numerous occasions and for many different reasons' (Wilbert 1987, 37), while in the northwest Amazon 'none of the tribes south of the Japura smoke their tobacco; *it is only licked*' (Whiffen 1915, 143).

Ambíl is often used alone, but more commonly while chewing coca leaves. Among the Witoto and the Bora,

> the preparation of ambíl is interesting, because it shows another example of the use of *alkaline ashes with a narcotic-alkaloid*, a custom widely spread in many parts of the world...Before the extract concentrates to make a thick syrup or in some cases a paste, the ambíl is taken out of the vessel and, while being carefully stirred, is mixed with *alkaline salts*. These *salts* are prepared by water which has been poured over and drained through the ashes of various plants commonly used for this purpose. A huge forest tree of the genus *Lecythis*...is probably the most used source of alkaline ashes. (Kamen-Kaye 1975, 58, translating from Schultes 1945)

In western Venezuela, a large portion of the non-Indian population still uses ambíl, the custom being derived from pre-Columbian Indian tribes (Kamen-Kaye 1975).

Snuffed Tobacco and Snuffed Hallucinogens

Tobacco was a familiar plant to ancient Peruvians, and they snuffed it: 'smoking was unknown' (Friederici 1960, 553).

Although the most widely-used snuff is made from tobacco leaves, a number of snuffed plants have hallucinogenic effects; among them are space-time disorientation, visual and auditory hypersensitivity, and enhanced color perception. Because these drugs offer their users a means of escaping mundane existence (and thus the experience of a different and, according to reports, a higher reality), their use in religious rites is common. For example, the Aztecs worshipped their deities through ingestion of an alkaloid, mescaline, from the peyote cactus, *Lophophora williamsii*.[3] (A peyote cult still exists among Indian tribes in northwest Mexico and the southwestern USA.) Under the influence of *teonancatl*—the name for their famous sacred mushrooms—the Aztecs 'communed with the spirit world...Divination, prophecy and curing rites likewise depended on the narcotic effects of these fungi' (Schultes and Hofmann 1973, 37). Indole appears in the two main alkaloids of *teonancatl* (Schultes and Hofmann 1973, 43). It also appears in strychnine, as well as the hallucinogens *paricá*, *cebil*, *yákee* and *caapi*, to be discussed shortly.

In the 16th century, the instruments used in taking snuff were much the same from the West Indies to Amazonia. The following accounts of snuff among two tribes of the middle Amazon were written in the early 20th century:

> The [Kachinaua] Indians claim their tobacco is much stronger than ours. [They may have used *Nicotiana rustica, rather than N. tobacum.*] It is hardly ever used except as snuff. It is inserted *into the nose with a v-shaped bone tube*. They introduce one prong into the nostril and the other into the mouth and exhale deeply. (Tastevin 1943, 70)

And among the Amazonian Pano tribe:

> Every evening, before retiring, the men gather in a circle and regale each other with roasted and pulverized tobacco. *To do this, the user places the precious powder in two bamboo stalks joined at right angles with black resin. One of the loaded tubes is introduced into the nose of the guest and with a vigorous puff through the other tube, the contents are blown into his cranium.* The operation is repeated many times, first in one nostril then in the other. (Tastevin 1943, 97)

Following is a description of the instruments used in snuffing tobacco (as well as the hallucinogen *paricá*) by the Piro tribe of Amazonian Peru. Leaves of tobacco are dried and pulverized, and inhaled through:

> a V-shaped instrument made of two leg bones of a heron. The end of one bone is *decorated so that it may be distinguished from the other*. The snuff is placed in the decorated end, while the other end is placed in the nose, and *an assistant blows the snuff with a sharp puff into the nostril*. Sometimes the arms of the V are made so short, that while one end is placed to the mouth, the other reaches the nostril and allows the operator to do his own blowing. This same instrument is used for taking the pulverized, roasted seeds of *Acacia niopo* [*Anadenanthera peregrina*] as a stimulant and narcotic. The hunter *administers the same powder to his dogs*, believing that both he and the dogs will be more alert and have clearer vision. (Farabee 1922, 56-57)

This is the only instance that I know of in which a hallucinogen is administered to animals—except, of course, in modern laboratory procedures. The last few lines of the above passage refer to paricá snuff which is obtained from the leguminous tree, *Anadenanthera (Piptadenia) peregrina*, formerly known—as above—as *Acacia niopo*.

[3] In Peru the columnar San Pedro cactus (*Trichocereus pachanoi*) is used for its similar effects; the principal alkaloid is also mescaline (Sharon and Donnan 1977, 381).

Paricá snuff is variously called niopo and cohoba, depending on the area where it is made. It is used by tribes in the drainage basin of the Orinoco River and, in the West Indies, the Taino Arawak people of Hispaniola also made a snuff (there named *cohoba*) from *Anadenanthera peregrina*. The seeds—often mixed with tobacco leaves—were ground to a powder and snuffed through forked tubes. The custom—and probably the tree itself—was introduced to the Caribbean islands from South America. Although extensively cultivated on the South American continent, and often present in regrowth vegetation, the tree is also found in a wild state there. Indole appears in several alkaloids from *Anadenanthera peregrina*; among them is the alkaloid bufotenine. (Although bufotenine was first found in the skin of a toad, *Bufo* sp., it is widely distributed in plants as well.)

In 1802 Alexander Von Humboldt described how the Otomacs, a tribe living near the Orinoco River, made *paricá* (*niopo*) snuff—a procedure which included a stage of fermentation:

> They gather the long pods of mimosacea, which we have made known by the name of *acacia niopo* [now, *Anadenanthera peregrina*], *cut them into pieces, moisten them, and cause them to ferment. When the softened seeds begin to grow black, they are kneaded like a paste, mixed with some flour of cassava and lime procured from the shell of a helix* [a snail], *and the whole mass is exposed to a brisk fire, on a grate of hard wood. The hardened paste takes the form of small cakes.* When it is to be used, it is reduced to a fine powder, and placed on a dish five or six inches wide. The Otomac holds this dish, which has a handle, in his right hand, while he inhales the *niopo* by the nose, through a forked bone of a bird, the two extremities of which are applied to the nostrils. (Humboldt and Bonpland 1972, 5:662-63)

As evidenced by artifacts in the archaeology of south-central Andes (around Tiahuanaco), a distinctive snuff-taking paraphernalia appeared there about 600 A.D.:

> This artifact complex consisted of a usually cylindrical container of wood, bone, or shell, a small delicately engraved spoon made of wood or bone, used to extract a *measured quantity* of the powder from the container. The powder was deposited on a wooden or stone tablet from which it was inhaled by means of a decorated wooden or bone tube. (Pochettino et al. 1999 127)

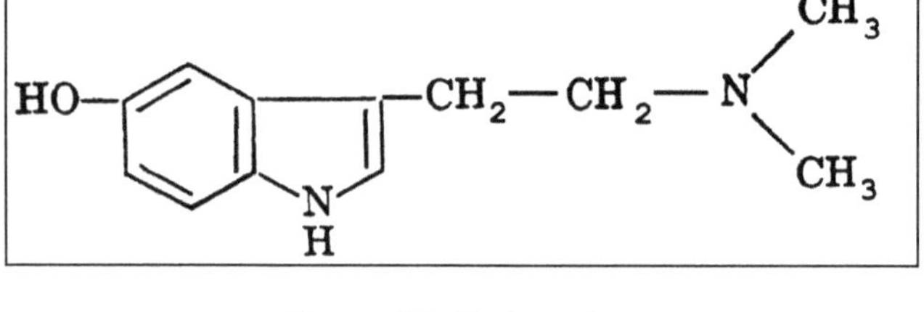

Figure 26. Bufotenine

This set of utensils also appeared in northwestern Argentina several centuries later (around 900-1000 A.D.), where it persisted until Inca occupation of that area in the fifteenth century. The principal source of the snuff has been identified as *Anadenanthera colubrina* var. *cebil*, a close relative of the tree from which paricá is made. In the same general area, in addition to the use of cebil snuff, a variety of narcotic materials has been smoked, and the 'use of pipes to smoke cebil is very old.' In whatever way they may have been used, cebil seeds have been found in preceramic levels that are dated at approximately 2130 B.C. (Pochettino et al. 1997, 128).

Yákee or epená snuffs are obtained from several trees of the genus *Virola* in the nutmeg-family, Myristicaceae. Likely some of the paricá snuffs reported in the literature are actually derived from the genus *Virola*, not from *Anadenanthera* (Schultes and Hofmann 1973, 89). Snuffs from Virola trees contain at least six indole-based alkaloids, four of which also contain the methoxyl group (Schultes and Hofmann 1973, 79).

Several different methods are used in preparing yákee (epená) snuff. In one example, among the Waiká Indians inhabiting the upper Rio Negro drainage of Brazil, the snuff is made with parts of three principal plants: *Virola theiodora* (the chief ingredient), *Justicia pectoralis* var. *stenophylla*, and *Elizabetha princeps*.

First, the bark is stripped from the *Virola* tree. Then,

> the soft, inner layer [of the bark] is carefully scraped off; the shavings are dried—first in the sun, then by gentle toasting over a slow fire—and stored until needed. When a new batch of snuff is to be made, these...crudely ground shavings are collected...and placed in a hollow lecithydaceous fruit as a mortar (*Bertholettia excelsa* H.& B.) [i.e., the fruit of the familiar Brazil nut] and triturated with a heavy wooden pestle until the material is almost completely pulverized. This powder is then sifted through a small basket...The resulting powder is very fine, homogeneous, chocolate-brown and highly pungent. (Schultes and Holmstedt 1968, 135)

Next, the cultivated herb *Justicia pectoralis* var. *stenophyllaa* is dried and stored until needed. Then a powder is made from its leaves, and the 'fine greenish dust is added to an equal amount of the brown *Virola*-powder' (Schultes and Holmstedt 1968, 135).

The third ingredient of yákee or epená snuff is ash, made from the bark of *Elizabetha princeps*. These ashes 'are then mixed, in approximately equal amounts with the *Virola-Justicia* powder. The resulting *epéna* snuff is greyish and extremely fine' (Schultes and Holmstedt 1968, 139). Indians living near the Vaupés River in eastern Colombia and neighboring Brazil use a somewhat different, but even-more painstaking, procedure to make their yákee snuff.[4]

[4] I quote, almost verbatim, from the meticulous descriptions of Schultes and Holmstedt for two reasons: (1) to show how detailed the native snuff-making recipes are; (2) because small details concerning the substances used are often important to a successful chemical outcome. The bark of the *Virola* tree is stripped off early in the morning. Almost immediately

In at least one of their otherwise pitiless human sacrifices, the Aztecs gave the victims an anesthetic which perhaps can be classified as a snuffed substance:

> after having performed many ceremonies with them, *they cast into their faces some powder* which they call *yiautli, that they might lose their sense of feeling* and not suffer greatly... (Sahagún 1950-82, 3:17)

The identity of the substance called *yiautli* has not been ascertained.

Psychoactive Drinks

A hallucinogenic drink was also obtained from *Anadenanthera colubrina* var. *cebil*; it is used in southern South America, especially in Argentina and Peru.

Since distillation was unknown in pre-Columbian America, as mentioned before, inebriating drinks with high alcohol content were unavailable. Psychoactive substances, however, were sometimes added to alcoholic beverages. For instance, the Inca added a narcotic plant, called *vilca*, to the chicha they drank in religious ceremonies (Cobo 1990, 169); vilca may have been *Anadenanthera colubrina* var. *cebil*.

In Mexico, too, certain roots were added to pulque:

> When it is fermented in a jar, as one ferments wine, with the *addition of some roots* which the Indians call ocpatli, meaning an agent which produces fermentation in wine, it becomes so strong that it *intoxicates violently*... (Motolinía 1950, 272)

Such substances may have been added for their effects on bacteria—that is, to control the rate of fermentation: pulque spoils quickly.

Caapi (*ayahuasca*, *yajé*), a narcotic drink, is made from the bark of a number of vines belonging to the genus *Banisteriopsis* (Malphigiaceae), chiefly *B. caapi* and *B. inebrians*. It is drunk throughout large areas in northern South America, from Amazonia to coastal Ecuador. Among the active constituents in *Banisteriopsis* are harmaline and harmine (Rivier and Lindgren 1972, 120). Both contain indole and the methoxyl group—i.e., the so-called 'methoxyindole alkaloids.'

Since prehistoric times narcotic plants of the genus *Datura* have been used for hallucinogenic purposes in both the Old World and the New. The leaves, bark, seeds or sap were all used. (In this genus, the active alkaloids contain neither indole nor the methoxyl group.) The daturas present a cultural parallel in the two hemispheres similar to that offered by fish-poisoning plants of the genus *Lonchocarpus* (*Derris*) and dye plants in the genus Indigofera. (In pre-Columbian times, daturas may have been the most widely recognized of all hallucinogenic plants.) For instance, *Datura metel* was used in China, India and the East Indies,[5] while in the New World daturas were used in the American Southwest, in Mesoamerica and in South America. The Jimson weed, *Datura stramonium*, is famous in American colonial history, because of its part in the so-called Jamestown Rebellion in which soldiers, who had eaten greens with which datura leaves were accidentally mixed, became disorderly. The origin of Jimson weed is disputed:

> Although there is *disagreement as to whether D. stramonium is native*, evidence seems to indicate that it was indigenous to the New World. It is *employed as a narcotic in both hemispheres.* (Schultes and Hofmann 1973, 167)

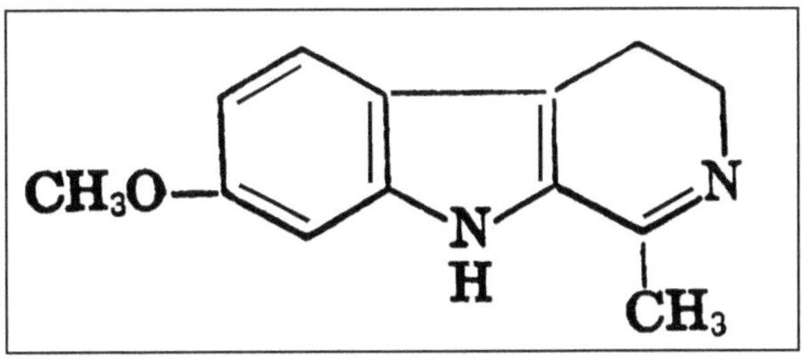

Figure 27. Harmaline

upon separation of the bark from the tree, a profuse exudation or 'bleeding' of a thick reddish resin-like liquid, which soon becomes viscous, oozes forth from the inner surface of the bark in small drops...According to the natives, this exudation is greatly reduced in quantity and is weaker in its narcotic effects when the trunk of the *Virola* tree has received the warmth of the sun's rays.

The bundles of bark are placed in water for a short time. When they are taken out, 'the soft inner layer, on the surface of which the red exudation has congealed, is rasped off with a knife or machete'; the raspings...are thrown into a large earthenware pot or enamel tray, and the rest of the bark is discarded.' (Both the making of machetes and enamel trays are techniques introduced by the Europeans.)
When enough shavings have been accumulated, a small amount of water is added, and the mass is thoroughly kneaded and squeezed by hand. The water becomes muddy and assumes a brownish or tan hue. This turbid liquid is strained several times, usually through a piece of finely hammered bark-cloth...into a small-mouthed earthenware pot. The residual shavings, when as much of the water has been expressed as possible, are thrown away. Enough water is added to the strained liquid to fill the pot, which is then set to simmer over a slow fire. From time to time, a sordid foam, which rises to the surface, must be scraped off with a piece of bark. The boiling is allowed to continue for three or four hours, more water being added if evaporation is too rapid, until nothing remains except a thick, dark brown syrup at the bottom of the pot. This syrup must not be dried rapidly over a fire; the pot is set in the sun, and the syrup is permitted to solidify slowly. When nothing but a dry, brown crust is left, the residue is scraped free from the pot and is ground into a fine powder with a water-smoothed stone as a pestle, and the pot or an enamelware tray as a mortar. It is then ready to be mixed with ashes which have been made from the bark of a small wild cacao tree (*Theobroma subincanum* Martius). Usually, equal amounts by volume of ashes and ya-kee powder are used (Schultes and Holmstedt 1968, 121-122).

[5] Recent genetic analysis has updated the knowledge of some plant origins: 'Although originally thought to be from Asia (Linnaeus, 1753), *D. metel* [a domesticated species] is now considered to be of New World origin' (Luna Cavazos et al. 2000, 494).

The tree-daturas, however, are native only in South America. In the Andes, drugs from tree-daturas were taken most often 'in the form of *pulverized seeds dropped into fermented drinks* or as an infusion of leaves and twigs' (Schultes and Hofmann 1973, 172).

Wound-Injected Hallucinogens

Whether the following substances from amphibia are best described as drugs or medicines is unclear; even in modern medicine there is no clear distinction between the two:

> The Kachinaua, Kurina, and Kanamari Indians [of middle Amazonia] fasten it [i.e., a toad or frog] down alive with its legs spread out like a cross and place it over a hot fire. A sort of glue oozing from its body is smeared on small sticks and kept in reserve. The animal is then freed, for should it die the glue would take vengeance on its executioners. When an Indian falls ill, becomes thin, pale or puffy, or when his hunting has been persistently bad, it is believed that an evil spirit has entered his body and must be expelled. Before dawn, after they have fasted, the sick man and the unlucky man nick their arms or bellies with the tip of a red firebrand. They then vaccinate themselves with the toad 'milk' as they call it. Soon they are seized with violent nausea and diarrhea. The evil element leaves the body...[After this] Not an animal escapes [the hunter's] piercing glance. His ear catches the slightest sound and his weapon shoots true. (Tastevin 1943, 72)

Along the border between Brazil and Peru a similar preparation with the same effects continues to be used among the Mayoruna and Matses tribes. A skin secretion, previously scraped from a live frog (identified as *Phyllomedusa bicolor*), 'and stored dry on a stick, is mixed with saliva and introduced into a line of fresh burns on the arms or chest' (Daly et al. 1992, 10960).

Narcotics and Religious Ritual

In the absence of written records, ritual and ceremony may serve as mnemonic aids, marking the successive steps toward completion of a recipe. An example is the following contemporary account of coca preparation among the Makús in the northwest Amazon:

> The coca plant is cultivated in large quantities in the Makú fields, since it is used daily to mix with their food [an unusual use]...Coca leaf is harvested by the Makús and placed in a large flat pan to toast, until it is crisp and dry. At the same time, a fire is made, and green banana leaves are burned. The dried leaves are placed in a wooden bowl, ground into a powder and mixed with the banana leaf ash. There is a ceremony attached to the pulverising, and a rhythm is often beaten out with the wooden grinding stick while other Indians chant. *The extremely deep, long, hollow pestle which they use makes a loud drum-like noise when the wooden pestle is knocked against the side. The different rhythms which they beat tell the rest of the Indians how the preparation is progressing.* When the ashes and leaves are ground into a fine powder, they are ready for use. The powder is mixed with cassava, either farinha flour or tapioca flour. This powder from coca forms a part of the daily diet of the Makús... (Prance 1972, 228-229)

Poisons

The plant poisons are mainly alkaloids, glycosides, and saponins.

Fish Poisons and Insecticides

Fish poisoning is practiced in many parts of the world. In the Americas, fishing with chemical substances was practiced both by hunters and gatherers (like the California Indians who used wild plants) as well as by agriculturists who grew plants domesticated for that purpose.

Fish poisoning was probably most complex in Southeast Asia and tropical America. (It was practiced in intermediate areas as well; for example, *Tephrosia piscatoria* and other plants were used in Samoa.)

In Spanish, *barbasco* is the common general term for fish poison; it was used in the Inca Empire:

> On the rivers, fishing with *barbasco*, which is a plant like a liana, is very common; when *barbasco* is pounded and put out in the water, it stupefies the fish in such a way that they float on the water as if dead. (Cobo 1990, 242)

Probably the discovery of the fish-poisoning potential of plants long antedates agriculture—and, likely, the first immigrants to the New World brought the knowledge with them. Among the oldest of bathing and washing sites are the edges of streams. And so, not surprisingly, there appears to be an old association between soap-substitute plants and fish-poisoning plants: probably, new ones would have been discovered by women who, while sudsing their garments or utensils at the streamside, observed the behavior of fish. (True soap was unknown in pre-Columbian America.)

The California Indians, who lacked agriculture, used the herb called soaproot (*Chlorogalum pomeridianum*) both as a cleansing agent and as a fish poison. Farther south, agricultural people also knew that some plants—for example, *Sapindus saponaria* of the soapberry family, which they cultivated—could be used both for poisoning fish and washing clothes. The modern Guaymí use more than a half dozen plants as fish poison (Gordon 1982, 46).

Many plant genera contain fish-poisoning substances, especially those of the soapberry and legume family. Of these, the most effective and widely used come from the closely inter-related (and thus a hard-to-distinguish) group of leguminous species sometimes called the '*Derris-Lonchocarpus* complex.' These plants are found in both tropical Asia and America. Certain 'species of *Lonchocarpus* have been propagated by cuttings for centuries' (Krukoff and Smith 1937, 573)—and have been domesticated so long that they are no longer able to reproduce from seeds. For example, *Lonchocarpus utilis* (formerly *L. Nicou*) is used as fish poison in Peru and is found throughout the Amazon basin. All collected specimens have been 'either of cultivated plants or of plants growing in secondary forest on the sites of old Indian plantations. It is doubtful whether the plant is found in a wild state' (Krukoff and Smith 1937, 580).

Other principal fish poisons in tropical America come from the genera *Sapindus* and the leguminous *Tephrosia*—some of which, having lost their ability to flower or fruit, can only be propagated vegetatively. These genera are widely distributed; for example, *Tephrosia toxicaria* is used as a fish poison by both the Colombian Chocó (Wassén 1935, 103) and tribes of northeastern Peru (Tessman 1930, 71).

The following is a description of fish poisoning by Indians living in the early 20th century in the middle-Amazon area:

> There are many kinds of poisonous plants in *their fields...when they throw the leaves of these plants into the water after they have been crushed and kneaded into pellets with earth soaked in their juice*, they see the fish rush madly to the surface, make prodigious leaps as if to escape from some pursuer, and then fall immobile, bellies up, asphyxiated but still edible. (Tastevin 1943, 70)

Curiously, both insects and fish are susceptible to some of the same toxic materials, and today species of Lonchocarpus are the source of rotenone, a commercial insecticide. The property for which the plants were originally domesticated may also be employed in industrialized countries: rotenone is sometimes used for ridding lakes of unwanted exotic fish.

Insect depredations were no doubt a major problem, as they are today. In their markets, the 16th century Aztecs sold various herbs to kill or drive away 'bedbugs, lice, fleas, flies and mosquitoes' (Hernández 1986, 105)—but information on indigenous insecticides is scanty.

Inorganic substances were also used as poisons. The natives of Peru took from their ancient copper mines a green mineral called coravari, said to be strongly poisonous if ingested by humans. The smoke from burning it with a kind of 'clover' was used to kill fleas and other insects (Cobo 1956, 1:127-128).

Arrow and Dart Poisons

The ingredients of the arrow poison used by the Indians who lived near the Guajira Peninsula (Venezuela) are poorly known. Indian informants have given conflicting (and, in the interest of secrecy, possibly purposely misleading) reports.

> I have been told by Indians that the poison that they use to tip their arrows is made from the sweet-smelling apples [i.e., manchineel, *Hippomane mancinella*] and certain *large ants*...and the venom of vipers, scorpions, and other poisonous ingredients which they mix. It appears to be a very black pitch. In 1514 in Santa Marta, in a village about two leagues or more inland, I burned a great quantity of this poison, as well as many arrows and the house in which it was stored. (Oviedo 1959b, 27)

The ants may have been added only to increase—through their formic acid—the pain of the wound.[6]

Although various toxic and pain-causing materials may have been mixed in its preparation, the poison's lethal effects were probably caused by the tetanus bacillus. Arrows, impregnated with the poison, were obtained from the present Goajiro Indians in 1921:

> The points were made of sting-ray tail fins [probably the serrated bony spine at the base of the tail] possessing little natural barbs that had been steeped in the substance claimed to be poison...[It was said] that this poison was obtained from decomposing cadavers of animals, snakes, toads, and such poisonous creatures. Since this poison acted much too slowly, the arrows in question were used only in war, not in the chase...Professor Santesson examined the supposed poison and found that the desiccated material contained germs capable of multiplying—the test animals died of tetanus...[Moreover, a bacteriologist] cultivated the germs located and found that they were still virile despite a lapse of at least five years since the impregnation. (Linné 1957, 153)

Curare is derived from extracts from plants of the genera *Strychnos* and *Chondrodendron*, though other plants are occasionally used. Variants of the word curare are found in Carib, Tupi, and Arawak, major language families of the South American mainland (Friederici 1960). Alkaloids are the active ingredients in curare, and in most dart and arrow poisons. An example is tubocurarine, the deadly Amazonian neurotoxin which causes paralysis by blocking the transfer of nerve impulses. (Despite its potential lethal effects, tubocurarine—derived from the large liana, *Chondrodendron tomentosum*—is used in modern surgery as a muscle relaxant, but in very small doses.) Acculturation of the tribes who once made curare, and the demise of elderly tribal informants, makes it unlikely that a complete knowledge of curare poisons can now be attained.

In their account of curare manufacture—from *Chondrodendron tomentosum*—on the upper Orinoco River, Humboldt and Bonpland gave one of the few eye-witness descriptions of an indigenous chemical laboratory (Humboldt and Bonpland 1972, 5:516-532).

---

[6] Along the coasts of Tierra Firme grows a tree called manchineel, *Hippomane mancinella*, which has a milky and highly caustic sap. 'When the wood of the [manchineel] tree is burned one cannot bear it, because instantly it gives one a splitting headache' (Oviedo 1959b, 91). Although the tree's effects are often exaggerated, 'travelers who through ignorance have used it for firewood have had their eyesight badly affected...contact with any part of the living tree produces inflamation and even ulcers on the skin' (Pittier 1908, 113).

The Chocó are exceptional in that they have discovered both a tree and a frog as sources of their dart poison. The tree, *Perebia* sp., is a member of the mulberry family. Its latex, tapped by making slanting cuts on the tree's lower trunk, 'contains a *non-nitrogenous glucoside* which is a powerful cardiac poison' (Wassén 1957, 79). As mentioned above, when ingested orally, glycosides and glucosides are generally not poisonous, but when introduced directly into the blood stream they may be quite toxic. By contrast, in the curare used by the tribes of Eastern Ecuador and the upper Amazon, the active agents are alkaloids which act on the central nervous system (Wassén 1935, 92).

The Chocó tree poison is prepared as follows:

> In order to get at the poison, they scratch a spiral groove in the bark...From this there immediately exudes a rather great amount of a fairly clear sap. In order to collect this, rags of *bark cloth* (woven of bast threads) are stuffed into the groove. The sap which further flows along the groove is collected in a bowl...The rags are then picked out carefully and allowed to dry. In this state the poison would seem to keep for an unlimited period. In order to use this dried poison to smear arrow [dart] heads with, such a rag is softened in a little water. The sap that has flowed out and been collected in the bowl is evaporated carefully over a slow fire...[An accompanying figure] shows an Indian 'boiling' poison. The sap is in a little bowl which, with the help of three fairly long wooden rods, is placed a good distance above the fire. The Indian stirs the liquid with a wooden stick. (Santesson 1931, 161-162)

Thus the poison seems to be obtained from the tree sap by two different methods: (1) Bark cloth rags are used to separate the poison from other sap constituents by preferential adsorption. (The method is reminiscent of one much used in twentieth-century laboratories, the separation of compounds by paper-chromatography.) (2) The poison is concentrated by stirring the sap while slightly increasing its temperature. Thus the poison is left undecomposed, while water and other liquids that volatilize more readily are driven off. The frog poison is extracted by suspending a frog, outstretched on a stick, over a fire. The secretion exuded from its skin contains the poison, and is scraped off (Wassén 1935, 100). The proper identification of the frogs, and of the tribes who used their poison, follows:

> Three extraordinarily toxic species of *Phyllobates* from rain forest on the Pacific versant of western Colombia are the only frogs known to be used for poisoning blowgun darts (not arrows). The only Indians known positively to practice this geographically restricted custom are the Emberá Chocó and the Noanamá Chocó. There is no evidence that other trans-Andean Indians or any Amazonian tribes ever tipped darts (or arrows) with frog secretion as a primary poison. (Myers and Daly 1993, 262:1193)

The Chocó use their poison darts for hunting game; they seem never to have used them against people. Whereas the Chocó are the only people known to have used frog poison for blowgun darts, as noted above, it is used for other purposes in the Amazon area.

Poisons Ingested by Mammals

As in other parts of the Americas, armadillos are eaten in the Brazilian Amazon. The Makú tribe, who live along a tributary of the Rio Negro, use scrapings of the bark of *Carpotroche amazonica* (Flacourtiaceae) to poison armadillos. The bark of another plant in the same family, *Ryania speciosa*, was 'employed by old people for suicide until rather recently, a type of euthanasia acceptable in this tribe. The poison was also utilized to kill enemies, but these uses have now stopped.' The species is 'a well-known toxic plant containing the alkaloid ryanodine, much used as an insecticide' (Prance 1972, 232-33).

Far to the north, there is a relatively recent example of the transfer of cultural traits between Asia and America: the Aleutian people probably carried the practice of whale poisoning with aconite (Heizer 1943). The leaves and roots of herbs of the genus *Aconitum* (Ranunculaceae) contain powerful toxins.

Noxious Gases and Burning Arrows

In the early 16th century, native people living in the Orinoco area used a sort of 'poison gas' in their warfare against the Spanish. When the Indians went into battle two young men went ahead of their squadron, each carrying a couple of basins, 'like earthen cooking-pans.' In one hand, the basin was full of glowing coals (probably of charcoal); in the other, the basin contained powdered pepper ('*ají molido*'—that is, cayenne pepper, or some other species of *Capsicum*). When the wind blew in the right direction, the pepper was scattered over the burning coals. The smoke blew into the Spanish soldiers' noses, causing a good deal of sneezing and much confusion in the ranks (Oviedo 1959, 2:395).

Apparently this stratagem was quite widespread, since native people of the mid-sixteenth century in eastern Brazil also used the gas of pepper when 'laying siege to villages fortified by palisades' (Nordenskiöld 1931a, 493). They

> drive their enemies from the forts with pepper...*They make great fires when the wind blows, and then they throw thereon a quantity of pepper*: if the fumes were to strike their huts, they would have to evacuate them...[Once we were left] lying dry, with a ship in a river, for the flood had left us; and many savages came, thinking to take us, but they could not. Upon this they threw heaps of dry underwood between the ship and the shore, also *intending to drive us away with pepper fumes*, but they could not light the wood. (Stade 1874, 154)

In Mesoamerica, too, Aztec couriers are said to have been 'suffocated [by their enemies] with the smoke of chiles' (Durán 1994, 195). (The active ingredient in the now-famous 'pepper spray,' e.g., MACE®—used in law enforcement and for personal protection—is also an oily extract derived from plants of the genus *Capsicum*.)

When attacking enemy settlements, the Indians of southeastern Brazil 'take cotton-wool, mix it with wax, tie it to the piles [heads] of the [arrow] shafts, and set fire thereto; these are their burning arrows' (Stade 1874, 154). In ancient China as well, flaming arrows, made with 'rosin or other inflammable substances,' were used in warfare (Li 1948, 114).

CHAPTER 5

TREATMENT OF PLANT FIBERS

The world flora has been thoroughly canvassed for valuable fiber plants. Many hundreds have been found; probably few have been overlooked.

Making cords and strings, as well as using knots, is among the earliest human technical inventions. Long before the first human settlers came to the Americas such skills were known in the Old World. In dry areas where their preservation is possible, artifacts made of plant fibers are among the oldest organic materials discovered.

Basketry came later, but still well before any knowledge of agriculture. Although the first immigrants coming from northeastern Asia probably had some knowledge of basket making, more complicated weaving and dyeing techniques were either independently discovered in the Americas or introduced from elsewhere well before the 15th century. In making baskets, withes, strips of palm leaves, sedge stems, etc. are used, and fibers are usually not separated from the rest of the plant body.

The techniques used in making some baskets and weaving a cloth are quite similar. It is only a step, though a major one, from basket-making to the weaving of textiles: a major step, since it depends on extracting the fibers themselves from the surrounding softer, non-fibrous plant tissues.

Before Europeans arrived in the Americas, dogbane or 'Indian hemp' (*Apocynum cannabinum*, or its close relative *A. androsaemifolium*) was the most widely used bast plant north of Mexico. Its strong, silky fiber was used from the Pacific to the Atlantic coast. In pre-agricultural times, dogbane (a perennial weed) was traded from tribe to tribe. In agricultural areas, it was displaced when domesticated cotton was introduced. Usually the stems of dogbane were simply beaten to extract their fibers. The fibers of many other plants were used at an early date by pre-agricultural peoples.

A next step, more strictly a chemical art, was the extraction of plant fibers by retting.

Retting

Plants of the genus *Agave* (maguey), in which the fibers are usually imbedded in fleshy leaves, were probably among the first plants to be retted in the New World. In the retting process plant stems or leaves are soaked for some time in water, in order that their fibers can be easily separated from the other plant tissue. Anaerobic bacteria of the genus *Clostridium*, almost ubiquitous in soils and stagnant water, probably play a large part in decomposing the plant's soft tissue and thus facilitating the separation.

The retting process often includes roasting, pounding, rinsing and wringing to remove non-fibrous material. Nowadays among the Otomí and their rural neighbors in highland central Mexico, the green maguey leaves are heated, butt-end first, while being constantly turned in an open flame. Old maguey stumps are used as fuel; firewood is scarce in this area. The maguey leaves are then put in a damp retting pit. The pit is covered with old leaves and rocks, and water is occasionally poured over it. After four or five days the leaves are taken out and beaten with a wooden hammer on a large pounding stone to loosen further the soft, partially decomposed membrane from the fiber. One end of each leaf is then fastened to the top of a slanting scraping board, leaned against a pole, and all non-fibrous material is alternately scraped and washed away. Next the fibers are placed in the sun to dry and bleach. A ground-stone implement, which may have been used for scraping maguey leaves before the introduction of iron tools, is common in archaeological sites throughout the area (Parsons and Parsons 1990, 300). In earlier times the large spine or thorn that grows at the end of each maguey leaf was used in carding (combing) the fiber (Hernández 1888, 237); a tool was probably made by setting the spines in a wooden base.

Henequen and *cabuya* are general terms often used in Latin-American Spanish for fibers extracted from such monocots such as *Agave*, *Furcraea*, *Yucca*; *ixtle* or *pita* generally refers to the bromeliad, *Aechmea magdalenae*. (In contrast, tree bast is usually called *majagua*; such fibers are used for making ropes, nets, bags, hammocks, sandals, and so forth.) Oviedo describes how the fleshy leaves of *Agave* and *Furcraea* species were retted to extract their fibers:

> The Indians take these leaves and submerge them near the banks of a river or stream, weighted down with stones (much as linen is retted in Spain); and after the leaves have remained in the water a number of days, they take them out and put them in the sun to dry. After the leaves are dry, they beat them with a cudgel (much like a hemp beater) to rid them of pulp and chaff, thus exposing the fibers within. (Oviedo 1959a, 1:237)

The Spanish were familiar with the retting procedure, because it was very like that used to prepare linen fiber in their homeland.

Bleaching and Its Use in Counterfeiting

Textile fibers were bleached not only by drying in the sun, but also by treatment with lye, minerals, and plant juices.

The Aztecs frequently used lyewater as a bleaching agent: a variety of products was 'washed in the water of leached ashes' (Sahagún 1956, 3:135). For example, the sandals (*cotaras*) made from maguey fibers sold in Aztec markets were bleached with lye and then 'dyed in various colors' (Sahagún 1950-82, 4:73-4; Sahagún 1956, 3:140). Cloaks or capes (*mantas*) were, in today's terms, both 'starched and ironed':

> The seller of coarse maguey fiber capes...treats them with maize dough...He sells capes...clean ones, white...*with dough applied, burnished with a stone*... (Sahagún 1950-82, 4:73)

Among the Aztecs, too, a mineral called *tizatlalli* ('white earth') was used both to whiten cotton fiber and facilitate its spinning (Hernández 1888, 309).

In northern Colombia, the acid-sweet juice of *caña agria* (a species of *Costus*, a plant of the ginger family) serves as a bleaching agent among descendants of the Zenú people: fibers from the leaf midribs of a domesticated variety of *Gynerium saccharoides*, used in making hats, are dipped in *caña agria* juice before being placed in the sun to dry (Gordon 1957a, 81).

As in so many times and places, almost the same amount of ingenuity was devoted to making adulterants as to making genuine products. At the markets, second-hand cotton capes were fraudulently sold as if they were new.

Old, worn and spotted cloaks were first 'boiled in lye to bleach them.' They were then starched, either with 'ground tortillas' or with the white maize-gruel known as *atolli*, applied 'to give color and weight to the cloaks.' Finally they were burnished—that is, 'ironed' (Sahagún 1956, 3:134; Sahagún 1950-82, 4:63).

An inferior grade of red-dye cakes made from the cochineal insect was sometimes sold, mixed with potter's clay or flour. Or another inferior product, 'ashy-red cochineal' (made from a different insect which also grows on cactus plants) was covered with genuine cochineal (Sahagún, 1956, 3:342).

Another example of such counterfeiting is the sale in Aztec markets of substandard cacao beans. Unscrupulous traders soaked shrunken beans in water so that they would swell and look genuine. They also bleached cacao beans in hot ashes, then smeared them with potter's clay or moist earth, so that the shriveled beans looked plump and fresh. Since the surfaces of mature cacao pods are naturally spotted with a black mold, the traders discolored immature pods with a black wax. Sometimes vendors even filled emptied cacao pods with the seeds of avocados, cut into rounded pieces (Sahagún 1956, 3:135; also Sahagún 1950-82, 4:65).

Making Bark-Cloth

The inner bark of certain trees, when beaten into sheets thin and pliable enough to be used for making apparel, is called bark-cloth. No trees are more suitable for this purpose than those of the mulberry family.[1] Such plants have been used by the Sumu Indians of Honduras. For instance, a tree is cut down and

> deprived of its bark; the latter is *then soaked in water for a few days, after which the sticky gum or milk adhering to it is scraped off.* The bark is then dried in the sun and kept in the hut until the women find time to pound it into cloth...As it becomes hard and shrinks considerably, it has to be submerged in the neighboring stream for a short time before the pounding of it begins. The latter operation is performed on a small log...with the aid of a wooden mallet having the shape of a short thick club, into which longitudinal ridges have been made at the head part...The bark extends gradually upon being pounded, and it becomes soft and flexible. *After being washed and dried, it is ready for use*, and has a brownish color...A similar cloth, but almost white in color and of superior quality, is obtained by the same process from the inner bark of a species of *Ficus*...and likewise from the rubber tree (*Castilla* sp.). But in both cases the manufacture of the cloth is more laborious... (Conzemius 1932, 48)

Another mulberry-family tree, the *sándi* (*Pseudolmedia spuria*), is especially common around archaelogical sites in Guaymí territory. It is used for making bark-cloth by the Guaymí Indians of Panama, as well as by their eastern neighbors, the Bókata. Sometimes pieces are placed diagonally on a log and beaten and meshed together, forming sort of a felt.

For some purposes, Guaymí women still prefer bark-cloth for undergarments to the cotton goods they buy from traders. They say the barkcloth is warmer than cotton and that it stretches more readily to fit the contours of the body. It is also preferred for sleeping pads, tump-lines and pack straps. Piles of bark-cloth, made in forests of the Atlantic slopes, are carried in packs over the mountains to the savanna-Guaymí who live on the much-deforested Pacific side of the Isthmus.

Some fifty years ago, when I first visited this part of Panama, Guaymí women made voluminous 'mother hubbard' dresses from printed cotton cloth obtained from traders. (The style itself is an old Spanish influence.) However, some Bókata women were still making their clothing entirely of bark-cloth in much the same style. Bark cloth is not a very satisfactory material for such expansive clothing: being thick, stiff, and easily shredded,

[1] Probably the best-known of bark-cloths is tapa, an especially thin, pliable product, made in Polynesia from the paper-mulberry, *Broussonetia papyrifera*. The equipment and procedure used in its manufacture are remarkably similar to those used in Mesoamerica.

it is difficult to wash. Thus Bókata women often washed their ample dresses without removing them; they stood at a streamside and scrubbed at the fabric with pieces of bark, dipped and softened in water (Gordon 1982, 54 and 149). Bark-cloth is used by many South American Indians, among them the Chocó in Colombia and various tribes living in the upper drainage of the Amazon River. For example, the Witoto make it from a mulberry-family tree, yet another species of fig (*Ficus* sp.). However, it is said that bark-cloth was 'used to a greater extent among tribes of eastern Bolivia than in any other part of America' (Métraux 1963b, 67). Fragments of bark-cloth excavated in Peru indicate its use there before 2000 B.C.

Although many New World tribes apply colored decorations to their bark-cloth, sometimes by means of wooden or baked clay stamps, few chemical changes are involved in its manufacture, aside from a partial retting.

Mesoamerican Bark-Paper and Its Uses

In the pre-Columbian western hemisphere, bark-paper was made only in Mesomerica, where, as in parts of the Old World, it had a special religious significance. Few Aztec ceremonies and sacrifices were conducted without painted paper and incense—and many of them with liquid rubber.[2] The use of paper, rubber, and incense was commonly linked in religious ceremonies; for example, in one Aztec ceremony the celebrants used ears of corn wrapped 'in red paper and painted them with liquid rubber' (Sahagún 1950-82, 3:61). When a person recovered from some illness, he paid his debt to the gods with an offering of paper spotted with liquid rubber and incense (Sahagún 1950-82, 3:185). (In China, too, paper has special ceremonial significance: note the use of red paper and joss-sticks, incense, in Chinese religious ceremony.)

At one Aztec sacrificial ceremony, the priest descended from a pyramid,

> thereupon likewise descended *the fire serpent*, looking like a blazing pine firebrand. *Its tongue was made of red arara* [parrot] *feathers*, looking like a flaming torch. *And its tail was of paper, two or three fathoms long*. As it descended, *it came moving its tongue*, like that of a real serpent, *darting in and out*. (Sahagún 1950-1982, 3:136)

Long before arrival of the Europeans, the Aztec and Maya peoples kept written records on bark-paper. Numerous manuscripts and pictorial books were made, and libraries were kept in Texcoco, Cholula and other cities. The largest, however, was at Tenochtitlán (Lenz 1973, 40), where a great building held the books on which a chief accountant kept all records of the tribute owed to Montezuma (Díaz 1960, 1:273).

Of pre-Hispanic (or at least early) Maya literature, only a few pictorial books (now called codices) survive: One is kept in Dresden (the *Dresden Codex*); another, in Madrid (the *Tro-Cortesianus*); and the third, in Paris (the *Peresianus*).[3] In addition, several codices of Aztec or Mixtec origin are in existence.

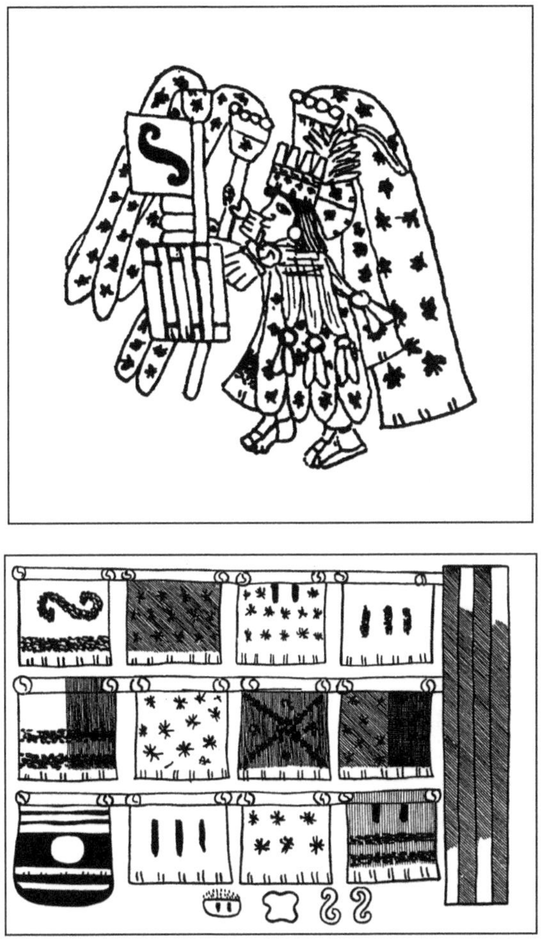

Above: Aztec priest wearing ornaments of painted paper (Codex Borbonicus, unpaginated).

Below: Various designs on painted sacrificial paper (after Danzel 1922, vol. 1, plate 6—from Codex Magliabecchiano). These two codices were probably created after the Spanish conquest of Mexico.

Figure 28. Aztecs and Painted Paper

Around the mid-sixteenth century, Diego de Landa wrote,

> these people [the Maya] also *made use of certain characters or letters*, with which they wrote in their books their *ancient matters and their sciences*, and by these and by drawings and by certain signs in these drawings, they understood their affairs and made others understand them and taught them. *We found a large number of books in*

[2] In his book, *The Aztec and Maya Paper Makers*, Victor von Hagen (1977) appropriately names one chapter 'The Paper-World of the Aztecs' because of the extensive use of paper in their ceremonies. They made enormous sheets of paper for their rituals; for instance, they spread before an image of one their gods a piece of paper, 'white paper and not yellow paper, a finger thick, a fathom [six feet] wide and twenty fathoms [120 feet] long' (Sahagún 1982, 3:69).

[3] A fourth, the oldest of all, is the Grolier Codex; it was published recently, in 1973, and is kept in Mexico City (Coe and Kerr 1997, 175 and 232).

> *these characters and, as they contained nothing in which there were not to be seen superstition and lies of the devil, we burned them all, which they regretted to an amazing degree, and which caused them much affliction.* (Landa 1941, 169)

Thus most of Mesoamerica's pre-Hispanic written accounts were destroyed. It is not surprising that the Maya and Aztec peoples felt 'much affliction' when their literature was burned: the books included their laws, the order of their ceremonies, their computations, astronomical notations, the manner and seasons of planting crops, and more (Mártir 1964, 1:426). Nor is it surprising—considering this appraisal of native literature—that the first European reports on paper and book-making in the New World lack detail. Certainly, little attention was paid to the chemical particulars.

Early Accounts of Paper- and Book-Making

Questions about the composition of Mesoamerican codices and manuscripts have been argued for a long time. Some are still incompletely answered. These include (1) the principal plant sources used; (2) the identity of the material used as a paper filler; and (3) the composition of the gums and glues used as sizing on the paper and in binding the books.

Although much like making bark-cloth, the process of making bark-paper involves significant chemical changes. As with bark-cloth, the tree sources are mainly in the mulberry family, especially the fig genus. In Aztec Mexico, such trees were known by the Nahua name *amatl*, and today bark-paper itself is generally called *amate*.

In Mesoamerica, bark-paper was still commonly used in making manuscripts and books throughout the 16th century (Lenz 1973, 76).

Of the earliest Spanish accounts of Mesoamerican book-making, perhaps the most accurate are those of Pedro Mártir. Among the gifts sent by Hernán Cortés to the King of Spain were certain books, some of which Mártir had inspected. The following (translated from the Spanish) are Mártir's comments:

> The material upon which the natives write are sheets made from a thin membrane which grows beneath the outer bark of a tree...This netlike tissue is smeared with a sticky gum. When sufficiently softened, the sheets are stretched to the shape desired. Later, when hardened, they *are covered with gypsum, it seems, or with some similar substance*. Likely Your Holiness has seen these sheets, *covered with gypsum as fine as flour*, upon which one can write whatever comes to mind and *then rub it off with a sponge* or small cloth so that they can write again. Tablets are also made with these sheets from a fig tree, which the stewards of major households carry to market in order to note with *a metal stylus* what they have bought and erase it after it has been entered in their accounts. (Mártir 1964, 1:425-426)

A few decades later, Bishop Landa also noted that Maya paper was made from a tree, and added comments on the format of their books:

> Their books were written on a large sheet doubled in folds, which was enclosed entirely between two boards which they decorated, and they wrote on both sides following the order of the folds. And they made this paper from the roots of a tree and *gave it a white gloss* upon which it was easy to write. (Landa 1941, 28-29)

(Landa may have been referring to *the aerial roots* of strangler figs.) Another early observer, Toribio Motolinía, wrote that

> *They make a very good paper from the metl, or maguey.* The sheet is about twice the size of ours; a great deal is made in Tlaxcallan and circulates through a large part of New Spain. (Motolinía 1950, 274)

Later, Alexander von Humboldt, who traveled to the Americas in the late nineteenth-century, took back to Europe a number of (mainly post-Columbian) manuscripts from Mexico. Humboldt wrote that these manuscripts were made of maguey (agave) fibers. Disagreement soon arose as to whether fig trees or maguey were the source of the paper, though Motolinía said that both were used, and that the fig tree was a source in the 'tierra caliente.'[4]

There can be little doubt that pieces of maguey fabric were commonly used for written communication before the Conquest. For instance, Montezuma had received news from Tabasco that the Spanish were approaching Tenochtitlán well before their arrival: 'All this news had been brought to him on pieces of painted cloth made of *henequen* [i.e., maguey fiber] which is like linen...:' 'y todo se lo habían llevado pintado en unos paños que hacen de henequén, que es como de lino' (Díaz 1960, 1:70). Likewise, envoys from distant countries painted their dealings on henequen cloth before showing them to Montezuma (Díaz 1960, 1:295). And Montezuma presented Cortés with an accurate map of the area 'from Pánuco to Tabasco' which was painted on henequen cloth (Díaz 1960, 1:317). Clothes of agave fibers are frequently mentioned in early documents—not paper, but painted 'cloth' made of henequen.[5]

---

[4] The two most comprehensive publications on pre-Columbian Mesoamerican paper making are those of Victor von Hagen (first published in 1944) and Hans Lenz, *El Papel Indígena Mexicano* (first published in 1948). Victor von Hagen was of the opinion that only fig bark was used in making the paper. Hans Lenz made microscopic observations of 44 manuscripts, most after the Spanish conquest, belonging to the National Museum in Mexico. While 36 were made of *Ficus* bark, Lenz (like Schwede) found that four were made of maguey; the remaining four were made of the old world plants, linen or hemp. However, Lenz 'does not give us a satisfactory description of how paper could have been made of maguey fiber in pre-conquest times' (Christensen 1963, 362 and 363).

[5] Feathers sewn to woven goods of both cotton and henequen were sold in sixteenth-century Aztec markets. That Otomí women spun henequen with spindles, is mentioned specifically: 'hila nequén con huso, con que hilan las mujeres otomitas' (Sahagún 1956, 3:155 and 198). A spindle is a slender, straight stick—some 20 to 45 cm long—which is twirled to twist a fiber into thread.

In 1912, the results of a chemical and microscopic analysis of the *Dresden Codex* were published (Schwede, 1912). This research showed that Mártir was right—that the pages are made of fibers obtained from the bark of a species of *Ficus*, not from a monocotyledon such as maguey. (Microscopic examination of the fibers in the pages showed the presence of sap-carrying tubules, characteristic of inner bark in trees of the mulberry family.) Later, however, an analysis of twenty-one other manuscripts showed that, while twenty were of fig-bark, one was made of agave fiber (Schwede, 1916). While there is little doubt that in pre-conquest Mesoamerica most paper was made from the bast fiber of wild fig trees, apparently some of lesser quality was made using maguey fiber.

Both the identity of the plant used for making the paper and the composition of the powder that covered the paper were disputed. As noted above, Mártir was uncertain whether this sizing was made of *yeso* (gypsum, a hydrated calcium sulfate) or 'some similar substance.'[6] Regarding sizing, Bishop Landa, as noted before, said only that the Maya 'gave it a white gloss upon which it was easy to write' (Landa 1941, 29). Other accounts suggested that the material was chalk, talc, or, as in ancient Chinese paper, alum (Li 1948, 162).

Schwede's analysis proved that the page surfaces are coated with a chalky substance, a carbonate of calcium—not gypsum, a sulfate of calcium. This chalky substance was mixed with some material of vegetable origin that turned blue when tested with iodine, indicating that it was a starch (Schwede 1912, cited in von Hagen 1977, 62). According to a later examination, the sheets are '*covered with a fine white substance, probably a bicarbonate of calcium obtained from the ashes of a plant* called *tizate, Zexmenia frutescens*' (Christensen 1963, 361).

Covers of the books were painted in various colors, as were the figures on their pages after the sizing had been applied. The painter first drew the image he desired with either charcoal or soot (Sahagún 1956, 3:115: 'dibujar o señalar las imágenes *con carbon*'). Various colorants were used: red from achiote, and purple from mixing cochineal with alum (Sahagún 1956, 3:343-44). Some of the blue was probably made from indigo: among the Aztecs, indigo was a common '...medium for painting black, for painting in colors' (Sahagún 1950-82, 12:242). Inorganic pigments such as red ocher were used as well. Figures on the sheets of paper were painted in the various colors, usually with a thin brush.

It is possible that cylindrical baked-clay seals were also employed to make designs or symbols on book covers. However, whether paper itself was stamped is unknown. (Stamping is both a mechanical and chemical process, and, since the materials used are partly perishable, the chemical aspect of the process is often lost.) Because excavated pottery often bears stamp imprints, it is frequently assumed that most stamping was done on clay—but flat and cylindrical pottery stamps were commonly used for printing cloth and for body painting.

> Cylindrical stamps are found over the greater part of Central and South America. Two generalized types are used, in one of which there are projections at each end on which the stamp spins, and in the other of which there are depressions at the ends in which the fingers are inserted. (Lothrop 1926, 181)

One author finds a pottery stamp directly related to printing:

> Since stamping as practiced in Middle America and over a large part of the western hemisphere was actually a form of printing, then the inventors of the pottery-stamp were the developers of at least one form of printing—centuries before that form was introduced in Europe following the Renaissance. (Ries 1932, 415)

> When the aboriginal potter employed a stamp to impress a design into the wet clay in his hand, he was putting into practice the identical principle of printing with movable type. *He was printing.* (Ries 1932, 417; italics in the original)[7]

Glyphs, pictorial symbols used to represent words or sounds, appear on a cylindrical seal, recently discovered near the old Olmec center at La Venta on the north side of the Isthmus of Tehuantepec. This ceramic seal, dated to about 650 B.C., suggests that the Olmec initiated New World writing well before the Maya (Pohl, Pope and von Nagy 2002).

Numerous excavated spindle whorls, made of baked-clay or stone (formerly on the end of spindles, presumably for spinning the coarse fiber of maguey), have been found in the Valley of Mexico; they are heavier and more than three times the size of cotton spindle whorls (Parsons 1975, 210). Such whorls, acting something like flywheels, are used to steady the motion of the spindles.

[6] In his book, *De Rebus Oceanicis et Novo Orbe*, Mártir (Martyr) comments:

To render it [the sheets of tree bark] pliable they fill up those porous membranes with bitumen and stretch them to whatever form they please, and setting and becoming hard again, they cover them with gypsum. I, however, presume that *the paper which they...*[other observers] *have seen preparing*, was made with a substance that is *only similar to gypsum*, beaten and sifted into fine flour, and thus a substance is prepared upon which one may write whatsoever would occur to one, and erase it with a wetted sponge or cloth, and so prepare to use it again. (Pedro Mártir, as translated by Victor von Hagen 1977, 29)

Nevertheless, unlike the Maya codices which were made of bark-paper, some pre-Conquest Mesoamerican books were made of dressed animal hides, 'but were also coated with a fine white layer as a paint base; tests on the *Selden Codex*, a Mixtec manuscript in Oxford's Bodleian Library, showed this substance to be a mixture of calcium sulphate (*gesso*), calcium carbonate, *and animal glue*' (Coe and Kerr 1997, 144). This seems to contradict Gilmore's assertion (in the Introduction) that there was no 'glue from animal gelatin' made in pre-Columbian America. (Now, the glue in ordinary use is common, or impure, gelatin.) The people of ancient Arica also made 'animal glue' (Linné 1957, 150-151).

[7] According to Ries, a

flat pottery-stamp is intensely similar to a piece of movable type in any print-shop of today. (Ries 1932, 425; see Figure 2 on page 418 for a 'comparison of Maya flat stamps with a piece of modern movable type')

Aztec painter and son (Codex Mendoza 1992, 4:147)

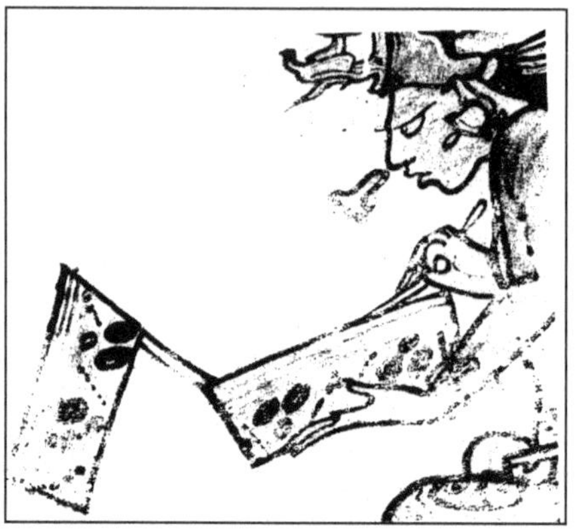

Maya book painter (Coe and Kerr 1997, 170; see also Bishop 1994, 43)

Figure 29. Aztec Painter and Maya Book Painter

A flat and a cylindical terra-cotta stamp from the Uloa Valley, Honduras (Gordon 1898, 117), and below, a cylindrical stamp from Kaminaljuyu; the roll-out design is extended to the left (Kidder et al. 1977, 215)

This remarkable seal 'depicts two speech scrolls that emanate from the beak of a bird' (Pohl, Pope, and von Nagy 2002, 1985). The seal is some 7cm tall; the roll-out extends to the right

Figure 30. Ceramic Stamps and Seals

Not all Mesoamerican manuscripts were pictorial and vari-colored, however. There are numerous references to a black ink manufactured from soot (lampblack—i.e., fine bits of amorphous carbon) that was made by burning resinous pine wood, copal, or other materials. This ink was of better quality than some made later. For example, on certain post-Conquest manuscripts 'the outlining of the pictorial forms has tended to fade, an indication that it is not the ink of native pre-Columbian or even early colonial manuscripts which still retains its dense blackness. A European ink subject to fading was used...' (Robertson 1975, 254). Lampblack was also the chief constituent of ancient Chinese ink (Li 1948, 119).

Maya ink was used for a system of conventionalized rebus writing:

> The characters they use are very different from ours and consist of blocks, hooks, loops, *limas* [?] and other objects arranged in a line like ours...Between the lines are drawn figures of people and animals... (Mártir 1964, 1:426)

Sheets in the books were 'fastened together with *a glue so strong and flexible* that, bound between covers of wood, they appear to have come from the hands of an accomplished bookbinder' (Mártir 1964, 1:425).[8] Not all constituents of the glue have been identified, but one was obtained from the root of an herb of the orchid family, *tzacutli*, identified as *Epidendrum pastoris* (Lenz 1973, 162; also Christensen 1963, 361). This glue was also smeared over the figures painted on paper to make their colors adhere firmly and prevent their

[8] These are generally called screen-fold books, and appear to have been a widespread type:

> The date of the first manufacture of bark paper and of manuscript books in screenfold format in Mesoamerica is unknown. Both of these traits may be the result of *diffusion from Asia*. (Glass 1975, 14:3)
>
> The *screenfold format* has also been reported as a cultural trait from Europe, China, and southeastern Asia, where it is still used as a vehicle for Buddhist literature. (Glass 1975, 14:8)

fading.[9] Glue on pre-conquest books was made, as well, from animal sources.

Bark-Paper in Ethnography

Though details of how the paper was manufactured are scanty in the early Spanish accounts, practices of Indians who still live in the area shed much light on the subject. At the beginning of the 20th century it was discovered that bark-paper was still being made in isolated Otomí villages in the Mexican state of Puebla, and in much the same way as it was in the 16th century. The fibers used by the modern Otomí for making paper come, as in earlier times, from bark of wild trees of the mulberry family. The process of making paper and the alkali treatment of maize are, and probably were, closely associated. Among the Otomí,

> at San Pablito two kinds of bark are used: *moral* gives a whitish, *xalama* a purplish paper. The bark is best gathered when full of sap, but is kept after drying. A board is used for a foundation on which to beat. A stone approximately rectangular and generally with the corners grooved for convenient grasping is used for a beater. The bark is carefully washed in *lyewater* [or limewater], *taken from maize that has been prepared for tortillas;*[10] it is then washed in fresh water and finally *boiled until it shreds readily into slender strips.* These are arranged upon the board...They are *beaten with the stone until the spread fibers are felted together.* The sheets are dried in the air, folded, and done up in packages...The work is done by women[11] and usually in the houses *with a certain degree of secrecy.* (Starr 1901-1902, 1:81-82)

[9] Francisco Hernández, writing in the 16th century, described how the roots of *Epidendrum pastoris* (*tzacutli, tzauhtli, tzavctli*) were cut in small pieces, put in the sun to dry, then ground to a powder which was mixed with water. The prepared glue was 'used by painters to fix their figures strongly, thus the images which they paint are not obscured or erased so quickly' (Hernández 1888, 246).
Hans Lenz conducted further experiments on *Epidendrum pastoris* and added to the information provided by Hernández:

> We were able to find the plant in the mountains of Tepoztlán, Morelia, and investigated the glutinous material produced by its tubers. This is a true natural glue, similar to gum arabic and cherry-gum. Possibly we are dealing with a mixture of complex polysaccharides like pectin...The gum of *tzauhtli* is slightly soluble in cold water but quite soluble hot water. It is insoluble in alcohol and *contains no starch.* If the tubers are cut in small pieces and dried in an oven at fifty degrees centigrade—and afterwards pulverized in a mortar until one obtains a yellow gray powder which is moistened with water—one obtains an adhesive of very good quality. (Lenz 1973, 162-64)

[10] There seems to be an intimate association between the making of bark paper and the alkali process, or 'nixtamalization,' of maize (Coe and Kerr 1997, 143; Christensen 1963, 363).

[11] A more recent report on Otomí procedure adds this detail:

> The men collect the bark, and *the women do the actual paper-making.* After the peeling, the inner bark is separated from the outer bark and sold to the women. It may be dried and stored away as it is for later use; but before it is used, it must be boiled in ash-water, or lime water, in which the corn for the tortillas has been soaked [that is, the liquid remaining after making nixtamal]. It must boil for several hours, generally from three to six; then it is rinsed in clean water and is finally ready for use. While making the paper, the women keep the fibers in a wooden bowl filled with water to keep them soft. The paper is made on a wooden board, the size of which depends on the size of paper wanted...A woman spreads a layer of fibers on the board and beats it out with a stone until it is felted together. The stones are either grooved or smooth on the pounding surface with fluted [grooved] sides...The boards with the wet fibers are placed in the sun to dry, and after a while the paper can be lifted off the board. (Christensen 1963, 363)

Such procedures have the following benefit: paper of better quality contains less lignin and resin.

Though present Otomí methods are essentially the same as those practiced by the ancient Aztecs, no doubt they have become simpler.[12] Moreover, paper had a more practical application among the Aztecs (e.g., book-making), and was omnipresent in their religious ceremonies. In contrast, the paper made by present-day Otomí has few practical uses; it is 'not used for writing or printing, but for witchcraft' (Christensen 1963, 364; also Starr 1900-1902, 308).

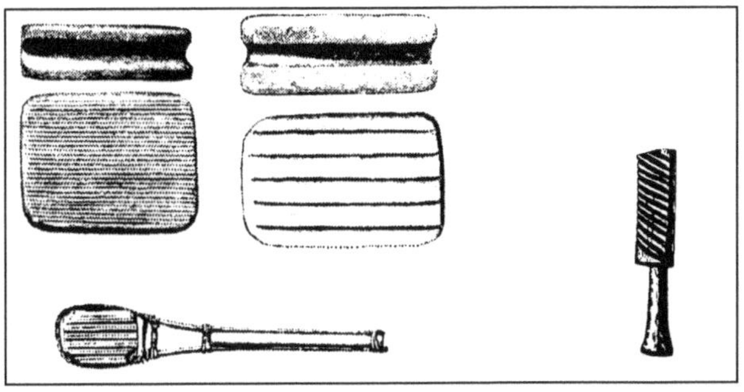

On the left: Stone barkbeaters, one with a racket-type handle (from Linné 1934, 203). On the right: A Cayapa wooden barkbeater from Ecuador (Barrett 1925, 106). In both South and Mesoamerica the barkbeaters used for making bark-cloth are mainly wooden mallets. In contrast, beaters used by the Aztecs in the 16th century and by present-day Otomí for making paper are made of stone. Stone beaters found in the Basin of Mexico are typically rectangular with more-or-less rounded corners; a deep groove running around their edges makes it possible to fasten a handle. The flat beating-surface of the stones is usually 'fluted' (i.e., grooved), often on both sides. Their size and weight, as well the prominence of their fluting, vary considerably.

Figure 31. Barkbeaters

'True Paper'

Having been written upon and bound into books, Mesoamerican bark-paper clearly functioned as 'real' paper. Technically speaking, however, in manufacturing true paper the raw plant material is almost completely reduced to a dough-like mass of cellulose fibers, which are then collected on a rectangular screen to form paper sheets. True paper (i.e., the paper of today) was first made in China in the first century A.D. (Li 1948, 159).

[12] Certain flat-iron shaped artifacts (*planches*) made either of stone or baked clay are commonly found in the Valley of Mexico. These have flat, smooth working surfaces, and are thought to have been used for smoothing the absorbent surface of paper sheets before the latter were coated with sizing (von Hagen 1943, 65, Plate 36). Some of these stones are 'on their sides heavily incrusted with lime,' suggesting 'that they must have been used for some different purpose, e.g. for smoothing plastered walls and floors' (Linné 1934, 198). Other stones, however, have no such encrustation (Lenz 1973, 84). Whether the planches were actually used for 'ironing' bark-paper, is an unsettled question. The stones may have had multiple uses.

Although making true paper in ancient China involved a more complete disintegration of plant material than was the case with Mesoamerican bark-paper, the basic processes were quite similar. In both areas, the plant material was washed, soaked, cooked, pounded, and treated with alkaline materials, generally with lime or wood ashes (lye). In China after bamboo was pounded and washed, 'the flax-like soft mass is mixed with *milk of lime* and put into a cooking cask to be *boiled over a fire*...After this, it is transferred to a strainer and strained with the addition of the *hot liquor of wood ashes*' (Li 1948, 162). (Milk of lime, composed of quicklime and water, has the consistency of milk; it is synonymous with whitewash.)

From China, knowledge of this procedure was diffused westward by the Muslims, and reached Egypt around 900 A.D. Thence, it was carried to Spain by the Moors in the 12th century. Until that time, Mesoamerican peoples may have had writing materials superior, or at least of equal quality, to those used by inhabitants of Europe who wrote on parchment made from the skins of sheep or goats.

CHAPTER 6

COLORANTS AND DYEING

That colors often have special religious significance is well known; for instance, the sacred color of Buddhists is yellow-orange, and of Muslims, green. For the ancient Maya, that color was blue: 'Blue is the color symbol for virtue, for chastity, for religion, and, by extension, heaven' (Thompson 1932, 186). Blue was for priests, whereas black and red were for warriors. Blue was also the color chosen for holy objects. The balls of incense used in Maya religious rituals were colored a bright turquoise blue; sacrificial victims were painted blue (Landa 1941, 11); and in preparing

> for a general festival...in honor of all the gods...their purpose was to *anoint with blue bitumen* [*betún*]...all the appliances of all their pursuits, from *the priest to the spindles* of the women, and the *wooden columns of their houses.* (Landa 1941, 159)

In contrast, red was a sacred color for the Inca: even their stone tools were often painted red (Lothrop 1950, 161). Bernabé Cobo states that a llama ('which we call sheep of the land') was sacrificed: '...a smooth-furred sheep was sacrificed to the Sun *every day*. This animal was dressed in a red vest before it was burned' (Cobo 1990, 113 and 129).

For their sacrificial fires, the Inca went to great lengths to find wood of the appropriate color:

> since there was in the province of the Chicha Indians a *red firewood* that was excellent for carving, *even though it was two hundred leagues from Cuzco*, the Chichas brought the wood themselves, all carved and *prepared to be burned in the sacrifices.* (Cobo 1979, 234)

Colorants

Archaeological evidence indicates that Old World hominids were using colored mineral pigments for decoration some 300,000 years ago. A knowledge of such uses (and, no doubt, the knowledge of various vegetal colorants, too) was brought from the Old World by the first immigrants to America.

Organic Sources

Achiote (*Bixa orellana*), genipap (*Genipa* sp.) and carajura (*Arrabidaea chica*)—the latter also widely known as *chica* or *bija*—were among the most important organic sources of color in pre-Columbian times. Widely distributed, and frequently found together, they are planted around many tropical Indian households, and are still much used by the inhabitants.

Although scarcely known among non-Indians, carajura (this English word comes from the Tupi name) was of great importance in indigenous tropical America. Being a liana, it is associated with wooded areas and is either planted near trees or grows as an escape in regrowth. In the rainforest of Colombia, the frequent occurrence of carajura is 'the certain indication of the site of an old aboriginal settlement, destroyed after the conquest' (Vezga 1936). As a result of its semi-domestication and being traded about, carajura ranges from Mexico to the West Indies and Argentina and grows in a great variety of terrains.

Soot or lampblack, '*cisco de teas*' (literally, the 'dust of resinous wood'), was sold in Aztec markets for painting pottery and as writing ink. Where available, they used 'the lampblack of pine pitchwood...[as] a medium for blackening, for dyeing, for tracing lines' (Sahagún 1950-82, 12:242). Soot, the principal organic source of black coloring, consists mostly of carbon particles. Its pigment, almost unchanged by exposure to light and moisture, is extremely long lasting.

Special pottery pieces, similar to the upper part of a still, were used for collecting the soot from which they made a fine ink: 'they have certain receptacles, called *tlicomalli*, something like *alquitaras* [alembics]' (Sahagún 1956, 343). These pottery pieces, however, were not true alembics—albeit shaped similarly to the condensation chamber of a true still. They were simply ceramic pieces meant to act as collection chambers for the smoke.[1]

A few years earlier than Sahagún wrote, Oviedo had remarked on soot made from the smoke of burning pine wood by the 'indios chondales' (a group of Chontal-speaking Maya); the soot was formed into small loaves, wrapped in *bihao* leaves, and sold in native markets (Oviedo 1959a, 1:177). The Lacandones, a Maya-speaking group, still use a specially-formed piece of earthenware to collect soot from burning copal for making paint:

> [An] earthen cover...[is] placed over the burning *copal* to *collect the soot* for the manufacture of the *black paint*. This process has a ceremonial object as well. The rounded interior of the cover represents the dome of the heavens and the soot collected in it is symbolic of the black rain cloud. (Tozzer 1907, 71)

---

[1] In China soot was made in a sort of vertical 'kiln'—and, similarly,

> made either by *burning pine wood* or vegetable oils...Instead of the chimney, at its [the kiln's] top there was a jar of a capacity of fifty liters, on which five others were put one upon another. They were of different sizes, each having a hole at its bottom...When the lampblack produced in the jars gradually became thick, the fire was stopped and the lampblack was swept out by means of a feather brush. (Li 1948, 123)

The Aztecs also made an ink or a black paint from the fruits of the leguminous tree, *Caesalpinia coriacea,* or another species of the genus:

> It is the fruit of a large tree which grows in the hot lands...It is small and twisted...which one leaves in turpentine unguent...It is sticky...It is a medium for dyeing things black...It is a medium for painting, a medium for writing. (Sahagún 1950-82, 12:241)

The Aztecs made ink from yet other materials, including an unidentified tree, called *nacazcolotl*, which also grows in the *tierras calientes*: from the tree's fruit mixed 'with aceche and other substances is made a fine ink for writing' (Sahagún 1956, 3:342). *Aceche*, a Spanish term, is hydrous ferrous sulfate, $FeSO_4 \cdot 7H_2O$, known in English as copperas or green vitriol. (Somewhat confusingly, 'blue copperas' or blue vitriol is a name for bluestone, hydrous copper sulfate, $CuSO_4 \cdot 5H_2O$.) Aceche was sold in Aztec markets and served, as it did in the Old World, 'for many things, from dyeing to the *making of ink*' (Sahagún 1956, 3:343).

Mineral Sources

Chief among the minerals used as pigments are the ochers. When these minerals are ground to a powder, their color varies somewhat according to the fineness of pulverization. Ochers occur in three principal forms: (1) *Hematite*, or red ocher, is an iron oxide, Fe_2O_3. (Although hematite means 'blood-like,' the unweathered mineral is black. If finely ground, however, the powder is bright red.) (2) *Goethite*, or yellow or orange ocher, is an iron hydroxide, $Fe_2O_3 \cdot H_2O$. (3) *Limonite*, or brown ocher, is also an iron oxide-hydrate, $2Fe_2O_3 \cdot 3H_2O$ (Schmandt-Besserat 1980, 145).

It was probably these three ochers that the Andean Aymara recognized in Spanish colonial times under the general name of tacu:

> *Tacu* is found in the rich mines of Potosí and in other metal mines, especially in iron mines, in three different kinds: the first is an earth red like blood [probably hematite ocher] *which is used by painters and principally by gilders*; the second kind is the color of liver [probably limonite ocher], and the third and most common and used by the Indians to cure various illnesses, is yellow [probably Goethite ocher], *which the Indians sell in the plazas in small loaves and cakes*;... they drink it, powdered, in chicha which is their wine. This mineral is valued, and especially if it has been burned [calcined], to dry up any type of ulcer, and *unburned it is mixed with vinegar*...used against all kinds of inflammations. (Cobo 1956, 1:116)

In order to change the color of some minerals, calcination was carried out long before the human species migrated to the New World. Calcination of 'natural ochers intensifies their color and transforms yellow ochers into red and red ochers into deeper hues.' The process involves

> loss of water of the iron hydroxides and their conversion into iron oxides: $2Fe_2O_3 \cdot 3H_2O \rightarrow 3H_2O + 2Fe_2O_3$.
> The degree of temperature required is about 260-80°C. (Schmandt-Besserat 1980, 129)

In Mesoamerica, the Aztecs mined hematite (red ocher) and used it as 'a medium for beautifying, for reddening' for various objects (Sahagún 1982, 12:243). The Aztecs also used vermilion (that is, cinnabar or mercuric sulfide, HgS) for its scarlet pigment (Sahagún 1956, 3:343). In addition to cinnabar, two arsenic minerals occasionally replaced the ochers (especially in Peru) as a source of pigments: orpiment, As_2S_3 (lemon or golden-yellow) and realgar, As_2S_2 (orange to red). When ground, two copper carbonates, azurite or $Cu_3(OH)_2(CO_3)_2$, and malachite or $Cu_2(OH)_2CO_3$, furnish blue and green pigments respectively.

The Process of Dyeing

Before discussing dyeing itself, it should be noted that application of colorants does not, in general, constitute 'dyeing.' Countless colored organic compounds are present in nature, but the great majority cannot be used as dyes; that is to say, most cannot be attached to a fabric so that they remain fast. ('Fast' means that a colorant or dye is resistant to repeated washing or long exposure to light.)

Dyeing is a quite a different process from merely wetting a fabric with some colorant mixed with water. And it should be noted that the painting (or printing) of colors on fabrics—or on bark-cloth, bark-paper, hides, and gourds—is also a different process. Moreover, painting is often done with materials that are not water-soluble.[2] Also prehistoric Pueblo peoples scrubbed cotton fabrics with colored mineral substances (e.g., ochers and bluestone). Although these substances are called 'dyes' (McGregor 1931, 3), clearly their application is not a dyeing process.

Dyes are soluble compounds which when deposited on a fiber become insoluble. Thus, dyeing involves a mutual chemical reaction between the fibers and the applied color.

In pre-Columbian America, there were at least four kinds of dyeing: mordant-, direct-, saliva- and vat-dyeing.

Mordant-Dyeing

Most of the natural dyes require a fixing agent (mordant) to become fast, especially on vegetable fibers; these are the so-called mordant dyes. Mordants are usually mineral

[2] Although many of the famous textiles of the ancient Paracas culture (preserved until the present time because of the extreme aridity of coastal Peru) were dyed, some were only painted. These include fabrics of brilliantly painted cotton cloth, perhaps the oldest examples of their kind (Linné 1953, 114-115).
Even recently, on the other side of the Andes, cotton fabrics were often coated with paint:

> Painted cotton-cloth is to be found in South America chiefly on the Upper Rio Amazonas, on the Rio Ucayali in Colombia, and was formerly used on the coast of Peru. (Nordenskiöld 1924, 209)

substances that combine chemically with an organic dye to form insoluble compounds.

Although both vegetable and animal fibers were dyed in the Americas, the fibers were overwhelmingly vegetal, except in parts of western South America where the hair of domesticated camelids was available. In those parts of the New World where weaving was at its finest, cotton was by far the most important vegetable fiber. Maguey fiber (henequen) was also used in both Peru and Mexico to make clothing and footgear, especially among the poor. Although the footwear of the Andean people was often made of animal hide, the Inca, like the inhabitants of Mesoamerica, also used maguey fibers: In sixteenth-century Peru, the Indians wore '... sandals made of a root or plant they call cabuya' which yields 'white fibers, like hemp' (Cieza de León 1959, 54).[3] The Aztecs made 'sandals (cotaras) from maguey fiber, dyeing them diverse colors' (Sahagún 1956, 3:140).

Vegetable fibers are, of course, more plentiful and easier to obtain than animal fibers. The carbohydrate cellulose is the principal component of vegetal material.[4] Most cellulose fibers cannot be dyed without the use of a mordant for the dye to become fast. (A few exceptions, such as saliva-dyeing and vat-dyeing, are noted below.)

As well as minerals, parts of various plants may act as mordants, probably because of the presence of tannin in their wood, leaves or fruit. Tannin, a mildly acidic polyphenol, was formerly an important mordant because it allowed basic dyes to be applied to such vegetable fabrics as cotton. The pods of the tropical American tree, *Caesalpinia coriaria*, long used to make a black dye, 'contain 25 to 30 per cent of tannin' (Standley 1926, 423).

Dyeing thin twigs black for use in basketry was also noted in the Yurok-Karok area of northern California. The sticks were put 'in a mass of ground rotted acorns' where they were left 'for about five days' (O'Neale 1932, 27). As said before, oaks and their fruit have an unusually large amount of tannic acid.

Perhaps another example of tannin acting as a mordant is the following:

> in Guatemala an infusion of the leaves [from a species of *Jatropha*] is *used commonly by some of the Indians for setting dyes of cotton*... (Standley and Steyermark 1946, 6:127)

Mordants may be substances that are either inadvertently incorporated in the dyeing process (as in mud-dyeing) or consciously chosen beforehand.

Mud-Dyeing

Many tribes dyed vegetable materials black by burying them in wet, soft earth. For instance, in northern California among the Yurok-Karok basket weavers 'the most common dyeing method employed was to bury [thin] hazel or willow lengths in mud' (O'Neale 1932, 27). Iron tannate, a salt of tannic acid, (or other likely fixing agents) is often present at the bottom of a former pond where tannin-containing vegetable material is decomposing. Thus, no particular mordant was consciously selected.

> The most primitive form of mordant dyeing is the combined use of tanning and ferrous substances, e.g. in mud dyeing. In mud dyeing the iron *salts work as mordants*, and combine with the tanning substances to form indissoluble complex salts which draw on the fibre. It may also happen that the tanning substances themselves work as mordants and the ferrous mud as auxiliary mordants. Mud-colours are dark, ranging from buff-brown to black shades. Primitive as these methods are, the resulting colours are usually fairly fast...The submersion of the fibres or fabrics in ferrous clay or mud was no doubt the earliest method of mordant dyeing. *But to obtain a fast dark shade, it is essential that the material to be dyed* (bark, wood, bast) *should contain tanning substances*...Mud is an important factor in dyeing dark shades or black even in places where the craft has reached some degree of perfection. (Bühler 1948, 2496)

Whether the Maya, with a sophisticated technology, also mud-dyed their semi-domesticated sedge, the following passage does not say:

> They have a certain plant which they raise in their wells and in other places, three cornered like the rush but much thicker, from which they make their baskets [presumably a species of the genus *Cyperus* which has triangular stems and generally favors wet ground], and they are accustomed *to dye it in colors* and they make them wonderfully pretty. (Landa 1941, 195)

Consciously-Selected Mordants

In more complex processes than mud-dyeing, specific mineral mordants are deliberately chosen. In the 15th century, both in the Old World and the New, the most frequently-used mordant was alum; it 'was used by the Asiatic Indians, American Indians, Egyptians, Chinese, and Greeks' (Bender 1947, 7). In ancient times much of this alum may have been alunogen[5] (also known as

[3] The word *cabuya*, another name for fibers of agave-like plants, comes from the native languages of the West Indies, whence it—like the words *maguey*, *pita* and henequen—passed widely into mainland use as the Spanish conquest proceded (Sauer 1966, 61).

[4] In accordance with its natural functions in the plant world, forming the supporting structure of terrestrial plants, cellulose is chemically inert. Probably, this inertness is one of the reasons that vegetable fibers are harder to dye than animal fibers. That cellulose does not react readily with most chemicals is illustrated by its use in the modern chemical laboratory for making filter papers.

[5] Where such selected mordants as alum were used by peoples with a relatively simple technology, they have 'generally been taken over [i.e., adopted] from more developed civilizations' (Bühler 1948, 2485). For instance in the American Southwest, alunogen (and wood ashes, another commonly reported mordant) was used in dyeing by the Navajo before arrival of the Spanish (Amsden 1964, 74). The Navajo probably learned

'feather-alum' and 'hair-salt'), a hydrous aluminum sulfate, $Al_2(SO_4)_3 \cdot 18H_2O$—rather than what is called common (potash) alum, $K_2SO_4 \cdot Al_2(SO_4)_3 \cdot 12H_2O$. Thus, alunogen differs mainly from common alum in that it lacks potassium. (Alunogen is occasionally found in volcanic vents or as white, feather-like efflorescences on quarry or mine walls.)

Alum was a common item in Aztec Mexico: 'The mineral alum is well known...; there is much trade in it because dyers use it a good deal' (Sahagún 1956, 3:343). The following is a sixteenth-century description of the purified alum, decanted and concentrated, used in Aztec Mexico:

> The alum which is called tlalxoctl or sour earth...before anything else they grind up the luminous earth, and put it in certain large vessels made of clay with spouts, and later dissolve it in water, and that which is decanted off is useful; the decanted water is cooked for as long as seems necessary...and [the sediment] made hard enough to be easily divided into small pieces...and after being well thickened, it is sold in the markets and tiangues, white and lucid and transparent, of sharp taste and astringent; like ours brought from Spain with not one difference; dyers use it as mordant for their dyes;...it has the same effects and serves for the same purposes as our alum of Spain. (Hernández 1888, 310)

One can only make reasonable assumptions as to the identity of minerals mentioned in sixteenth-century literature. Hernández had for his time an unusual knowledge of mineralogy, but most of the early chroniclers were by no means so expert. Moreover, the uncertainty of identifications is occasionally accentuated by translators of their works.

Various other mineral substances were used as mordants in pre-Columbian times. Iron tannate (as in mud-dyeing) and wood ashes are commonly-reported mordants among American tribes; aceche, iron sulfate, was a common Aztec mordant (Sahagún 1956, 3:343), and is still widely used as such.

When ancient textiles found in Peru were sent to France for analysis, 'it was discovered that silicate of chalk, aluminium, silicate of aluminium, and oxide of iron were all used in ancient Peru as mordants' (Means 1964, 467). But the mordants used by prehispanic Peruvians are still poorly known.

Direct-Dyeing

That vegetable and animal fibers differ in their dyeing qualities was well known to early sixteenth or seventeenth-century Europeans, as is illustrated in the following comment on Indian clothing worn in highland Peru

> Cotton...is also abundant in this kingdom...Moreover, the Peruvians enjoyed a plentiful supply of wool from their llamas, and vicuñas, from which they made the majority of the cloth for their clothing. Both the wool and the cotton cloth are decorated with fine colors...Cotton does not hold its colors as well as wool. Newly dyed cotton has bright colors, but with use they fade. This does not happen to wool clothing, which always holds its original colors fully without fading. (Cobo 1990, 223-224)

Animal fibers, such as hair and feathers, can usually be dyed without mordants—that is, dyes can be applied directly to them from an aqueous solution. Animal fibers are proteins and amphoteric in character; that is to say, they are capable of functioning either as an acid or a base. Thus they can form salt-like combinations with either acidic or basic dyes. Although most animal fibers can be dyed without mordants, the fixing agent often changes or intensifies the dye's color.

The principal components of animal fibers, of hair and feathers, are keratins. These fibrous proteins, like cellulose fibers, are insoluble in water and relatively inert (although more active chemically than cellulose). Like other proteins, animal fibers such as llama and alpaca hair are amphoteric. Thus they are more easily dyed than vegetable fibers.

In Mesoamerica and southward to Colombia, there was little use of animal fibers in weaving, no doubt because abundant sources were not available there. Mesoamerica had no domesticated mammals from which hair or wool could be woven—aside from the dog. Except for feathers (which the Aztecs dyed yellow) and the pelts of wild mammals, few animal fibers were dyed there. One of the items traded in markets of the ancient Mixtec city of Cuixtlahuac, however, was 'thread made of [wild] rabbit hair, dyed in many colors' (Durán 1994, 184).

In Andean South America, where some of the world's finest fabrics were woven, the hair of several large mammals was available, especially the four native camelids: llama, alpaca, vicuña, and guanaco. The first two being domesticated, their hair was readily obtainable. The llama has 'relatively coarse hair.' In contrast, the alpaca 'has been bred for its long silky hair...The hair ranges in color from white, the most highly valued, to all shades of brown, as well as grey and black.' On the other hand, the wild vicuña 'has quite short light brown hair, but the fiber is of exquisite softness and was reserved for the finest textiles by the Incas' (Rowe 1884, 25).

The hair of at least two smaller mammals was also used in Peruvian fabrics:

> The very rich cloths made for the Inca and the great lords were made entirely or partially of vicuña wool, and they mix in viscacha wool, which is very thin and soft, and bat fur was also added, which is the most delicate of all. (Cobo 1990, 225)

much of their dyeing methods from the Pueblos, and the latter may have gained their knowledge of alunogen as a mordant from native peoples living farther south.

The 'viscacha' referred to in this passage is probably the mountain viscacha, Lagidium sp., a chinchilla-like burrowing rodent. The species of bat used is uncertain.

Saliva-Dyeing

Baskets (made using the stems of grasses and rushes, as well as the bast or withes of shrubs or trees) are among the oldest of dyed materials. Among peoples of simple technology, dyeing was often a cold process. Where pottery was unknown, other containers may have been used to hold the dye—for instance, water-tight baskets or bark containers.

In both the Old World and the New, saliva acting as a solvent has been used in the dyeing process:

> Saliva is an ingredient of dyeing among almost all primitive people...mainly used for the dyeing of red and brown hues...There are numerous records of saliva dyeing in America.' (Bühler 1948, 2511)

> Presumably *with its slightly acid effect* it [saliva] breaks up the substances of the plant which contain the colouring matter, *while by its enzymes it forms the emulsions which promote the dyeing process*...It must not be presumed however that saliva acts as a mordant. (Bühler 1948, 2486)

In California, the Yurok-Karok tribe dyed red or orange the stems of the chain fern (*Woodwardia fimbriata*) in making their basketry.

> The method of dyeing woodwardia most often mentioned...is that of drawing the strands through the mouth after chewing alder bark. Or..., the juice might be spit out into a basket and water added for the dye solution...Some alder trees give richer, darker colors than others. Bark from too young an alder will make a weak, light dye...The bark [after being peeled off the alder tree and cut] is about a half inch thick. It turns brilliant orange in about twenty minutes. (O'Neale 1932, 27-28)

Vat-Dyeing

Vat-dyeing often involves a lengthy fermenting process. It is discussed below, under 'Alizarin and the Vat-Dye, Indigo.'

Dye Sources

Where the early Spanish did not discover precious metals, they often found, instead, a treasure-house of dyes. In the two centuries following the Spanish conquest, dyes became a large-scale commercial product and, next to precious metals, the most valuable item carried back to Europe.

When he arrived in Mexico, Hernán Cortés spoke of 'spun *cotton, in all colors*' (Cortés 1908, 1:258). Both dyeing and weaving were done with much skill. The fine textiles woven by the Inca peoples were appreciated by the first Spanish to arrive in Peru: 'For these garments they had such *perfect dyes*—red, blue, yellow, black, and other colors—that *they truly excel* those of Spain' (Cieza de León 1959, 177).

In fifteenth-century America, many substances were used in dyeing. Only the most common are discussed here.

Alizarin and the Vat-Dye, Indigo

The remains of people possessing a Mexican variant of Basket Maker culture have been found in a cave on the Rio Fuerte in southwestern Chihuahua. Found with these remains were loom-woven dogbane (*Apocynum*) fibers. (No pottery or cotton fabrics accompanied the bodies.)[6] The dogbane fibers were dyed with several different substances, including the organic compounds alizarin and indigo.

The prehistoric Chihuahuans also applied mineral colorants. In the cave were found orange fibers which had been colored by an ocher in what seems to have been at least a staining, if not a dyeing, process. (The difference between dyeing and staining is by no means clear.) As Kasha notes,

> Direct application of the insoluble pigment to the textile [scrubbing it with ocher] is a possibility, although chemically less plausible...But the use of ferric oxide [red ocher] as a textile pigment is more complicated, and *involves dissolution of the mineral in some* [organic] *acid as a first step*, then application of the solution to the textile, followed by *precipitation of the insoluble ferric hydroxide within the fiber*...A textile dipped into such a solution of ferric iron and *then dipped into a boiling solution of plant ash leachings (as a source of alkali)*, would give the desired reaction, precipitating ferric hydroxide *within the fiber by hydrolysis*. (Kasha 1948, 155)

To return to organic dyes, blue-black and brown fibers found in the cave were dyed with alizarin; these fibers were 'mordanted by ferric hydroxide and (chiefly) aluminum hydroxide, respectively' (Kasha 1948, 156). Alizarin is especially common in plants of the madder family, Rubiaceae; the best known alizarin-producing plant in the Old World is madder, *Rubia tinctorum*.

[6] Unfortunately, the dyed materials from Rio Fuerte cannot be dated more precisely; the 'only information obtainable as to [the materials'] provenience was that they came from a cave in the Sierra Madre of southwestern Chihuahua...' (Kidder 1948, 99). This Mexican variant of Basket Maker Culture has been called 'Rio Fuerte Basket Maker' (Kidder 1948, 100).
In the southwestern United states, itself, the Basket Maker people belonged to a culture which arose around 100 B.C. and preceded Pueblo culture. Unlike the Pueblos, the Basket Makers did not possess cotton; instead, they wove dogbane and yucca fibers to make baskets, bags and sandals (Amsden 1964, 70).

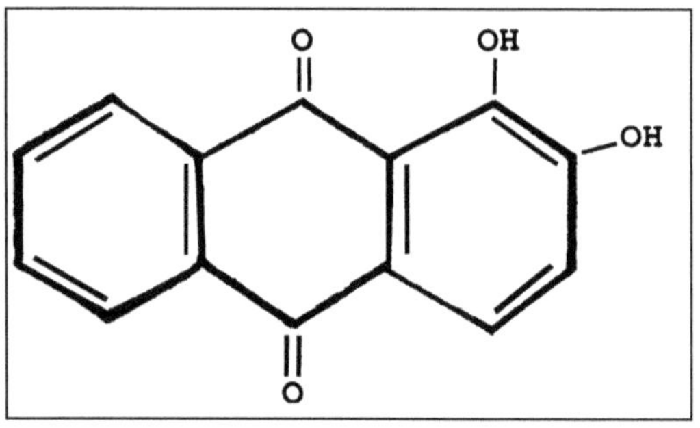

Figure 32. Alizarin

When madder roots are crushed and allowed to ferment, a glucoside in them undergoes decomposition with the production of alizarin. Alizarin is a well-known example of the so-called polygenetic dyes, which yield different colors with different mordants. Itself a reddish-orange compound, it dyes fabrics various shades of red, the actual shade depending much on the mordant used.

The Chihuahuan cave also yielded fibers which had been dyed with the vat-dye, indigo. In the fabrics, both dyes (alizarin and indigo) 'were modified somewhat by the yellow-brown color of the *Apocynum* fiber' (Kasha 1948, 156). The probable plant-sources of these dyes are unknown.

In the fifteenth-century, indigo was the world's most extensively used blue dye, being known to dyers in all of the inhabited continents, except Australia.[7] In Eurasia, indigo was 'already in use by the third millennium B.C.' (Balfour-Paul 1998, 12). Though perhaps not so old in the New World, its use is of considerable antiquity.

Although indigo is found in more than fifty plant species, distributed in several plant families, the dye in its natural form has come mainly from the leguminous genus *Indigofera*. Several species of this genus have been cultivated since antiquity in both the Old World and the New.[8] Before arrival of the Spanish, the common indigo of Central America was *Indigo suffruticosa* or *I. guatemalensis*. The green leaves of indigo-containing plants give no hint that they contain a blue dye.

Indigo is a highly insoluble compound and needs fermentation and a reducing agent to make a form that is soluble. It is applied principally to vegetable fibers, and no mordant is needed. The word 'vat,' obviously derived from the fermentation container, refers to 'the deep wooden barrels used by medieval woad dyers' (Balfour-Paul 1998, 116). The Old World woad, *Isatis tinctoria*, is also a vat-dye plant, long used for dyeing flax.

Basket Maker times seem surprisingly early for such a complex process. (Southwestern Chihuahua was at, or beyond, the extreme northwestern edge of Mesoamerica.) This early appearance has prompted some to speculate that vat-knowledge in the New World actually originated in Chihuahua and spread southward into the technologically more-complex cultures. On the other hand, as mentioned earlier, 'such evolved forms of vat and mordant dyeing as may be found among primitive people are almost certainly derived from highly civilized peoples' (Bühler 1948, 2505).

Vat-dyeing depends chiefly on hydrolysis and reduction-oxidation reactions. Steps in the process may be summarized as follows:

(1) The leaves of an indigo-containing plant contain a precursor of the dye in the form of a water-soluble glucoside, namely indican. When the leaves of such a plant are crushed and soaked for a few days in warm water, an enzyme present in the leaves themselves causes fermentation, and the indican is hydrolyzed, or 'split,' into glucose (a simple sugar, often found in ripe fruit) and indoxyl—the latter being 'a reduced and colorless form of indigo' (Kasha 1948, 155).[9]

(2) Atmospheric oxygen is added to the indoxyl by vigorously agitating (stirring or beating) the fermenting liquid. As a result of this oxidation, the indoxyl is converted to the insoluble compound, indigotin, sometimes called 'indigo blue.' In vat-dyeing, this blue color is developed in the fabric only after it is lifted from the yellowish-green dyebath and exposed to air.

The insoluble indigotin in the bath, when treated with the alkaline substance, is converted temporarily into a soluble 'leuco-form'—or, as it is sometimes called, 'indigo white.' The reducing agent prevents the oxygenation of the leuco-form. The process is reversible.

(3) When the fabric to be dyed is immersed in the fermenting liquid, the leuco-form is

> deposited on the surface of the immersed fibres so that when they emerge they too are yellowish-green in colour and only turn blue after several minutes exposure to the air, when oxygen converts the dye back to its permanent blue form again. (Balfour-Paul 1948, 116)

[7] In China, 'indigo is the oldest and most extensively used blue dyestuff. It was used as a blue color at a very early time in the Orient, especially in India and China...Indigo occurs in nature in considerable quantities and in various species of plants, such as *Isatis tinctoria* [woad], *Polygonum tinctorium, etc...*' (Li 1948, 140).

[8] It is sometimes said that a single species of the genus *Indigofera*—namely, *Indigofera tinctoria*—was used in both hemispheres before the European discovery of America: '*Indigo tinctoria*, indigenous to many parts of Asia, Africa, the East Indies, the Philippines, *and the Americas*' (Bender 1947, 3). It is likely, however, that *I. tinctoria* was confused with closely related species, native to the western hemisphere. (Identification of indigo source-plants by means of chemical analysis of the dye on ancient fabrics is rarely possible.)

[9] In a primitive version of the dyeing process, plants that are used as a source of indigo are 'left to ferment in a watery solution, which is then used as dyebath'; this 'may be classified technically as direct dyeing method' (Bühler 1948, 2494). This is, however, not a complete vat-dye. A fabric soaked in the dyebath is colored but a weak shade of blue.

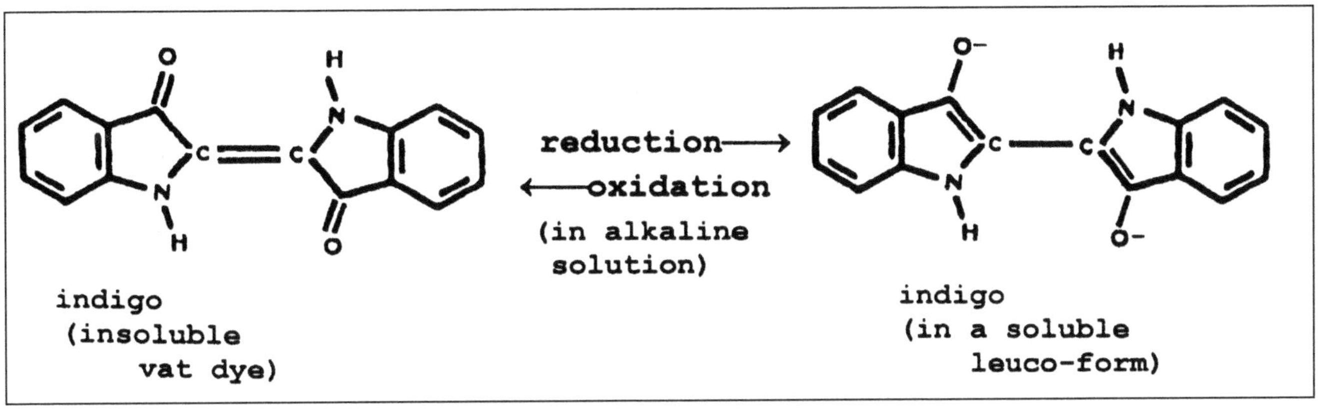

(Kasha 1948, 155; Goffer 1980, 182; Balfour-Paul 1998, 234)

Figure 33. Reduction and Oxidation of Indigo

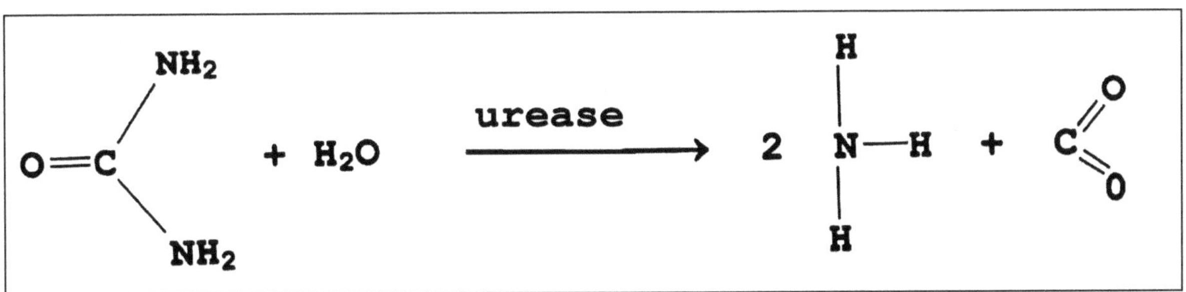

Mammalian urine is over 90% water. In addition to sodium chloride and other substances, it contains small amounts of urea. The nitrogen in the urea molecule constitutes most of the nitrogen in urine. During fermentation, the urea in mammalian urine (in contact with bacteria which supply the enzyme urease) hydrolyzes, releasing ammonia and carbon dioxide.

The carbon dioxide is released into the air.On solution in water, ammonia becomes ammonium hydroxide, NH_4OH, which is strongly alkaline

Figure 34. Urea and Ammonia

Thus, the leuco-form is reoxidized to the insoluble, blue-colored indigo when the textile is repeatedly dipped in, and lifted from, the dyebath. The dye, which is deep blue, then becomes extraordinarily fast to washing and exposure to light.

(4) In places, an additional (and more-sophisticated) step was taken: Instead of immediately immersing the fabric in the ferment liquid, sometimes

> ...the ferment liquor, which varied in shade from yellow-orange to olive-green, was drawn off and aerated so that oxidation took place and indigo blue precipitated out. The top liquor was decanted and the sludge heated to stop fermentation; more liquor was filtered out, and the paste, after drying in blocks, was ready to ship. (Bender 1947, 3)

When this step of extracting the paste-like blue pigment is taken, the indigo can be stored for long periods; it can also be easily traded—especially if dry cakes or balls of the hard, rock-like indigo are prepared. Whether, as in the Old World, indigo cakes were prepared in pre-Columbian America is uncertain, but very likely they were. Certainly, other coloring materials were treated in this manner: cochineal, achiote, and carajura were dried and made into cakes which were commonly sold in native markets or traded from place to place.

When dried indigo is extracted, it must be made soluble again if it is to be used as a dye. Since indigo is soluble in alkaline liquids, commonly some basic substance, such as urine or small amounts of lime or plant ash, was added to the dyebath as a reducing agent: 'Stale urine, whether human or animal, is for obvious reasons the most easily available ingredient to provide gentle alkalinity for the vat' (Balfour-Paul 1997, 86). Very likely urine was used in dyeing with indigo in pre-Conquest Mexico and other parts of the Americas; but, although Bender comments that urine is 'known to have been used by the Aztecs' (Bender 1947, 7), I was unable to find early references to that practice.

Probably the indigo pigment was extracted from the dyebath in prehispanic times; a description written soon after the Conquest describes the way in which Aztecs extracted indigo from an *Indigofera* species:

> It is an herb. Its growing place is in the hot lands. It is pounded with a stone. *The juice is squeezed out.* It [i.e., the pulpy mass of leaves] is wrung dry. (The juice) is placed in a bowl. There it *becomes thick*; there the *tlaceuilli* [i.e., the pigment] gathers. This color is *dark blue, gleaming, greenish.* It is a *dyeing* medium, a medium *for painting black*, for painting in colors. (Sahagún 1950-82, 12:242)

The fact that the juice became 'dark blue' shows that the pigment was oxidized, and the fact that it was placed in a bowl where it became 'thick' indicates that it was concentrated to something like a paste-like consistency. Whether the concentrated pigment, or the plant's dry leaves, was imported is not said. (Since the plant is described as growing in the 'hot lands' of Mexico, it seems not to have been grown at Tenochtitlán itself which is at an elevation of 2247m above sea level.)

How much the dyeing activities of Mesoamerica were influenced by Old World custom in the first few decades after the Conquest is an open question. The Spanish, too, were well acquainted with vat-dyeing. Indeed, in the American Southwest the use of indigo vat-dyeing with urine was introduced among the Pueblo and Navajo peoples by the Spanish (Amsden 1964, 73 and 90). Day-to-day dyeing operations were left for some time in Indian hands, but the Spanish soon became involved both in the trading of dyes and the transplanting of dye plants, because of their commercial potential. For example, the Spanish introduced an Old World plant, *Indigofera tinctoria*, in the last half of 16th century. In some areas it replaced local species of *Indigofera*.

Like the Aztecs, the Maya used indigo and harvested cotton 'in wonderful quantity'; most of their textiles were made of cotton fiber.[10] In Peru, also, 'some of the blues were derived from indigo' (Means 1964, 467). In Chancay times (i.e., between 900 and 1200 A.D.), indigo and cochineal were used to dye cotton fibers and alpaca hair. Alum and some iron mineral were used as mordants (Lothrop and Mahler 1957, 33-34).

A description of indigo processing in the last part of the 16th century shows that Aztec dyers at that time were indeed extracting solid indigo from its aqueous solution. Chopped leaves of indigo plants were put in an earthen pot or '*un baso de cobre*' (a vessel made of copper), full of either lukewarm or cold water, to ferment, and then *shaken* vigorously. Next the liquid was poured into a large earthen jar which had a hole or spout near its top. Through this aperture the upper liquid containing the dissolved dye was decanted. The leafy mass which settled to the bottom was wrung in a coarse cloth. All of the liquid that was gathered was 'placed in the sun' to evaporate, and the concentrate was 'shaped into 'tortillas'.' These tortillas were hardened and dried by being placed on plates 'over live coals' (Hernández 1888, 90-91). A Spanish influence may be indicated by the reference to a '*baso de cobre*': the Aztecs made many copper artifacts, including copper cups, but there seems to be no record of their making a metal vessel of that size.

Another way of making the extracted indigo soluble again has been suggested:

> To solubilize solid indigo it is necessary to reduce it to *indoxyl*. Perhaps the most common and available reducing agent was glucose or a mixture of fructose or sucrose found in honey, sweet fruits, sugar cane [the latter, however, not being available before the 16th century], and similar plant materials. By the simple process of *boiling indigo with solutions of these natural sugars in the presence of a little lime, the indigo is reduced and solubilized.* (Littmann 1982, 405) (Littmann believes, however, that it was boiled fresh plant material, rather than the insoluble extract, that was used in making the paint called Maya Blue, discussed below.)

Dyeing was, of course, a smelly business. However, one should not assume that use of urine in vat-dyeing is merely one of those unpleasant practices characteristic of 'primitive' peoples: the history of urine vats extends from ancient Egypt to twentieth-century Europe. In New England, descendants of those who crossed the Atlantic in the Mayflower also used human urine in dyeing. This urine

> was often called chamber dye and since the dye had to stand to allow penetration of the fabric or fibre the offensive smell was offset by many bunches of sweet herbs set around the room. James Franklin, elder brother of Benjamin, dyed and printed calico in the 1720s. (Robinson 1969, 32)

Indeed, in the first half of the twentieth century, in 'the Highlands and Islands of Scotland housewives still kept a *urine-based* woad or *indigo dye pot* permanently on the go' for dyeing tweeds (Balfour-Paul 1998, 125). (Urine has, of course, lost most of its place in the chemical arts. But from such humble beginnings, one of the most important of modern sciences arose: In 1776 the Swedish chemist, Karl W. Scheele, discovered uric

---

[10] With regard to the history of weaving in South America,

> the earliest evidence for textile fabrication in Peru surfaces [was] *almost two millennia before the cultivation of cotton...From approximately 5000 to 3000 B.C.*, looped and twined constructions are fashioned exclusively from *bast fiber*...During the cotton preceramic period, *a limited number of dyes also were developed*... (Doyon-Bernard 1990, 88-89)

In Peru cotton was 'first cultivated along the coast perhaps as early as 3500 B.C.' (Doyon-Bernard, 1990, 88-89). It has been suggested that the earliest use of the wild cotton plant was for its bast fibers—that is, the fibers in its stem rather than its seed hairs, which are unusably short in undomesticated cotton. Plants of the mallow family, to which cotton belongs, often contain useful basts.

Weaving cloth requires the use of some sort of loom—that is, a frame to hold the warp system taut. Using a heddle and long loom poles, broad pieces of fabric can be woven. Some of the Chihuahua cloths 'were woven on stationary looms. The backstrap type...does not plausibly explain [the] wide materials' (O'Neale 1948, 119).

The heddle is a thin rod from which hang loops that encircle selected warp threads so that the latter can all be raised together. This ingenious device was found in both the Old World and the New. In the New World, the heddle first appeared surprisingly early in Peru: 'Between 2000 and 1800 B.C., the sudden proliferation of woven pieces, their size, and their uniformity *suggests the introduction* of the heddle loom' (Doyon-Bernard 1990, 71).

acid in urine. Organic chemistry began when Friederich Wöhler in 1828—after uric acid had been extracted from urine and analyzed—synthesized urea from inorganic materials without the help of a living mammalian intermediary.)

In summary, the first steps toward dyeing fibers (for example, the dyeing of baskets) may have been based on plant knowledge that was available to the first immigrants in America. (Little is known about the fund of technological ideas that were stored in their cultures.) Surprisingly, as evidenced by the Chihuahua Basket Makers, pottery was unnecessary for the vat-dyeing process to be carried out successfully. Moreover, vat-dyes (and, probably, extracted-indigo pigment) were known in the Americas before the European conquest.

Shellfish Dye

Shellfish dye was used both in the Old World and in pre-Columbian America. Chemically, the dye of the New World is the same as that used in the ancient Near East and Mediterranean area—the celebrated Tyrian purple. The dye is produced mainly by several edible genera of marine univalves—namely, *Murex*, *Purpura* and *Thais*—which occur in all oceans.

In coastal areas, marine shellfish have been a major source of human subsistence since the human species arose. Thus the shells' characteristics were intimately known, and the coloring properties of these dye-producing genera were likely noticed long ago and in many different places.

The structural formulas of marine shellfish dye and the indigo obtained from plant sources are much the same, except that shellfish dye contains bromine: 'It is of course not surprising that an indigo of marine origin should be halogenated, since the salts of seawater include halides [including bromine] in considerable amounts' (Kasha 1948, 153). Through oxidation, when exposed to air (and to sunlight) the fresh, frothy milky-white fluid changes to green and finally to purple. Like indigo, shellfish dye remains fast and requires no mordant. Unlike indigo, it requires no lengthy fermenting process and can be applied immediately.

The first written report of shellfish dye in the Americas comes from around 1529 from the Gulf of Nicoya in Costa Rica, where the natives 'dye mantas, and cotton thread and other things they want to color, and there are various shells or oysters for making purple' (Oviedo 1959a, 4:424). Throughout the 17th and 18th centuries in Costa Rica, a Spanish dyeing-industry used Indian laborers to gather shellfish dye.

The dye was usually (but not always) applied to cotton. In the mid-seventeenth century, around Nicoya Peninsula, it was also used to dye pita thread:

> [The Indians make] a kind of thread called *pita*, which is a very rich commodity in Spain, especially of that *purple color* wherewith it is dyed in these parts of Nicoya. The Indians...work about the seashore to find *certain shells wherewith they make this purple dye*. There are also shells for other colors... (Gage 1958, 322)

In this area, pita was probably not an agave fiber, but the saltwater-resistant fiber of the 'wild' pineapple plant, *Aechmea magdalenae* (Gordon 1982, 72).

In the last years of the 18th century, at the port of Tehuantepec in Oaxaca,

> the Indian women collect the purple liquor [of the mollusk], following the course of the shore, and rubbing the cloak [i.e., the interior?] of the murex with *cotton stript of its seed.* (Humboldt 1966, 3:83)

At the end of the 19th century shellfish dye was made at the town of Huamelula on the south side of the Isthmus of Tehuantepec (Starr 1900-1902, 1:68). A few years later, use of the dye was again reported from Huamelula:

> [Fishermen] started in boats laden with cotton skeins and went northward along the coast...Slipping a skein over his left wrist, the fisherman wrenches one sea-snail after anther from the wet rocks, *blows on it, causing it to exude the dye-stuff which resembles a milky froth, and then dabs the cotton thread with numerous shells in succession, until it is thoroughly saturated.* When each shell had yielded its small supply of liquid dye, some fishermen pressed it to the rock and waited until it adhered thereto, but others laid the shell in a pool. When treated thus the same shells yielded a second though diminished supply when the rocks were visited on the return journey. (Nuttall 1909, 369-370)

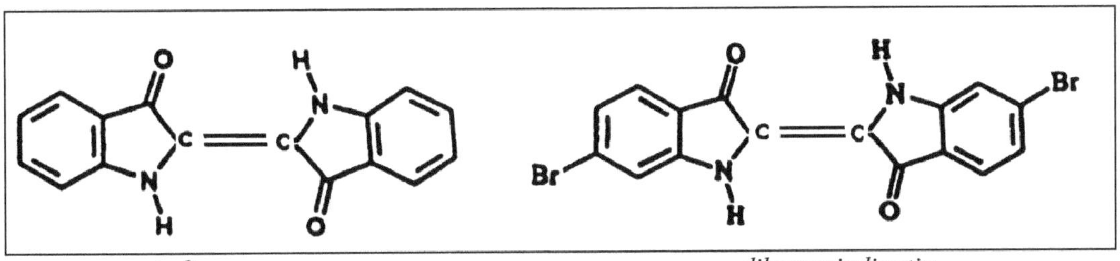

indigo *dibromoindigotin*

Indigo and dibromoindigotin (the main colorant of shellfish dye) have similar structural formulas

Figure 35. Indigo and Dibromoindigotin

The colorless fluid emitted by the mollusk acquires color when exposed to the air:

> After having been moistened by the dye fluid, the cotton thread turned green, and, when exposed to and dried in the sun, the green turned purple and became permanent...A *disagreeable* feature of the dye...[is] the strong fishy smell, *which appears to be as lasting as the color itself.* (Nuttall 1909, 370)[11]

The Boruca (Brunka), on the Pacific coast near the border of Costa Rica and Panama, are probably the only Central American tribe that continues to make shellfish dye:

> Here a mollusk (*Purpura patula* Gould), called la morada, is gathered at low tide off the rocks...On obtaining the mollusk, they first turn over the shell so that the sea water runs out. Then they blow into the shell, *holding the thread in front...* Another, not so strong, purple dye is gathered from a small mollusk (*Purpura kiosquiformia* Duclos) which lives on the roots of the mangrove trees...The soft part of the shell is removed, and the liquid it contains is put on a thick stick and *drawn* on the material in the same manner as using a colored pencil. When first applied, the liquid is black, but turns purple on drying. (Stone 1949, 17)

Although use of shellfish dye has largely disappeared in South America, the Cayapa Indians who live near the coastal border of Ecuador and Colombia recently got 'lavender from an ocean univalve' (O'Neale 1963, 125). Whether shellfish dye was used at one time in ancient Peru was, for a time, unsure. In 1962, however,

> ...a selection of pre-Columbian purple fabric samples from Peru was subjected to spectrophotometric analysis...Seven of them were of alpaca or vicuña wool in *which purple came from a single unidentified source*. The remaining two samples, *dating perhaps from the first century B.C.*, were *white cotton* mummy wrappings from the Ica valley, in which purple designs in the shape of hands and dots *had been painted onto the woven fabric*. In these the source of the color was positively identified as shellfish. Thus it appears that shellfish dye was at least occasionally used in ancient Peru as a pigment on cotton... (Gerhard 1964, 179)

Although the shellfish dye-fluid was usually applied to cotton 'wool' from the boll or cotton thread, according to this report, it was sometimes applied to woven fabrics as well.

[11] It would be surprising if the strong fishy smell lasted so long, since 'the amino group in the molecule is tightly bound....It indicates that some kind of volatile amine is formed somehow—possibly by hydrolysis under the influence of water and sun' (personal communication, Dr. Hanns Ahrens).
Zelia Nuttall may have been wrong about how long the odors on such fabrics last. In fact, a piece of cotton fabric—dyed purple with this mollusk dye (and blue with 'indigo, of which the woof consists' Nuttall 1909, 369)—is enclosed in Zelia Nuttall's book in the Richmond library of the University of California, Berkeley. Although the colors are well-preserved, as far as I was able to detect, the fabric has no odor. To be sure, however, the fabric has remained protected there from the weather, and mostly from the sunlight (a principal cause of deterioration of dyes), for almost a hundred years.

Sacatinta

In Central America several species of *Jacobinia* shrubs, called by their Spanish name, *sacatinta*, are used (or were recently used) by the Boruca (Brunka) on the Pacific slopes of Costa Rica and in highland Guatemala. In Guatemala it is often mixed with indigo; in Costa Rica the sacatinta is used alone. In Costa Rica, handfuls of leaves and twig-tips of Jacobinia tinctoria or J. macrantha are lightly heated over a flame, then thrown into boiling water. The mixture produces a dark bluish-purple dye that is used to dye white cloth (Pittier 1908, 132). (Since Pittier mentions no mordant, this dye may not be fast.)

The following account describes the dyeing of cotton in highland Guatemala and the preparation of blue dye in, and around, the towns of Totonicapan and Salcaja. Although the two principal dye materials are indigenous, a number of introduced traits are apparent. (Synthetic aniline dye powder is sometimes used as a supplement; not only cotton is dyed, but sheep wool as well; moreover, while some dyeing is still done in large earthenware pots and by women in the home—probably, as it was originally, most dyeing now is done on a commercial basis using concrete vats with much participation by men.) Of the two native shrubs used in the dyeing process, the better known is indigo from an Indigofera species. The other is *Jacobinia spicigera*. The sacatinta supplies 'come from the coast and...[are] brought to the markets in the form of great bunches of green leaves'; it is 'a plant of great importance to the dye industry in the Totonicapan area' (O'Neale 1945, 24):

> The initial dyestuff comprises a blend of natural indigo, powdered, and the cheaper imported aniline dye to which is added about eight to ten times its weight in sacatinta leaves and stalks. The ground indigo, dye powder, and green leaves and stalks are put to steep in one of the vats. After a couple of weeks *the dyer tells by the green color on his stirring stick and the bubbles rising from the fermenting sacatinta at the bottom of the vat, by the smell, and by the surface foam that the mixture has aged properly*. The mass of greenish black foliage is dragged out, and the dye is ready for use.
>
> Mordant for indigo dye consists of wood ashes from Cajola, the village producing charcoal, and lime from San Francisco el Alto. These are mixed together...Over them [are] poured...cans of water. The filtrate is ready for immediate use, or it may be stored. With age the liquor becomes stronger and better, *turns to purple, and develops the odor of ammonia.* (O'Neale 1945, 25)

(There is no mention of urine here, but the odor of ammonia probably came from decomposing urine used in the vat.)

Cochineal

Cochineal, one of the first dyes brought to Europe from the New World, is produced by the domesticated scale insect, *Dactylopius coccus*, which feeds principally on leaves of the prickly-pear cactus (*Opuntia* sp.). Related wild scale insects contain a similar or identical dyestuff, but not in equal quantities. Both the insect and the dye itself are known as cochineal.

The Aztec name of the cochineal insect was *nocheztli*; the common Spanish name was *grana*, or grain, because of the seedlike appearance of the dried insect bodies. Many Mexican place names are associated with the insect's culture: Nochistlan, Autlan de grana, and so on.

The insects were brushed from the prickly-pear cactus into bowls; then, to make what was considered the best dye, the insects were either spread on mats and dried slowly in the sun or killed by hot water. If dried, the insects were ground on a metate and put to soak. This nocheztli or grana 'is sold in the [sixteenth-century Aztec] markets made into loaves, so that it can be bought by painters and dyers' (Sahagún 1956, 3:341).

After discovering the use of cochineal dye in Mexico, the Spanish began an extremely profitable trade in the product. They established plantations throughout tropical America, including Peru, for cultivation of the insect and its cactus host. As a result, it is unclear whether the domesticated Mexican variety of cochineal insect was also cultivated in prehispanic Peru, or whether a less-productive wild species of *Dactylopius* was used as cochineal dye. The question is still unsettled.

The Inca's preference for red has already been noted:

> Both the [camelid] wool and the cotton cloth are decorated with fine colors and excellent weaving. For dyeing it they have excellent colors of blue, yellow, black, and many others, *and especially crimson or scarlet*, which are known to be better than the dyes from many parts of the world and compete favorably with the best that can be found anywhere. (Cobo 1990, 223)

Cochineal is a mordant dye, and the mordant most commonly used with it was probably alum. In ancient Peru, however, both alum and an iron compound were used as mordants to obtain fast red shades on cotton and alpaca textiles (Lothrop and Mahler 1957, 33-34). Since cochineal is also a polygenetic dye, it may yield either red or black depending on the mordant used; 'it is interesting to note that the Indians in the area of Cajamarca [northern Peru] still dye their products black with this product' (D'Harcourt 1962, 6).

In addition to cochineal, it was found that some of the reds on Peruvian ancient fabrics came from a plant of the madder family, a *Relbunium* species (Fester 1953, 15). People of the Chimú culture, which flourished in northern Peru between 1000 and 1470 A.D., before being incorporated in the Inca Empire, were using the Relbunium plant to dye both cotton and camelid fibers:

> A few Chimú textiles contain cotton yarns dyed black. Probably an *iron mordant* was used to achieve this color since the fibers are in disintegrated condition, an effect well known to be caused by iron. Some Chimú cotton yarns are a *pinkish* color. This color has never been analyzed, but it could have been produced by using either *relbunium*, a madder-related plant, or cochineal...Both of these dyes are known to have been used in Peru on animal fibers to produce red, although in late times, cochineal was probably preferred. Cotton does not take up dye as readily as animal fibers, a fact probably accounting for the faint pink found on these examples, for *although cochineal can be used to dye cotton a true red, the process was never perfected* in [ancient] Peru. (Rowe 1984, 19-20)

In southern coastal Peru both cochineal and relbunium dyes seem to have been used in funerary practices long before the Chimú or the Inca people existed. It was in the Paracas culture, which existed in Peru between 400 B.C. and 100 A.D.,

> ...that wrapping of *brightly dyed camelid fiber yarns around cotton warps began*. Also, the dull red color believed to have been derived from a plant (possibly *Relbunium nitidum*) was completely replaced by *the intense red derived from cochineal*. (Shimada 1999, 3[1]:393)

Carajura

Carajura (*Arrabidaea chica*) is mainly used for dyeing other plants, rather than cotton thread. In highland Guatemala, for example, carajura is used for dyeing sleeping mats and baskets (Osborne 1942, 90). It is used similarly by the Sumu in Honduras and Nicaragua (Coe and Anderson 1999, 369).

Depending on how they are processed (boiled in water, fermented in water, or buried in mud), carajura leaves yield colors ranging from red through purple to black. The Guaymí ferment carajura leaves to make a reddish-purple dye for carrying bags. In the Colombian Sinú, descendants of the Conquest-time Zenú people still cultivate carajura in the shady spots of their tree gardens. From its leaves, which have been fermented in mud, they prepare from its leaves a black dye for hats and basketry (Gordon 1957a, 81; also 1982, 85). The dye remains unfaded even after years of exposure to rain and sunlight.

Logwood and Brazilwood

Several genera of leguminous trees that are called in English either logwood or brazilwood were used as sources of dye. Logwoods are generally—but by no means always—from the genus *Haematoxylon*, and brazilwoods from the genus *Caesalpinia*.

Two closely related tropical American trees, *Haematoxylon campecheanum* and *H. brasiletto*, still supply commercial dyes made from their heart wood. Logwoods make especially good blacks and are, for that reason, one of the few natural dyes still used on a large scale.

In 16th century Mexico solutions of splintered brazilwood and water, sometimes mixed with alum as a mordant, were used in dyeing. This wood comes from a small tree, called cuhuraqua by the people of Michoacán and brasil by the Spanish; it is identified as *Haematoxylon brasiletto* (Standley 1926, 419):

> Dyeing with this tree yields a better color than true cochineal, because it is very similar to red sandalwood [an east Asian tree]. The decoction is at first tawny and later red. If it is boiled longer the color changes to purple—or, if alum is added, a showier color than the vermillion [of cinnabar]. (Hernández 1888, 103)

The Maya used one of the logwoods, probably *Haematoxylon brasiletto*, to 'make black, blue and purple dyes' (Roys, 1931, 240).

The dark red sticks of campeche, *Haematoxylon campecheanum*, are still used for dyeing in Guatemala (but nowadays for dyeing wool, not cotton).

> Freshly cut wood from the tree is colorless; *as with indigo, the color probably exists as a glucoside.* The wood, in the form of chips or paste, *is fermented to prepare the color for dye application.* (Bender 1947, 3)

This is done by first splitting the sticks 'into small slivers, and finally...reducing these to shreds by pounding them between stones' (O'Neale 1945, 28).

The Aztecs made a red dye from a tree which they called uitzquauitl. The tree was either Caesalpinia echinata (as identified by Anderson and Dibble, translators of Sahagún's Aztec version) or Caesalpinia crista (as identified by Standley 1926, 422).

> This is a big tree. And a color is made from its wood, its trunk, with which to dye. Its peculiarity is that it is that which can be *cut, split, steeped*; it is for steeping; water can be extracted. It is only a little dark, a little blackish. Later they make it clear with *alum*; and besides they mix it with *other things*. It *becomes chili-red*...It is a dyeing, not a painting medium. (Sahagún 1950-82, 12:241 and 112)

Perhaps Sahagún's comment on 'steeped' refers, instead, to fermented—because (as noted above) fine chips of these woods, nowadays at least, are moistened and subjected to fermentation for some time (Bender 1947, 3).

Caesalpinia echinata, found from Mexico to Brazil, has a red heartwood;

> this wood, split into splinters, drenched and redrenched in water, tints the water making it red, and this is not a pure red but slightly blackened; but stirring it with *alum* and other materials that are red, makes it pure red. With this coloring, hides are dyed red; and *to make a black dye it is mixed with aceche* [copperas], and with this hides are dyed black. (Sahagún 1956, 3:342)

In Brazil, this tree 'named locally *pau brasil* eventually lent its name to the country because of the quantity of [purple] dye shipped to Portugal' (Gottlieb 1985, 224). A red dye was also made from *chacte, Caesalpinia platyloba*; when its wood is 'cut up fine and thrown into water, it turns to blood and the [Maya] Indians make use of it for dyeing their garments' (Roys 1931, 231; *C. platyloba* is erroneously spelled as *C. platyoba*). Logwoods and brazilwoods may produce different colors when different mordants are used; for example, *C. pulcherrima* is 'said to give a yellow dye with alum and a black dye with iron salts [e.g., copperas]' (Standley 1926, 425).

Tezuatl

The Aztecs combined cochineal with an infusion made from the leaves of a shrub or small tree called *tezuatl*, either *Miconia argentea* or *M. laevigata* of the melastoma family. After adding mordants and boiling, the Aztecs produced a concentrate which was either used as a liquid dye—or, through further boiling, extracted the dye as a solid.

> It is a shrub which grows in the hot lands. This *tezuatl* is converted into color. *Cochineal* is crushed; it is added to the tezuatl. It is provided with *alum*, with copperas. To purify [to concentrate?] it, it is *boiled*. From it comes the *dried color or a good colored water. With it* [that is, with the mixture either in a dried or liquid form] rabbit hair is tinted. (Sahagún 1950-82, 12:242)

In his Spanish edition, Sahagún identifies the mineral used with alum as *tlaliac*—that is, as aceche or copperas (Sahagún 1956, 343). As noted earlier, the 'copperas' mentioned in the above passage is hydrous sulfate of iron, also known as green vitriol. In early China, too,

> red wood, and [an unidentified] *log-wood* were largely used as dyes in early times. In ancient times these natural colors were used for dyeing *cotton* or silk by mixing them with either soda or vinegar according to their properties; in other cases, some *alum or green vitriol was added as a mordant.* (Li 1948, 141)

Although in the Aztec case, the fibers being dyed were animal (i.e., rabbit hair), much the same mordants were used. (As mentioned earlier, cochineal is a mordant dye.)

The Zapotecs, renowned dyers in prehispanic times, today know *tezuatl* (*Miconia argentea*) as *tejute*. In the Zapotec village of Teotitlán it is mixed with cochineal and lime juice to dye wool (Ross 1986, 70). (Both the lime tree and the sheep were introduced here by the Spanish.) The *tejute* plant itself yields no color, and probably functions with cochineal dye as a color intensifier.

Xochipalli

In Mexico a large herb called *xochipalli* (probably *Cosmos sulphureus*) was used by dyers and painters. *Xochipalli* is described as standing about six feet tall, bearing yellow flowers:

> It grows everywhere you walk in *tierras calientes* and is

known by everyone...Great *use of the flowers is made in dyeing*...and for painting images and other things a yellow color or tending toward red. To do so, *they are boiled in water along with salitre* [Spanish for saltpeter], and *finally the fluid is expressed and strained.* It is then used by painters and dyers as we have said. (Hernández 1888, 212)

Saltpeter had a number of uses among the Aztecs, some of them unexplained from a chemical standpoint; for example, its use as 'a medium for refining colors' (Sahagún 1950-82, 12:243).

Resist (Reserve) Dyeing

Resist-dyeing is one way of creating colored patterns on a piece of cloth. Parts of the fabric being dyed are protected with wax—or some other impermeable, dye-repellent substance—so that the dye does not affect them; then the wax is dissolved in boiling water.

Tie-dyeing (plangi), practiced in the Andean area,[12] is a simple form of resist-dyeing. The meaning of the term 'tie-dyeing' becomes obvious from the following description:

> In the more common [method] the finished fabric, usually a gauze, was systematically puckered up into a number of sac-like bulges so arranged as to form rows or groups. Each pucker probably contained *a small pebble or a pellet of clay* beneath which there was formed a sort of shank made of the gauze, a shank which was tightly bound *with waxed thread or with some other material impervious to dye*...When all was in readiness the whole fabric was placed in the dye-pot where all of it, *save the areas covered by the wax thread*, received the chosen color. (Means 1964, 490)

---

[12] In the Andean area,

> Reserve or resist-dyed textiles are recovered from graves of Perú, Argentina, and Bolivia. The simplest form of tie-dyeing is done by pinching up a bit of the fabric and winding string tightly around it. After the cloth is dyed, dried, and the string removed, the protected portions show as squarish white spots on a colored ground. This procedure was known to ancient Peruvians. A second method, also exemplified by archaeological finds, produces striped patterns. The piece of cloth is rolled up and tied with cords at regular intervals before dipping in the dye. These simple methods survive today among Mataco, who tie-dye bark cloth, and Indians of Calilegua in Argentina.
> Tying on cords to protect certain sections of one or both weaving elements before they are interwoven is the ikat process named from the Malay word to tie around. Rare finds of textiles patterned by ikat are reported from Peruvian graves. (O'Neale 1963, 122-123)

CHAPTER 7

ETCHING

Etching, another resist method, was a very rare process in pre-Columbian times. The Hohokam, whose culture flourished from 900 to 1450 A.D. in what is now southern Arizona, etched the interior surfaces of imported sea shells in 'intaglio patterns,' *probably using vinegar* (i.e., dilute acetic acid) derived from the fermented fruits of saguaro cacti. Among the shells so etched were those of the marine cockle, *Laevicardium* sp., a bivalve of Pacific coastal waters. 'Acid etching of shell continues to stand out as a technological achievement unique to the Hohokam' (Haury 1976, 318).

According to one writer, this 'was some 300 years before Europeans invented the technique of etching' (Rouse 1972, 30).

> [The process] is dependent on the following steps: (1) employment of a *wax or pitch as a resistant material*, applied on the shell surfaces in those areas to be protected against the corrosive effect of the acid and (2) use of an acid to chemically reduce the calcium carbonate of the shell *where not protected by wax or pitch*, thereby producing a pattern in relief. (Haury 1976, 318)

The source of the wax or pitch ('a heavy, dark brown substance that appears to have been tacky when applied') has not been determined. It is perhaps significant that the original home of the Hohokam was near the coast in northwestern Mexico, where marine mollusks are abundant.

CHAPTER 8

PROCESSING RUBBER, CHICLE, AND BEESWAX

In parts of both North and South America rubber and beeswax were produced in large quantities. The production of chicle, however, was probably confined to Mesoamerica.

Rubber

In pre-Columbian times, rubber was made in the West Indies, throughout Mesoamerica, and in much of South America. Rubber is probably the major contribution made by inhabitants of the New World to the world's technology:

> It is to Indians that credit is due for the discovery of rubber and its utilization in the form of rubber balls, enema syringes, waterproof fabric, elastic rings, etc. It is a matter of fact that in pre-Columbian times *the Indians were acquainted with all the qualities that make rubber so valuable in modern industry*. No corresponding discovery had ever been made in the Old World prior to the discovery of America, in spite of the fact that both in Asia and Africa there are found rubber trees of various kinds. (Nordenskiöld 1930, 12-13)

Latex of various plants can be made into rubber. At the time of Spanish conquest, the principal sources were two trees: *Castilla* species in the fig family (Moraceae) and *Hevea* species in the spurge family (Euphorbiaceae). The genus *Castilla* is found both in Central America and parts of Amazonia, whereas the genus *Hevea* is confined to South America itself. The natural distribution of *Hevea brasiliensis*, which yields Pará rubber, is in Amazonia—mainly to the south of the Amazon River.

In pre-Columbian America, despite rubber's utilitarian value, its most important use was to make balls for games. In Mesoamerica numerous special courts were made in which to play ball games.[1] Rubber balls were a major item in commerce; for instance, towns in the 'hot country,' between northern Oaxaca and the Gulf of Mexico, paid 16,000 rubber balls as tribute every year to be used in Aztec ball courts around Tenochtitlán (*Codex Mendoza* 1992, 4:96).

Shown on the left is a ball court and a player, with a protective glove and rawhide apron; the court was typically shaped like a horizontal letter 'I' (after Danzel 1922, vol. 1, plate 6—from Codex Fejérváry-Mayer). Shown on the right are the relative sizes of player and ball (Codex Mendoza 1992, 4:145). The balls were about the same size as the air-inflated balls used in Spain (Oviedo 1959a, 1:145)—but, being solid, they were heavier. The object of this game was to pass the ball through rings of wood set high on the parallel walls of the court.

Figure 36. Ball Court and Player

The balls used in Mesoamerica and the West Indies were solid, whereas most of those in South America were hollow.[2]

The Spanish, on first seeing these balls, were amazed at their elasticity. The following account is probably the first reference to rubber in European literature:

> the most important of games among those of our islands is a game played with balls. These are made *from the juice of a certain plant that climbs over trees* as the hop does over fences; they *cook this juice until it hardens on boiling*, after which one shapes the mass to the form he desires...but I do not *understand how these balls when they hit the ground spring into the air with an incredible bounce*. They are skillful at this game, striking the ball with their shoulders, elbows, heads, but not the hands, and sometimes their hips... (Mártir 1964, 2:547)

---

[1] Ball-courts were constructed in south-central Arizona, well to the north of Mesoamerica—further evidence of cultural relationships between the two areas (Haury 1937, 282). A ball, found there by archaeologists, was subjected to chemical analysis and found unquestionably to be made of rubber. The ball is 'identifiable as the product of the Hohokam culture...[and] may be roughly dated at 900-1200 A.D.' Neither the latex-producing *Castilla* nor the moonflower would have been locally available.

[2] Both in the Arawak Haiti and in Mesoamerica, the rubber balls were solid, having neither 'agujero ni vacuo alguno' (Oviedo 1959a, 1:145). In contrast, those in South America were hollow with one possible exception: the Guiana Indians (at least in the early 20th century) had 'a kind of ball made from the milky juice of a tree, which they boil and then pour into molds made of clay kneaded with sand' (Roth 1924, 705). Also in contrast to the processing of rubber farther south in South America, the rubber latex is boiled in Guiana, or at least heated in water in Nayarit, Mexico (Isabel Kelly, 1943), as it was in the West Indies, according to Peter Mártir.

As far as it goes, the account is reasonably accurate: Pedro Mártir was correct in writing that some kind of vine is critical in the rubber-making process, but he seems not to have been told about the principal ingredient, the latex of the *Castilla* tree. (Much faulty information circulated among the early Spanish about the constituents of this new substance.) The following description, written some ten years later, corrects this:

> When the [rubber] tree is punctured it exudes white drops which are collected and make a substance that thickens and turns black, like soft pitch. Of this substance they make the balls that the Indians use, which bounce more than the inflated balls of Castile, and are of the same size and a little firmer...*they are much heavier*... (Motolinía 1950, 68)

Other principal uses of rubber in Mesoamerica were similarly non-utilitarian. In close association with incense and bark-paper, rubber was among the more common items in Aztec ceremonies, and rubber figurines were used in some Aztec rites (Sahagún 1950-82, 3; Sahagún 1956, 3:21-22). Rubber bands were made, similar to those which have so many uses today. However, such bands seem to have been used mainly as adornment; for instance, 'Many Indians of the middle Guaporé...wore rubber bands under their knees or around their ankles' (Métraux 1963c, 228).

Rubber-tipped drum sticks were used in both North and South America. In Mesoamerica, the Aztecs struck a 'two-toned drum...with *rubber-tipped drum sticks*' (Sahagún 1950-82, 3:98), and the Maya had a

> drum made of hollow wood with a *heavy and sad sound.* They beat it with rather a long stick *with a certain gum from a tree at the end of it*... (Landa 1941, 93)

Similarly, in the northwest Amazon, signal drums are beat with a 'wooden mallet headed with a knob of rubber' (Whiffen 1915, 215). (Rubber-tipped sticks are used today on percussion instruments—for example, in playing the xylophone. Of course, such sticks were not used in the Old World before the discovery of America.)

Nevertheless, many utilitarian purposes for rubber were found. One example is footgear: Olmec and Mixtec men 'walked in rubber sandals.' Since these sandals 'were very precious,' they were probably only for nobles (Sahagún 1950-82, 11:188). In one Aztec ceremony, the celebrants wore 'sandals of liquid rubber' (Sahagún 1950-82, 3:43). Also repeatedly mentioned is a kind of footgear which, translated from the Aztec, was called 'foam sandals' (Sahagún 1950-82, 3:87 and 199). Whether these foam sandals were different from the 'sandals of liquid rubber' is unclear; they may have been something like modern sandals with crepe-rubber soles. (Nowadays, foam rubber is made by beating air into liquid rubber.)

Aztec sandals were sometimes rubber-soled (after Danzel 1922, vol. 1, plate 47—from Codex Borgianus)

Figure 37. Seated Drinker with Sandals

In another utilitarian function, Native Americans employed rubber as waterproofing, although only on small objects. For example, the Chocó use blowgun darts (the ends of which are wrapped with balls of balsa fiber), and these must be kept dry (Wassén 1957, 83-84).

Isoprene contains only hydrogen and carbon, and those elements (especially the former) release great amounts of heat upon oxidation. Thus rubber—of which isoprene is the principal constituent—is readily flammable, and much used by tribes of Amazonia 'for starting fires when the wood was wet' (Métraux 1963c, 228). 'Rubber is collected in lumps of coagulated latex and carefully kept to light fires. *The lump is drilled with the fire drill*, which produces a highly inflammable dust' (Lévy-Strauss 1963b, 476).

The oldest artifacts made of rubber, as well as the oldest remains of ball courts, are found in Mesoamerica; thus, games using rubber balls may well have begun there. The oldest known ball court, dated at some 1400 B.C., was excavated at 'in the Soconusco region of Pacific coastal Chiapas' (Hill et al. 1998, 878). The oldest rubber balls are even older:

> The oldest archaeological specimens [of rubber artifacts] are 12 *solid rubber* balls recovered at the Manati site in Vera Cruz, Mexico...The two oldest Manati balls date to 1600 B.C. on the basis of radiocarbon dates. Manati...is an *Olmec ritual burial site* located in a swampy region along the lower Coatzacoalcos River. (Hosler, Burkett, and Tarkanian 1999, 1988)

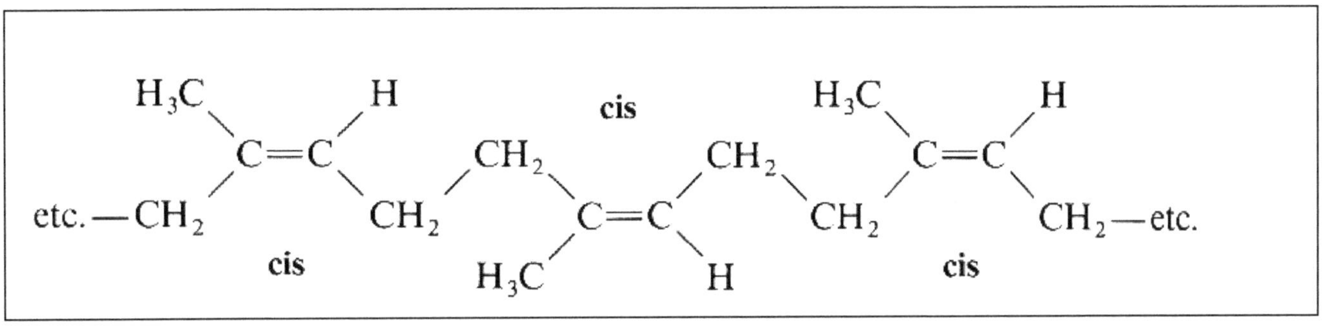

Natural rubber is an isoprene polymer: cis-1,4-polyisoprene. Explaining the meaning of 'cis' would take us into the rather-complex subject of stereochemistry. The interested reader will find clarification in a text book of physical or organic chemistry—for example, Solomons 1996, 1179-1182, 1106..

Figure 38. Natural Rubber

The Olmecs, who lived along the swampy coast of Veracruz, established one of Mesoamerica's first major civilizations; their name is said to mean 'Rubber People.' Despite the antiquity of these artifacts, whether rubber was first made in South or in North America is an unsettled question.

The latex drawn from rubber trees is an emulsion in which tiny rubber particles are suspended. To produce elastic rubber, the Indians mixed the latex of *Castilla elastica* with an extract taken from the vine, Ipomoea alba. (*Ipomoea* is the morning-glory or sweet potato genus. The binomials *Ipomoea bona-nox* and *Calonyction speciosum*—as well as *Ipomoea alba*—are simply synonyms. This twining vine, with a milky juice, is commonly known as *moonflower*; it is now widely cultivated for its white, night-flowering blooms.) This extract causes the rubber particles to coagulate or clump together. For example, a twentieth-century report from Honduras and Nicaragua states that,

> in former days the [Sumu and Miskito] Indians used to collect rubber from a wild-growing tree (*Castilla* sp.), which is found nearly all over the country. This industry started about 1860, first on the Rio San Juan...*the latex was strained and caused to coagulate by the addition of an alkaline* [?] *decoction made from the juice* of the *chajmol* vine (*Ipomoea bona-nox...*) or from that of a liana (*Calonyction speciosum*). *To prevent putrefaction coagulation must be effected within 24 hours after* the collection of the milk. (Conzemius 1932, 46)

In Nayarit, toward the west coast of Mexico, such solid balls were also made in the 20th century. To coagulate the rubber latex, *Operculina rhodocalyx* (a plant closely related to *Ipomea alba*) was used; its root was:

> diced and added to water...The solution was then strained into the latex, three parts solution to one of latex. A small quantity of this mixture was *heated* to lukewarm, tested with a small pointed stick, and, when of the proper consistency, removed from the fire. The rubber adhering to the stick was shaped into a pellet...The [solid] ball was thus built up, layer by layer, as small quantities of rubber were added to the pellet-like nucleus. (Kelly 1943, 164)

Unfortunately, the plant source of the latex in Nayarit is not named.

Castilla trees also grow in South America, and Indians of the middle Amazon harvest and process the latex. (In Spanish America, rubber latex is usually called '*caucho*,' which derives from the Tupi name for *Castilla elastica*, *caa uchu*. This is also the origin of the English word, *caoutchouc*.) As in Central America, to coagulate the latex, 'they must *mix it with the juice of a bindweed* called *batata rana* because it resembles the sweet potato' (Tastevin 1943, 29). Thus, an extract of the moonflower vine or its close relatives seems to have been required for processing *Castilla* latex; at least, it was used for that purpose in both Mesoamerica and South America.

Rubber balls were made, however, in large parts of South America where neither Castilla trees nor the moonflower vine are present. In South America, the principal rubber-yielding plant was *Hevea brasiliensis*, although other trees were used. There, as in the West Indies and North America, no elaborate masonry courts were constructed: ball games were played in open fields.[3]

In Brazil, 'rubber is known throughout the province [of Pará] only by the name *seringa*, the Portuguese word for syringe...it was in this form only that the first Portuguese settlers noticed it to be employed by the aborigines' (Bates 1962, 87). Hence the name *seringuero* is given to one who taps tree for rubber. In early historic times, detail is scarce on methods used by South American Indians for processing the latex of *Hevea brasiliensis*. Nevertheless, Portuguese rubber tappers (*seringueros*) first learned to prepare the latex from the Omagua Indians in the upper Amazon area (Métraux 1963c, 228); thus, the methods once used by Indians and tappers were probably much the same.

[3] However, what is believed to be an I-shaped ball-court (dated to between 1800-900 B.C.), 'architecturally very similar' to those known in Mesoamerica, has been found in Peru (Pozorski and Pozorski 1995, 274).

The latex of *Hevea* species was prepared differently from that of *Castilla*. While heating was still needed, heat alone was not enough: a fuel that produces much black smoke (apparently of a special kind) was required. The incorporation of carbon (soot) gives rubber an increased toughness, elasticity, and durability. The practice was to pour a cup or so of the white latex over a paddle or the broadened end of a stick. This was then rotated in a thick smoke made by burning certain palm nuts,[4] keeping the stick well away from flames. On making contact with the *smoke*, the latex dries, turns quickly a brownish color, and becomes elastic. Another cup of latex was poured over the stick which was again turned in the smoke. The layers of rubber accumulated until the whole lump reached the desired size; it was then cut from the stick with a knife (Woodroffe 1915).

The following is a description of the Indian method of making balls in the upper Amazon basin within the last century:

> A hollow rubber ball is made by the Indians by shaping clay into a round ball, which is then covered with a coating of rubber [latex]. This is then made to *coagulate in the smoke over a slow fire*, whereupon the rubber is pierced and the clay washed out. This leaves only the rubber shell remaining. The aperture through which this core of clay in this way has been removed is then closed with a clot of rubber so that the walls of the rubber ball become airtight. When once the Indians possessed hollow rubber balls they could find no difficulty in inventing the rubber syringe which is nothing more or less than a rubber ball with a hole in its wall. (Nordenskiöld 1930, 62-63)

In the mid-eighteenth century the Mojo—who live on the Mamoré River in northern Bolivia—made their hollow rubber balls in substantially the same way (Métraux 1963c, 227).

The custom of making hollow balls and rubber bands extends southward from the Amazonian rainforest into the seasonally dry savannas. In the early twentieth century, the Apinayé people who live between the Tocantins and the lower Araguaia Rivers made them ceremonially (Nimuendajú 1939, 61-62) from the mangabeira tree, *Hancornia speciosa*, of the dogbane family.

Rubber balls for a game were made in tropical South America (in the Department of Amazonas of southern Venezuela), not only from tapped latex but also from a natural deposit called *dapicho*—a white and corky, semi-fossilized substance, excavated from beneath hevea trees. This 'corky caoutchouc' is found in quantities beneath the roots and trunks of certain large hevea trees—'more especially when the trees have attained a great age.' It appears that the balls are no longer made, permanent courts were not constructed, and the game is no longer played.

An old Indian 'showed us, that, after digging two or three feet deep, in a marshy soil, this substance [*dapicho*] was found between the roots of two trees...'. It is found '...at the extremity of the longest roots, for we found masses of dapicho of two feet in diameter, and four inches thick, eight feet distant from the trunks' (Humboldt and Bonpland 1972, 5:263-265).

The preparation and uses of *dapicho* are further described:

> A Poimisano Indian, seated by the fire...was employed in reducing the *dapicho* into black caoutchouc. He had spitted several bits on a slender stick, and was roasting them like meat. The *dapicho* blackens in proportion as it grows softer, and gains in elasticity. The resinous and aromatic smell...filled the hut...The Indian beat the softened and blackened mass with a piece of brazil wood, ending in form of a club; he then kneaded the *dapicho* into *balls of three or four inches in diameter*, and let it cool...They are used at San Balthasar *in the Indian game of tennis*, which is so celebrated among the inhabitants of Uruana and Encaramada. (Humboldt and Bonpland 1972, 5:231)

And the local Indians used 'a drum, which was a hollow cylinder of wood...This drum was beaten with great masses of *dapicho, which served as drum-sticks*' (Humboldt and Bonpland 1972, 5:232).

As already noted, isoprene is the basic building block of natural rubber, in which the number of isoprene units, $(C_5H_8)_n$, approaches its high point: the polymer is composed of some five thousand such units, linked head-to-tail in long, kinky thread-like molecules. Pyrolysis—that is, heating in the absence of air—degrades natural rubber to isoprene.

Recently, the process of rubber making from the latex of *Castilla elastica* in Mesoamerica has been subjected to a thorough chemical examination (Hosler, Burkett, and Tarkanian, 1999). When this latex is mixed with an extract from the moonflower, it coagulates and leads to the cross-linking and entanglement between the isoprene chain-molecules. Like the carbon with *Hevea latex*, sulfur is a key element in this process, and moonflower-extract contains compounds with sulfur in their molecular structure (Hosler, Burkett and Tarkanian 1999, 1990).

---

[4] The tapper—that is, the...

> 'seringuero', in order to smoke his rubber, depends almost entirely upon the nuts of the Urucuri (*Attalea excelsa*), and when these are not easily obtainable uses small blocks of Massaranduba (*Mimusops elata*) or Carapanáuba [identity ?], *all three of which produce thick, oily smoke*, of a density and composition different from that given off by any other Amazon variety of nut or timber, except, perhaps *inajá* nuts...The *smoke* from the urucuri is more rapid in its effect than that from Massaranduba, while that of the Carapanáuba is the slowest and least suitable of the three. The finest quality rubber is that smoked by a careful seringuero using urucuri nuts; although it is said that the *inajá* nut [*Maximiliana regia*] is still better. (Woodroffe 1915, 198-199)

For over three centuries Europeans attempted to duplicate this process of turning crude latex into a tough, elastic material. This process had been known for over three millennia in pre-Columbian America.[5] It was not until 1844, that it was discovered that the qualities of gum rubber were greatly improved by cooking with sulfur—a process named vulcanization. Then rubber, which previously had had only a limited use in the outside world, suddenly became industrially important. (The damaging effects of the consequent search for rubber on upper Amazonian tribal communities were probably even greater than those associated with today's coca trade.)

Chicle and Chewing-Gum

Chicle, like other gums, has a semifluid consistency when moist, but when dry it is hard. Chicle is obtained by boiling down the white sap of the sapodilla tree, *Achras (Manilkara) zapota*. Aztec chewing gum, sold in the city markets, was made by mixing latex of the sapodilla tree with *axin* to soften it: 'They do not chew it alone...*They mix it with axin*. It cannot be chewed alone; it crumbles' (Sahagún 1950-82, 4:89; also Sahagún 1956, 142). Whether some Maya, like the Aztecs, chewed gum is uncertain.

Chewing gum is another New World contribution to modern commerce. Although chicle was commonly chewed in ancient Mexico, among the Aztecs approval of chewing gum was by no means universal. Except by young people and some unmarried adults, gum was not chewed in public. Women chewed gum to sweeten their breath, and even adult men chewed gum to 'clean their teeth,' but only in secret (Sahagún 1956, 3:152). According to a sixteenth-century Aztec superstition, a woman born under an inauspicious sign would do 'nothing but talk in a loud voice,' and go about 'snapping her chewing gum' (Sahagún, 1950-82, 5-6:95). A woman of questionable morals perfumes herself strongly; she 'constantly anoints herself with *axin*' (Sahagún 1950-82, 11:55), and chews gum so loudly that her teeth clack like 'castanets' (Sahagún 1956, 3:152).

Axin (also known as *aje*) is a waxy animal fat rendered from bodies of the scale insect, *Coccus* (*Llaveia*) *axin*. It was of much importance in pre-Columbian Mesoamerica. The insect is propagated on several cultivated trees, especially species of *Jatropha* and *Spondias* (for illustrations of the insect and associated paraphernalia, see Gordon 1957). *Axin* was employed in numerous ways: in lacquering and painting pottery; as a constituent of incense, of a depilatory, of body paint, of chewing gum, and perhaps of the paint used in illustrating books. In certain Aztec ceremonies, Montezuma 'had his body painted with a yellow pitch called *axin*' (Durán 1994, 424). *Axin* was even used in making something like a chap stick: 'In order that frost will not injure the lips, *axin* unguent is applied' (Sahagún 1950-82, 4:90; also Sahagún 1956, 153). Because of its numerous uses (it was also a part of a great many medicinal preparations), the *axin* insect may have been of greater importance in Mesoamerica than its better-known cousin, cochineal, *Dactylopius coccus*.

Although the major use of *axin* was in Mesoamerica, propagation of the insect reaches its southernmost extent among the Guaymí in Panama. Culture of the insect and preparation of its waxy fat are essentially the same wherever it is grown—the similarity often extending to details. Among the Guaymí, eggs of the insect are 'kept in calabashes or corn husks, bedded in dry balsa fiber, and commonly stored in attics. As the eggs begin to hatch, these containers are tied to host trees and the female insects swarm onto the branches' (Gordon 1957b, 38). When mature, the insects are 'picked and cooked in boiling water. The content of the container is then strained and as the fluid cools, a yellowish-tan fat, about the consistency of butter (which the Guaymí—like the insect itself—call *kŭrrón*) rises to the surface and is skimmed off' (Gordon 1957b, 42). Although mostly a fat, *axin* has a waxy consistency which apparently comes from the wax-based cottony fluff that covers the female insects. The Guaymí mold the waxy fat into bars or balls and store it in corn husks. On exposure to air, a hardened and darker crust forms on its surface. The Tarascans of Michoacán, Mexico, treat the insect similarly (Starr 1900-1902, 1:13). The Tarascans, well-known for their lacquerwork (and formerly for their metallurgy and excellent featherwork) live in the highlands near Lake Pátzcuaro. Their ancient capital was Tzintzuntzan.

Beeswax

As noted earlier, apiculture was a commercial enterprise in some areas. Beeswax had many uses, and was sold in the markets of Tenochtitlán. Quantities of wax were needed, both in North and South America, for the lost-wax casting of metals, the resist-dyeing of textiles, the decoration of gourds, and (perhaps) the resist painting of pottery. A wax tribute was exacted by Aztec rulers on their subject peoples.

As in the Old World, bees played an important part in the symbolism of Mesoamerican religions. Well known in the rituals of the Roman Catholic Church, bees were also important in early Maya religious ceremonies. Bees still have this religious significance for some Maya descendants:

> Among the Lacandones today the beehives are kept in the sacred hut and the honey [for making their ceremonial drink, balche] never used for secular purposes. (Landa 1941, 92; A. M. Tozzer's footnote)

[5] According to one modern polymer chemist,

> It's a marvelous example of technology demonstrated at an incredibly early stage...They probably had a pretty good R&D team. (Frank Bates, University of Minnesota, quoted by Erik Stokstad in *Science* 1999, 284:1898-1899)

The wax of stingless bees is usually darker than that of the common honeybee. Wax made by bees that feed on flowers of the sacred copal tree is said to be especially dark (Landa 1941, 197). This may explain the fact that black beeswax had a sacred character among the ancient Maya. In Chan Kom, a modern Maya village, 'beeswax is accredited a slightly supernatural character'; and at Roman Catholic funeral-ceremonies in this area, *black wax* candles are sometimes burned for the adult dead. 'The domesticated [i.e., the native stingless] bees and the hives...enter prominently into the religious beliefs and practices' (Redfield and Villa 1962, 49-50). Since candles are made in communities where native customs are well preserved, some believe that making candles—or at least making black candles—is a native custom. Nevertheless, it is very questionable that candles were made in pre-Hispanic times: 'We taught them how to make candles of the native [stingless bee] wax...for up to that time they did not know how to make this use of the wax' (Díaz 1960, 1:163); also the Aztecs 'lacked candles and lamps for which they used torches' (Hernández 1986, 106).

Apparently, even in the most technologically-advanced areas of pre-Columbian America, night-lighting was provided by torches.

CHAPTER 9

PERSONAL BEAUTIFICATION, PERFUMES AND INCENSE, AND CLEANSING AGENTS

Body Paints

The 'Red Indians' may have been so named because of their fondness for red body paint. For instance, the following note describes the Caribs in the Antilles:

> It was difficult to judge of the colour of their skin, because they were always painted with *roucou* [achiote], which gave them the appearance of boiled lobsters. (Sheldon 1820, 372)

Achiote and genipap were the plants most widely used for body paints in tropical America.[1] They are still used for that purpose, and applied in a variety of ways and patterns. Achiote is favored as a body paint because the reddish-orange colorant, bixin, although soluble in fats and oils, is not soluble in water. Therefore achiote paint does not readily wash off in the rain:

> The color is quite indelible and the Indian *can pass through a heavy rain* with the coating still adherent and little disturbed. Its removal from the body entails considerable soaking and bathing, and the clothing even after prolonged washing is never completely cleansed of it. (von Hagen 1939, 24-25)

Nevertheless, face paint made from achiote and fat is often partially rubbed off, and traveling Indians must occasionally reapply it.

According to an account written in the early 18th century, the Indians on the Orinoco prepared achiote (*onoto*) in the following way: They 'knead it with oil of turtles' eggs, and form it into round cakes of three or four ounces weight. When turtles' oil is wanting, some nations mix with onoto the fat of the crocodile' (Humboldt and Bonpland 1972, 4:512-13).

At about the same time, the Caribs in the Antilles prepared achiote (roucou) as follows: They rubbed the seeds

> ...with their hands, which *they had previously soaked in carapat*, or palma-christi *oil*. In this way they detached the pellicles from the seeds, and reduced them to a fine, clear paste. This they scraped from their hands on a leaf...and set it by in the shade to dry; for *they apprehended the sun would diminish the beauty of the colour*...When the roucou was nearly dry, *they made it into balls about the size of the hand*, and wrapped them in...leaves of the wild plantain... (Sheldon 1820, 401-402)

Carapat oil is rendered from the seeds of the tropical-American, mahogany-family tree *Carapa guianensis* or from the seeds of palma christi, *Ricinus communis*, a spurge-family plant introduced from the Old World—in short, castor oil. The Caribs were right when they recognized 'that the sun would diminish the beauty of the colour,' because recent investigations have shown that the tinctorial power of bixin is decreased by light and heat. Many tribes recognized that the achiote colorant has some affinity for fats and oils. For instance, among the Cayapa of Ecuador, ripe achiote seeds are soaked in water for a considerable time. Then the colored liquid is drawn off.

> This red or yellow liquid is then boiled down to a pasty consistence and mixed with any hard fat, such as lard or *the fat of the jaguar, the puma, or other wild animal.* It is said that the viscous juice of a tree called *sande* [unidentified] is sometimes used instead of one of these fats; but the paint produced with it is *not so good* as that in which fat is used. This strong greasy pigment is applied with...[a stick]. (Barrett 1925, 70)

Since *sande* juice is 'used in making torches' (Barrett 1925, 114), probably it contains terpenes (or another organic solvent) that will, like fats, dissolve bixin.

Genipap juice (sometimes called *caruto*) is also used to color the skin. When exposed to air, the colorless juice of the green genipap fruit is gradually oxidized, first turning blue and then black:

> From the fruit is obtained a clear liquid which the Indians use to bathe their legs, and at times even their whole bodies...They even paint themselves with it for the mere pleasure. The liquid...gradually turns to a jet black any part of the flesh that it touches. This stain will not disappear until twelve or fifteen days or more have passed... (Oviedo 1959b, 90)

Humboldt, too, notes that the black pigment of genipap juice 'resists a long time the action of water' (Humboldt and Bonpland 1972, 3:519). That is, its action is something like a dye or stain, rather than a paint which supposedly forms a superficial coating.

Carajura, although costly, was much used as red body paint[2] among tribes of the Orinoco drainage and the Caribbean Islands:

[1] Several authorities contend that such body paints afford some relief from the attacks of insects (Salazar 1941, 235; Hernández 1986, 102; Sheldon 1820, 372; and Standley 1928, 27: 270). Humboldt and Bonpland disagree (1972, 2:517).

[2] At the beginning of the 19th century, among tribes living along the Orinoco River, achiote was

> The red pigment of *chica* [carajura] is not obtained from the fruit, like the *onoto* [achiote], but from the leaves macerated in water. The colouring matter separates in the form of a light powder. It is collected, *without being mixed with turtles' oil*, into little loaves eight or nine inches long, and two or three high, rounded at the edges. The leaves, when heated, emit an agreeable smell of benzoin [i.e., the resinous juice of *Styrax Benzoin*, an Old World perfume]...The processes of infusion and maceration are in general very common among all the nations on the Oroonoko...The little trade in *chica* [carajura] is carried on chiefly with the tribes of the Lower Oroonoko, whose country does not produce the plant that furnishes this much valued substance. The Caribbees and Otomacks paint only the head and hair with *chica* [carajura], but the Salivas possess this pigment in sufficient abundance, to cover their whole bodies. (Humboldt and Bonpland 1972, 2:513-515)

In the Amazon area carajura is still used as body paint:

> Another red paint used for body and pottery decoration and varying from orange to purple *according to the technique of preparation* is *caraweru* [carajura]...which comes from the boiled or fermented leaves of *Bignonia chica.* (Lévy-Strauss 1963b, 478; also Tessman 1930, 40)

In the Old World, for instance in the Middle East, indigo is used as a body and hair dye. In most of the New World, however, indigo was not commonly used to decorate the body: '*we see no American tribe painted with indigo*' (Humboldt and Bonpland 1972, 3:518). Nonetheless, in 16th century Mexico it was a cosmetic among Aztec women who were considered immoral. Sometimes achiote was applied with stamps made by carving embossed designs in an absorbent wood:

> They go about staining themselves with *indigo*. They apply *yellow ochre* to their faces. *With a stamp* they apply red coloring to their faces...They *stain their teeth with cochineal.* (Sahagún 1997, 207)

(In modern industrial countries, cochineal has been used in preference to synthetic products to make such cosmetics as lipstick and rouge.)

Next to crude charcoal, soot is probably the oldest of black body paints. Aztec priests painted their bodies black, and before battle the faces of their soldiers 'were blackened with the divine soot, as they call it' (Durán 1994, 352). The Miskito of Nicaragua obtain this black paint from burning pinewood. Over the soot, they 'apply a 'varnish' of turpentine' (Conzemius 1932, 25). Among the Guaymí, a resinous material (probably a gum resin) from the *chutra* (*hutrá*) tree is used for making black face paint: Balls of this flammable exudate are placed in a hole dug in the ground about a foot deep, and set afire. Either an overturned pot or some boards are placed over the hole and the soot collected on its inner surface. The soot is then scraped off and mixed with the waxy fat of the *axin* insect for making face paint (Gordon 1982, 66). *Axin*, which has a melting point of about 120°F or 48.8°C (i.e, above the temperature of the human body) is easily applied. Being insoluble in water, it is not removed by rain showers. In the Southwest 'The balls of carbon [soot] which are occasionally found in [Pueblo] ruins were *intended for body decoration...*' (Hawley 1929, 735).

Of the inorganic pigments used as body paint, probably the oldest and most widely used are the ochers (in Spanish, *almagre* or ocre). They were used in both North and South America, and by both simple and complex societies. In the early sixteenth-century, Aztec women painted their faces with yellow ocher; the men painted their whole bodies before going to war with the same mineral (Hernández 1888, 309); the faces of sacrificial victims were painted with red ocher. The faces of the Otomí, neighbors of the Aztecs, 'were smeared with yellow ochre,' while among the Maçauaque 'old women paint their faces with yellow ochre, or with red.' (Sahagún 1950-82, 4: 179-85). In Peru, women used yellow ocher as a face paint.

Though less commonly available than ochers, cinnabar was also a face paint from California to Peru. The Aztecs used cinnabar for body paint, too (Sahagún 1956, 59). The Inca used the mineral in the same way: 'They would smear it [cinnabar] or other earth colors on themselves at the time of their festivals...' (Cobo 1990, 176). Salazar, who was in Mexico in 1560, claims that the Maya Indians were 'painted with minium [red lead]' (Salazar 1941, 237). He was probably correct, although he could have been mistaking minium for cinnabar, as was often done in early times.

When lead is heated in air, it forms litharge (lead oxide, PbO). And when, in turn, litharge is heated to a dull red heat, it is further oxidized to red lead or minium, Pb_3O_4; red lead was used in the Old World as a crimson pigment. The Peruvians may have used litharge for coloring their keros (Nordenskiöld 1931b, 96). On the other hand, though its manufacture was well within their capabilities, there is not much evidence that Peruvian artists used red lead as a pigment. Or that it was made elsewhere in pre-Columbian America.

The Cakchiquels, a Mayan group in Guatemala, used 'white earth' to paint their bodies and faces (Rice 1977, 231).

> ...so indispensable, that both men and women would perhaps be less ashamed to present themselves without a *guayuco* [a waist apron] than destitute of paint. (Humboldt and Bonpland 1972, 2:520)

Since little clothing was worn by Indians around the missions established there, two paints (achiote and carajura) 'may be distinguished among them, according as they are more or less affluent' (Humboldt and Bonpland 1972, 2: 512). Carajura was so expensive that

> ...a man of large stature gains with difficulty enough by the labour of a fortnight, to procure in exchange the *chica* [carajura] to paint himself red. Thus as we say in temperate climates of a poor man, 'he has not enough to clothe himself', you hear the Indians of the Oroonoko say, 'that man is so poor, that he has not enough to paint half his body'. (Humboldt and Bonpland 1972, 2:515)

Tattooing

In the Old World, tattooing was practiced long before historic times. Among hunting peoples crude charcoal was commonly used in tattooing. In tropical America, charcoal and genipap were among the more commonly used materials; for example, tribes in eastern Bolivia and the Mojos region tattooed by making incisions in the skin with fish teeth or thorns, then pouring genipap juice or charcoal over the scratches (Métraux 1942, 65, 102-103 and 163). Among the Maya, the method of tattooing was first to paint the body part and to cut or rub in the paint:

> Those who do the work first painted the part which they wish with color and afterwards they delicately cut in the paintings, and so with the blood and coloring matter the marks remained in the body. (Landa 1941, 91)

Tattoos often had special social significance beyond mere personal adornment. For example, an early report from Maya territory states that 'according to the custom of that country only the valiant may have their hands tattooed' (Salazar 1941, 236). In the territory of the former Nicaragua, the Chontal Indians (a Maya group) used soot, made out of the smoke of burning pine wood, to brand slaves. With a stone knife, a cut was carefully made in the face or arm, to a depth between the skin and the flesh; then the fresh wound was covered with soot (Oviedo 1959a, 1:177-178).

Dental Decoration and Dentifrices

Among many tribes, colored teeth were considered either appealing or beneficial. In sixteenth-century Mexico, the Aztecs used vinegar mixed with the red of cochineal to clean their teeth (Hernandez 1959, 1:315); to the north, Huaxtec women also colored their teeth red (Sahagún 1950-82, 4:186). On the other hand, many favored black—for example, the Otomí (Sahagún 1956, 3:197). Similarly, among the Witoto of the northwest Amazon, 'women cover their teeth and their fingernails with a black pigment' (Whiffen 1915, 88). The same is the case along the middle Amazon, though not—it is claimed—for cosmetic reasons alone:

> All the Panos *habitually blacken their teeth in order to strengthen them and prevent decay*. For this purpose they prepare a wild shrub with a slender stalk divided by nodes like the bamboo. There are several species bearing the name *nichpo*...All are not effective...[Only the right species give] a *black stain lasting about a fortnight*...[The plants are] not chewed as the stalk is soft, the oily section is tapped against the ivory of their teeth. (Tastevin 1943, 97)

The Citará Chocó around Quibdó in northwestern Colombia 'are noted for their jet black teeth'; this effect is produced by chewing young shoots of a madder-family liana, *Schradera* sp. Children 'are taught to chew the plant until a complete blackening of the teeth occurs'; chewing is repeated 'on the average of every six months, to restore the worn spots in the black film' (Archer 1934, 402-403). Chocó in the Sinú country chew either the new stems or the root of a licorice-flavored, anise-scented shrub, a regrowth species of *Piper*. They consider the purple-stained teeth which result, cosmetic (Gordon 1957a, 18).

On the other hand, some may have considered white teeth desirable: nowadays, descendants of the Zenú people (neighbors of the Chocó) chew stems of the plant called either *caña agria* or *limpia diente* (*Costus* sp.) to whiten the teeth (Gordon 1957a, 81). This may, however, be a European influence.

The Aztecs and other cultures of Mesoamerica practiced another form of dental decoration—namely, inlaying their teeth with jade or pyrite. In skulls found at Teotihuacán,

> x-ray pictures show the superbly bored cylindrical cavity and the perfect fit of the inlay, four millimetres in diameter, notwithstanding the extreme hardness possessed by both jadeite and tooth enamel. The cement by means of which the filling was fastened must have been of extraordinary quality...The fact that this cement has fulfilled its task even after 1500 years is astonishing indeed—that used by our dentists of today can scarcely be depended on to outlast a patient's life-time. (Linné 1957, 152)[3]

Archaeological finds show that such inlaying was also common (perhaps, even more so) in parts of the Maya area (Landa 1941, 126; A. M. Tozzer's footnote).

Hair Dyes

Aztec women used an indigo-like plant, the leaves of which yield a deep blue color, to dye their hair (Hernández 1888, 90-91). Furthermore, a certain valuable earth or mud, called *palli*, was mined and used by women to color their hair black (Sahagún 1956, 349). In contrast, the Huaxtecs 'wore their hair dyed diverse colors—some yellow, others red, and some varicolored' (Sahagún 1956, 203).

In Peru, as in most places, some people went to great lengths to make themselves presentable according to current custom:

> The Indian women of Peru all have long, loose hair...they are naturally very fond of having it very long and black...they boil a cauldron of water with herbs in it over the fire. One of the herbs must be the root of the *chuchua* [maguey],...but I have seen it made on more than one occasion *with several ingredients*, though I was too young to notice how many or what herbs they were. In order to get her hair in the cauldron which was boiling away with this

[3] Linné, who gives the constituents of the fixative, states that the 'essential elements in Portland cement agree surprisingly well' (Linné 1957, 153). Although the constituents of Portland cement vary, they are mainly certain mixtures of clay and limestone. These are heated in a kiln, almost to the point of fusion (to be sure, there is no evidence that such temperatures were attainable in pre-Columbian America), then cooled and pulverized. Portland cement has a very valuable property: unlike ordinary lime mortar, it hardens when immersed in water.

decoction on the fire, the woman would lie on her back with some protection so that she did not burn her neck. They also took precautions against the boiling water touching the head and scalding the flesh. The hair that was not covered with water was wetted so that it too should enjoy the virtues of the brew. They would submit to this voluntary torment for...nearly two hours...I did not however fail to wonder at the ordeal, which seemed to me a severe one...However in Spain I have ceased to wonder, after seeing what many ladies do to bleach their hair by perfuming it with sulphur [probably, with sulfur dioxide, a bleach], wetting in gilder's aqua fortis [a name given to weak and impure nitric acid], exposing it to the sun at midday...I do not know which treatment is worse...

The Indian women, after washing their hair again to remove the scum from the boiling, appear with their hair blacker and more lustrous than the feathers of a newly-moulted raven. This and much more will the longing for beauty induce people to undergo. (Garcilaso de la Vega 1966, 506-507)

Depilatories

Although Amerindians generally have scant body hair (compared to most European peoples), many desire to remove what little they do have. For example, Maya men 'did not grow beards, and they said that their mothers *burned their faces* in their childhood *with hot cloths, so as to keep the beard from growing*' (Landa 1941, 88).

Depilatory tweezers were almost obsessively valued before arrival of the Spanish. (In parts of Mesoamerica—for instance, among the Tarascans—tweezers were sometimes worn purely as ornaments, or as symbols of high social rank.) One of the first accounts of the Maya notes that the men were accustomed to pluck out hair in their beards 'with some things like pincers, as [Spanish] women do their eyebrows' (Salazar 1941, 235).[4] In the archaeology of both Mexico and Peru, tweezers are among the most common of metal artifacts. Tweezers were major trade items, as they are among the Chocó today.

In early times, the Maya obtained a gum 'resembling balsam' from a small tree (*Exostema caribaeum*). Mixed with axin and pine pitch, the 'Indians formerly employed it for shaving,' that is, as a depilatory (Roys 1931, 301, citing a sixteenth-century MS).

Among the Indians of the northwest Amazon area,

> hair is looked upon as dirt; therefore it is always removed, only the hair of the head being permitted to grow...The method of removal adopted is to cover the hirsute parts with rubber latex. This is allowed to dry, so that a grip can be obtained and the hair removed simply with the forefinger and thumb or by means of two small pieces of cane. (Whiffen 1915, 273)

In the Andean area,

> The Indians of Peru call, *macay*, little rolls the shape of coins but thicker, which they make *from a certain rock which resembles limestone*. It is made by mixing the powder of said rock with stale urine and cooking it in an oven. They take it out and slake it with urine mixed with the powder of rock sulphur, with which it acquires a very bad odor...if this powder is dissolved in hot urine, to form a thin watery mass and it is painted upon the hair of any part of the body, allowed to dry, and *then washed with warm water, the hair will later fall off.* (Cobo 1956, 1:128)

Perfumes and Incense

As mentioned earlier, most perfumes are based on essential oils which are usually fragrant. Apparently appreciation of fragrant substances is a universal human trait, and the application of perfumes to the body is widespread. For instance, the Chocó Indians—like the Polynesians and many others—rub themselves with the leaves of fragrant plants. Beyond this liking for simple perfumes, there was throughout most of Mesoamerica and the West Indies an unusual emphasis on scents—one reason being their strong association with religious ritual or belief. The Taino even recognized some metals, not only by their color but by their odor, which they attributed to some heavenly association. For example, they distinguished between pure gold and guanín (an alloy of copper and gold)—and valued the alloy more highly (Las Casas 1951, 1:281 and 304).

The Maya 'were great lovers of perfumes, and for this they used bouquets of flowers and odoriferous herbs, arranged with great care' (Landa 1941, 89). Moreover, Maya women

> ...had the custom of anointing themselves with a certain *red ointment* like their husbands, and those that could afford it added to it a certain preparation of a *gum, odoriferous* and very sticky, which I think is *liquid-amber.* (Landa 1941, 126)

(A fragrant resin is found in bark of the sweetgum tree, *Liquidambar styraciflua.*)

Many plants were used as perfumes by the Aztecs; for example, they 'fumigate their lodgings with this herb [an unidentified plant], throwing it into braziers to perfume bad air and stench...' (Hernández 1888, 144). In parts of Peru, a small sedge was harvested to perfume clothes, and another sedge was put in the mouth to remove bad breath (Cobo 1956, 1:157).

According to Bernal Díaz, when he and his fellow soldiers first arrived on the shores of Mexico, Aztec priests dressed in white cotton 'brought incense, a certain resin which they call copal, and baked-clay braziers [censers] with live coals and began to fumigate us' (Díaz 1960, 1:49 and 67). (As a rule, the use of censers is

[4] For similar purposes, mirrors (looking glasses) were made of obsidian (Hernández 1888, 306). They were also made from silver, copper, and bronze in the New World. Sahagún states that the mirrors sold in Aztec markets were cut, sawed, and rasped by lapidaries from 'mirror-making stones [probably obsidian or pyrites]' (Sahagún 1956, 149-150).

limited to peoples with a strongly ritualized religion: incense and censers have been used in Old World ceremonies since antiquity. For instance, in Chinese religious rites they have been used for more than two thousand years; in ancient Egypt their religous use is probably even older.)

Copal, a word of Nahuatl (Aztec) origin, is now the general name for a number of aromatic tree resins. In pre-Hispanic times, however, the term was only used for incense-yielding materials.

Resins should be distinguished here from gums: Gums are viscous, water-soluble polysaccharides which dry into solids; thus they are commonly used as adhesives. In contrast, resins (like waxes and fats) are insoluble in water. When they are taken from plants, resins rarely occur without being mixed with gums or essential oils. Soft and sticky when removed from the tree, resins harden with time on exposure to air. Most resins soften when heated, and harden again when cooled; thus they—like gums—are used as adhesives. Because resins burn with a fragrant, smoky flame, they are widely used for incense. Oleoresins contain both resin and essential oils; probably turpentine is now the most widely-used oleoresin.[5]

Resins are now sometimes divided into three categories: hard resins, oleoresins, and gum resins (a mixture of gum and resin). Copals are among the hard resins. Hard resins are either obtained from underground deposits (as fossilized resin) or prepared from an oleoresin whose essential oils have been removed by the application of moderate heat. For example, as one source of their copal, the Mayan Chorti Indians of Guatemala use Bursera trees. After it is extracted from the tree, the oleoresin,

> together with a great deal of bark, is then dried in the sun for a day, after which it is put *to boil with water* in a large olla [maximum heat 100°C]. Fresh water is added as the old boils away. The copal gum [that is, the resin] rises to the surface slowly and is skimmed off with a gourd dipper. After eight or ten hours of boiling...it is placed in cold water to harden. It is then shaped in the hands into round, elongated pellets...[They are] *extremely hard and brittle.* (Wisdom 1940, 182-183)

Among pre-Hispanic Maya, trade in copal was a major industry. Although copal-yielding trees grow wild in forests of Mexico and Central America, they were often cultivated in special plantations:

> They raised many trees of the incense for the idols and they got it out by wounding the tree in the bark with a stone so that the gum or resin should run from it. (Landa 1941, 197)

While the Aztecs applied the name to several other plant products used for incense, the principal constituent of Mesoamerican copal was resin obtained from the following plants: torchwood-family trees of the genera *Bursera* and *Protium*; pine trees; and the tree, *Hymenaea courbaril*. Even though other plant materials were used as incense (including chicle and rubber), none was so important as these four trees.[6]

Hymenaea courbaril is a large leguminous tree (found from southern Mexico and the Antilles to Amazonia) which secretes an oleoresin. From Mexico to Panama, the tree is often known as *guapinol*, a word of Nahuatl origin. In parts of tropical America, quantities of the exudate drip off these trees and accumulate beneath the ground. The resin, having lost most of its essential oils, is sometimes dug up in a semi-fossilized (hardened and amber-like) condition. From Amazonia, the resin is sometimes called 'Brazilian copal.'

Copal, burned as incense, was 'called the superodor of the center of heaven' (Roys 1931, 278). The Maya perfumed their idols with incense (Landa 1941, 146)—and lumps of copal were sacrificed in the cenote at Chichén Itzá on Yucatan Peninsula. Maya '...travellers even carried incense on their journeys and a little dish to burn it in' (Landa 1941, 107).

The symbol for copal appears frequently in Mexican and Maya codices. Among the Maya, '...copal as incense is shown by a black spiral' (Landa 1941, 75; footnote by A. M. Tozzer). 'It is believed that, as it burns, the smoke of copal carries the message of the people into the heavens' (Case et al. 2003, 192). Priests with incense burners appear frequently in Mesoamerican codices, and censers are among the most numerous artifacts found in Mesoamerican archaeological sites. These censers, often well-decorated, were sometimes made of stone, but more commonly of baked clay.

If, as suggested above, candles were not made in prehispanic Mesoamerica, the archeological artifacts called '*candeleros*' (candle-holders), common throughout much of Mesoamerica, are misnamed. These objects are generally now believed to be incense-burners, because some 'have shown traces of some organic substance. When analyzed, this was found to be copal' (Linné 1934, 113).[7] Within the last century, the Maya-speaking Zoque people made similarly ornate censers (Starr 1901-1902, 2:63).

[5] The term 'betún,' used frequently in early Spanish literature, can only by translated vaguely—depending on context—as a resin, gum, pitch, or an asphalt-like substance.

[6] From an archaeological standpoint, some progress has been made in distinguishing these copals on the basis of their chemical constituents, as well as their botanical and geographical sources. For example, resins of *Protium* and *Bursera* are triterpenoid, whereas resin of *Hymenaea courbaril* is diterpenoid (Lambert et al. 2002, 73). Pinene and limonene are the dominant essential oils from the resin of *Hymenaea courbaril* (Case et al. 2003, 189 and 198).

[7] On the other hand, since such artifacts (found in northern Honduras) were

> bearing traces of coloring matter in the interior, it is not unreasonable to suppose that these 'candelaria' with their vial-like shape, and rough, clay, undecorated exterior, might well have been used as some sort of *paint-container.* (Stone and Turnbull 1941-42, 45)

Above: Three crude clay archaeological 'incense burners' (candeleros) from Teotihuacan, Mexico (Linné 1934, 113). Below: An ornate ceramic incense-burner, recovered from Xolalpan, Mexico. 'The bulging portion of the walls is well polished, and the feet are provided with rattles' (Linné 1934, 111). Some censers have holes near the bottom for ventilation.

Figure 39. Incense Burners

Among the Aztecs, some fragrances were the prerogative of an upper class. Only the most important people were allowed to burn the woody roots of certain low-growing pine trees; when burned, the root smelled like incense (Sahagún 1956, 284).

The Guaymí of Panama harvest a resin called chutra which seeps from the scarred bark of a forest tree, *Protium panamensis*, closely related to the Maya's famous copal tree; the resinous gum accumulates in globular masses near the base of the trunk. (In the local Spanish, the material is sometimes called '*caraña*,' but that name is also applied to a similar gum obtained from *Trattininckia aspera*—like *Protium* or *Bursera*, trees in the torchwood family, Burseraceae.) When dry, the resin is a hard white substance with a pleasant, piney odor. *Chutra*, found in almost every Guaymí household, is burned as incense—and sometimes chewed to freshen the breath (Gordon 1982, 66).

The custom of burning fragrant substances at religious ceremonies was not confined to Mesoamerica. The Cuna of eastern Panama still burn incense and cacao beans on baked clay censers. In highland and coastal Peru where firewood is scarce, the Inca considered incense especially fitting for their sacrificial fires. However, 'this fire was not started and fed with *just any wood*; it was done with a certain kind of *wood which was scented*, carefully carved, and very colorful' (Cobo 1990, 117).

The chemical operation of an Aztec smoking and scenting implement remains something of a mystery: the '*cañuto de humo*,' or the 'tube of smoke or fumes' (Sahagún 1956, 3:151). Bernal Díaz refers to these instruments as *cañutos de olores*—i.e., 'fragrance tubes' (Díaz 1960, 1:279). The action of these tubes is difficult to picture from old descriptions. (Apparently they are not preserved archaeologically.) Although the tubes were primarily smoking implements, they appear also to have functioned as devices for perfuming and, perhaps, filtering the air. An essential element in the tube's construction was charcoal. Because of its porous and foam-like structure, a small mass of charcoal contains a disproportionately large surface area. Thus it is unusually capable of adsorbing vapors and filtering liquids.[8] (Charcoal is still much used for removing foul or noxious gases from the air.)

The following passage describes these tubes sold in Aztec markets:

> The smoking tube seller, the tobacco [tube] seller, is one who provides a covering [for the tobacco tube]—a maker of reed smoking tubes, a cutter of reeds. He strips them, he removes the outer surface; he prepares charcoal, he grinds charcoal. He covers [the reeds with moist charcoal dust]; he paints them, he colors them, he gilds them. He sells the tobacco [tubes] *destined for fondling in the hand*—long, of *an arm's span*, [covered with] a thickness of clay, [with] a thickness of charcoal [dust], whitened with chalk, gilded, painted, *painted with a hidden design*, mottled, painted with flowers; [he sells] tobacco tubes in the form of a blow gun, [painted with] fish, eagles, etc....[The smoking tubes] are filled with tobacco, the inside is filled, crammed, the opening is closed, *the end is closed.* [The seller] prepares tobacco, rubs it in his hands, mixes it well with flowers... (Sahagún 1950-82, 11:88, translated from Nahuatl into English by J. O. Anderson and C. E. Dibble; in this case, the brackets are theirs.)

From this passage, and others, we gather that the cane segments were about as long as an arm. After being well cleaned, the segments were covered with clay and moist pulverized charcoal, then whitened with chalk. The tubes may have been lacquered—at least they were varicolored, gilded, and painted with figures of flowers and animals. For instance, when Montezuma dined 'they also placed on the table three tubes (cañutos de humo) that were much painted and gilded ['*muy pintados y dorados*'], and inside was liquidambar mixed with an herb they call tobacco...' (Díaz 1960, 1:273). Since their ends were closed, how air passed through these tubes is uncertain; perhaps it was drawn in through perforations in the cane walls. The implements were smoothly shaped and eutactimorphic—that is to say, 'destined for fondling in the hand.' When

[8] Parts of a ceramic 'filter jar' with perforations in its bottom were found in eastern Santa Domingo, West Indies: 'several large pieces of charcoal were found in proximity to the sherd...The author considers it remarkable that the aborigines of Santo Domingo should have known the principle of filtering water through charcoal' (De Booy 1962, 89-90; figures 23, 24 and 25). It should be said, however, that since the charcoal was *not inside* the filter jar, filtering a liquid through charcoal is still unproven.

smoked, the tubes were filled with a mixture of ground tobacco, aromatic herbs and certain mushrooms, stuck together by chicle (Sahagún 1950, 4:88).

These enigmatic canes were said to have been fashioned so that some designs appeared only when the tubes were smoked (Sahagún 1956, 3:151). Sahagún's Spanish text differs somewhat from his notes in Nahuatl (translated into English, as quoted above). The Spanish text speaks of certain tubes made in such a way that their design-figures were concealed, and only visible when the implements were heated and in use: 'otros hay que tienen pintura encubierta, *que no se ve, sino cuando se van gastando con el fuego*' (Sahagún 1956, 3:151). In contrast, the translation quoted above mentions only tubes 'painted with a *hidden design*'—leaving open the question of *how, and why*, a painted design might be 'hidden.'

Some 8000 smoking canes are listed as the annual tribute levied by the Aztecs on 22 towns in the 'hot lands' of southeastern Mexico (*Codex Mendoza* 1992, 4:88). In Tenochtitlán, at about the same time, there is a reference to 'those skilled in preparing smoking tubes for tobacco' (Durán 1994, 319). Another early source adds the following, scarcely illuminating, comment, '...loads of smoking canes, which are perfumes *which the Indians use for the mouth*' (*Codex Mendoza* 1992, 4:88).

Personal Cleansing Agents

Since true soap (made by treating a fat with an alkali) was unknown in the Americas before arrival of the Europeans, in its place numerous 'soap' plants were used for washing. One of the many such plants used is the *jocote* (*Spondias* sp.):

> Considerable bark of jocote is gathered [by the present Chinantecs of Oaxaca]. This is taken off in long strips and done up in bundles...The bark contains a juice which can be *worked up in water so as to yield a suds* which is a refreshing wash for the head. (Starr 1900-1902, 1:71)

Cleansing agents were especially needed where textile-clothing was manufactured. (Because of their milder action, plant saponins may be less destructive than real soap to native dye materials.) In sixteenth-century tropical America a tree of the soapberry family, *Sapindus saponaria*, was cultivated and much used for this purpose. (Literally, the Latin name *Sapindus*, is created from *soap* plus *Indian*.) This small tree has a globular fruit a little smaller than a walnut. The fruit has a brown and shiny skin, an orange-brown flesh, and a black and bony pit which is often used to make beads. Having removed the pit, the 16th century Aztecs threw 'this fruit into *hot* water' and washed their clothes; the fruit forms 'suds in the same quantity as the finest Spanish soap' (Hernández 1888, 82). The same plant was used by the Maya:

> There is another [tree] which bears a small fruit with a stone like filberts from which are made good beads and *with the rind they wash their clothes as with soap*, and thus it lathers like it.'(Landa 1941, 197) (For other soap-making plants see Oviedo 1959a, 1:285.)

A scene from the Florentine Codex. The canes were an 'arm's span' in length (Sahagún 1950-82, 11:88). Since the far-halves of the canes are both thicker—and of a different color—than the mouth-halves, the canes' charcoaled and decorated coverings seem to have been on the far-halves.

Figure 40. Smoking Canes

Nowadays, too, in Guatemala and El Salvador,

> a vegetable product rather than soap as such is used by the Indians for washing their textiles. The one most commonly used is jaboncillo (*Sapindus saponaria* L.), the outer skin of whose fruit contains 4 per cent saponin. The wash water should be very cold [in contrast to Hernández's statement, above], *never hot.* (de Jongh Osborne 1965, 42)

In textile areas of the southern Andes, the soapbark tree (*Quillaja saponaria* of the rose family) plays much the same role as soapberry:

> The foaming properties of the bark were first recognized by the *indigenous people* of Chile (Mapuches) who used aqueous bark extracts *for washing their hair and clothes.* For this purpose the bark was placed in boiling water and allowed to settle all night before using the extract. (San Martín and Briones 1999, 303).

The Mapuche, an Araucanian tribe, keep herds of llamas; they still weave llama hair into clothing and blankets. The word *Quillaja* is derived from the Mapuche word 'quillean' which means 'to wash' (San Martín and Briones 1999, 303).

In Mexico agaves and yuccas, especially those that contain large amounts of saponins, are known by the name *amoles*. These act as cleansing agents by emulsifying the oils and greases on soiled fabrics. For example, the dried scrapings of *lechuguilla* (*Agave lechuguilla*) mixed with water, are still 'commonly used as soap throughout the Otomí-speaking area' (Parsons and Parsons 1990, 52).

Even outside the area in which textiles were made, plants were also important for their detergent qualities. Among California Indians the most important was the soaproot, *Chlorogalum pomeridianum*, a member of the lily family, which they used for washing their hair. The soaproot contains a saponin, called 'amolonin,' a word derived from *amole*.

Inorganic materials were also used as personal cleansing agents. For instance, the Indians of Peru used pasa, a kind of clay mentioned earlier:

> It serves to remove spots and *makes up for the lack of soap* because, making suds, it cleans clothes; for this reason the Indians wash their heads with this clay; and it kills lice... (Cobo 1956, 1:115)

Lye, ashes, and saltpeter functioned as cleansing materials as well (Sahagún 1950-82, 12:243). In Guatemala and many other places, pottery utensils are still cleaned with ashes and hot water (Reina and Hill 1978, 246).

CHAPTER 10

HIDE CURING AND FEATHER WORK

Tawing

After being dried, animal skins are stiff, unbending, and hard to work. In order to soften dried hides and make them workable, a process called 'tawing' has been employed from early times: like hide coloring, it was probably known to America's first human immigrants. The term tawing refers to the manufacture of a 'leather' from rawhide or skins without the use of tannin; other substances—fats, oils, alum, iron salts, etc.—are employed instead. For example, the Eskimos chewed fat into skins, and the softening process was aided by enzymes in their saliva. Similarly, the Plains Indians made clothing of painted bison hides which they cured by a prolonged *chewing*, rubbing and stretching—accompanied by the application of fat or bison brains rubbed in with a smooth stone.

In the southern hemisphere, the Tehuelche of Patagonia made cloaks of painted guanaco hides in a procedure that did not differ greatly:

> The skins are first dried in the sun, being pegged down with thorns of the algarroba tree. When dry they are taken up, and scraped with pieces of flint, agate, obsidian...They are then *smeared over with grease and liver* kneaded into a pulp, after which they are softened in the hand until quite pliable...The surface is slightly damped, and each woman takes a cake or piece of *red ochre*, if the ground is to be red, and, keeping it damp, lays the paint on with great care. (Lothrop 1929, 10-11, citing another source)

Alternatively, the skin is

> scoured with a coarse-grained stone, till it has a bright polish...then crumpled and twisted in the hands till it becomes perfectly soft and pliable...The fur being worn inside, there remains the work of outside decoration. With a due quantity of *clay, blood, charcoal and grease*, amalgamated for the purpose, the artist arms herself with a stick for a brush, and executes divers figures in black on a red ground. (Lothrop 1929, 11-12)

Among the hides offered for sale in Aztec markets were the 'skins of deer and other animals, dried and *treated*, with the pelt or *with the hair removed*, and also *dyed in various* colors' (Hernández 1986, 104).[1]

Tanning is a somewhat different process from tawing: in tanning, plant tannins (mostly from tree barks) are applied to the hide, thereby transforming its colloidal protein content into insoluble compounds. Thus rawhide is converted into true leather (a more pliable, more bacteria-resistant, and much less water-absorbent material) mainly by chemical changes in its collagen content. (An insoluble fibrous protein, collagen constitutes some three quarters of the weight of dry mammal skin. When wet, it is subject to bacterial decay.)

In the Old World, leather making is a craft of considerable antiquity: it seems to have been known in the Mediterranean area as early as 1500 B.C. It was, however, apparently unknown to native peoples in the New World before the 16th century: '*Neither in North nor South America are skins bark-tanned*, but only prepared by mechanical treatment and rubbed in with liver and fat' (Nordenskiöld 1931b, 80).

The animal hide used in making the sandals worn by the Maya was not tanned: 'They wore sandals of hemp [maguey] or of *the dry untanned skin of the deer...*' (Landa 1941, 89). Similarly, in pre-Hispanic Peru the footgear made of animal skins was mainly rawhide which, when drenched, soaks up much water, swells, and loses its shape:

> They [the llamas] have long smooth necks, the skin of which the Indians flayed and *greased with tallow* till it was soft and *appeared to have been tanned*, when it was used for the soles of their shoes. But as *it was not really tanned*, they used to take their shoes off to cross streams or when there had been heavy rains, otherwise they *became like tripe when they got wet*. (Garcilaso de la Vega 1966, 512)

Although references to 'tanning' and 'tanned leather' are common in the literature on Amerindian artifacts, the distinction between tawing and tanning is often overlooked. For example the Chichimecs, who collected the skins of deer, coyotes, gray foxes and squirrels, were described (in translation) by the Aztecs as tanners:

> ...they cured skins; they were *tanners* [tawers?]; for all the clothing of the Chichimeca was of skins, and the skirts of their women were of skins. (Sahagún 1950-82, 11:173)

On the Yucatan Peninsula, too, 'small bits of *tanned skins* have been recovered from the [Maya] Cenote of Sacrifice' (Landa 1941, 199; footnote by A. M. Tozzer). Yet the Aztec word, *cuetlaxtli*, is translated as '*dressed* leather' (*not as tanned leather*) by those who know Nahuatl well (Dibble 1971, 328).

---

[1] For some purposes, hair on the animal skins had to be removed. In North America, the Plains Indians removed hair from rawhides—characteristically bison hides—by soaking them in water mixed with wood-ashes (lye); a common name for skins prepared in this way (and various items prepared from them) was 'parfleche.' In the Andes, modern Aymara also use ashes: the 'hair is removed by applying hot ashes and then scraping with a stone' (Tschopik 1963, 535).

A minor advantage of tawed or dressed leather over tanned leather, is that persons facing starvation can cook and eat it: among the Chichimecs, when there was 'little food, they roasted it, broiled, boiled it' (Sahagún 1950-82, 11:173). This advantage was also recognized by the Jicarilla Apache (Gordon 1970, 55) and, no doubt, by other tribes as well. When heated in water, the insoluble collagen is converted by hydrolysis to gelatin, a protein which is soluble and easily digested. Although collagen is composed of many amino acids, it 'contains 25-30% glycine' (Belitz and Grosch 1987, 9). In contrast, the proteins in tanned leather are not digestible.

One note on the history of tawing remains something of a mystery. (In chemical discussions like this, one's normal tendency to withhold unpleasant details should be resisted.) At the market of Tenochtitlán and along waterways around Lake Texcoco, boat loads of human ordure were collected and sold. It is claimed that this material was used in curing hides:

> I must also state, with apologies, that they sold many canoe-loads of human excrement, which they kept in the waterways near the market, and this was for making salt [?] or *for curing hides, which they claim cannot be done well without it.* I am well aware that some gentlemen will laugh at this, but I declare that it is so. Furthermore, along all the roads they have the custom of providing shelters, made of cane or straw or herbs, where they cannot be seen by passers-by; there they can empty their bowels when they wish to do so, but also in order that this filth should not be lost. (Díaz 1960, 1:278)

Most of this ordure was likely used as fertilizer.[2] On the other hand—considering Bernal Díaz's generally acknowledged reliability—some use of it was probably made in treating hides. (Although special garments of deer and ocelot skin were favored by Aztec warriors and nobles, it seems unlikely that the people of Mexico, who did not keep herds of domesticated mammals, would have engaged in a mass production of cured hides. Deer and peccaries, drawn to rural gardens, were likely their only major source.) Since there is no mention of bark tannin, some form of tawing was likely involved. Bernal Díaz's account is made more credible by the fact that, in Europe at least, both dung and urine were once used in curing hides—as well as in vat-dyeing fabrics ('Stale urine and excrement are both alkaline' Balfour-Paul 1997, 84 and 86). Chicken droppings were used by sixteenth-century Italian dyers (Balfour-Paul 1998, 126). Curing hides and dyeing fabrics are historically related processes, and often in both similar materials were used. For instance, dog dung was commonly used by tanners in northern Europe until recently, and by dyers in Aleppo until the 17th century.

> In Britain in the 1930s dogs' turds were still commonly used by tanners, as some who worked in the business can even recall today. (Balfour-Paul 1998, 126)

In most urban societies the professions of tanner and dyer were lowly ones. Wherever it was practiced, curing hides was a messy procedure. For obvious reasons, specialists did not willingly publicize their trade secrets.

There is still something to be learned about the chemistry of curing hides: an authority on the subject of vegetable tannins notes that 'a scientific understanding of the traditional tanner's art remains, at best, incomplete' (Haslam 1988, 155). Included in this statement should be the tawing of hides by treating them with smoke, so typical of the tribes in northern North America: 'The smoke gave the skin a characteristic odor, a pleasing color, and was said to prevent the leather from hardening when drying after being wet' (Hough 1926, 71-72).

Either an inadvertent or deliberate partial tanning may have occurred in pre-Hispanic America because of the presence of tannin in the dyes used to color hides. Dyeing and tawing hides are probably historically related processes; as mentioned before, the 'most primitive form of mordant dyeing is the combined use of tanning and ferrous substances, e.g. in mud dyeing' (Bühler 1948, 2496).

Concerning the possibility that a partial tanning may occasionally have occurred in early America, a note on the chemicals used in the Old World for coloring hides may be pertinent:

> Originally, *vegetable dyes* with or without mineral salts were commonly used. *Black* was formed from *tannins mixed with* iron and *copper salts*. Vivid colours were produced on *tawed leather* by the action of the *aluminum salts* with many vegetable dyes. The dyeing of alum-tanned leather became a highly skilled technique among the ancient Phoenicians and Egyptians. (Robinson 1969, 12-13)

Much the same constituents are recorded in early-sixteenth century Mexico: When dyeing hides, the Aztec sometimes used a solution made by soaking chips of a leguminous tree (*Caesalpinia* sp.) in water. At first the solution turned reddish-black, an indication that the chips contained considerable tannin,

> but stirred with *alum* [an aluminum salt] and other red materials, the solution turns strongly red. *With this color hides are dyed red, and to give them* a black tint, *aceche* [copperas] and other materials are stirred into the watery solution, and with this the hides are dyed very black. (Sahagún 1956, 3:342)

As with most of the arts and skills under discussion here, the use of tannins (plant polyphenols) in curing hides has developed 'over centuries by the application of empirical methods and reasoning, and by the familiar processes of improvisation, trial and error' (Haslam 1988, 1). Occasional tanning has probably occurred from time to

[2] 'Manuring with human excrements, a method which to the Indian tribes east of the Andes not only is unknown but most undoubtedly would be highly offensive and disgusting, was used *in Mexico and Peru*, as well as in China' (Nordenskiöld 1931b, 46). Use of 'night soil' would be repulsive to agricultural tribes almost anywhere else in the New World.

time in many places, including in the Americas.[3] In the Near East and Europe, the benefit of applying the bark, or other tannin-rich parts of shrubs and trees, to hides was clearly recognized, and tanning there became standard practice.

Featherwork

Feathers have been used as decoration in many parts of the world, including areas where woven fabrics were unknown. For instance, the California Indians, who did no weaving, made tiny non-utilitarian baskets decorated with hummingbird feathers. Nevertheless, it was in areas where weaving was practiced that featherworking reached its high point. In the Americas, two areas in which feathers were used to decorate woven fabrics stand out, namely Mexico and Peru.

Making feather mosaics was a highly respected Aztec profession. Feathers were among the principal items in Montezuma's storehouse of treasures at Tenochtitlán, and huge quantities were collected as tribute from subject tribes by their Aztec rulers. The following items were among the annual tribute levied by the Aztecs on seven towns in the 'hot lands' of southern Mexico:

> Also *one thousand four hundred bundles of rich feathers of blue, red, green, turquoise-blue, red and green...Also eight hundred bundles of rich yellow feathers. Also eight hundred bundles of rich, long green feathers, called quetzalli.* (*Codex Mendoza* 1992, 4:98; see also 4:96)

For a local supply, Montezuma also kept a special aviary where numerous species were bred, hatched at the proper season, and plucked—each species being fed according to its need (Díaz 1960, 1:274). Featherworkers were housed in a special part of the city, and children were trained to make a special glue, an essential part of the featherworking process.

Figure 41. Aztec Featherworker and Son (Codex Mendoza 1992, 4:145)

[3] In the Maya area, and elsewhere, grows a small tree called nance (*Byrsonima crassifolia*), 'the bark of which is very good for tanning skins' (Landa 1941, 199). Landa does not say, however, that the tree was actually so-used by Indians themselves.

A detailed account of the featherworking craft, as practiced by the Aztecs in sixteenth-century Mexico, was given by Bernardino de Sahagún. As was the case with many native American crafts, Aztec artisans, despite their simple tools, accomplished much by skill and patience. Tools of the featherworker were little more than the following: a paint-vessel with some sort of black crayon to outline patterns; a copper or obsidian blade for cutting the feathers into the required shapes; a cutting-board made from bald cypress wood; and a bone spatula used in gluing and smoothing the cut feathers.[4]

Sahagún's description shows that the Aztec featherworkers manufactured with their vegetable glue a translucent colloidal sheet which they used for tracing. This cut-out, translucent sheet[5] was used 'after the manner of a stencil...The stencil was used as a check throughout the whole process...' (Joyce 1970, 145-147;

[4] According to Sahagún, the following steps were followed in making the Aztec feather mosaics:
(a) The featherworkers painted to scale a pattern for the desired finished product, their scribes outlining every detail.
(b) A broad section of a perfectly smooth maguey leaf was chosen whose skin had a 'smooth, shiny surface, with no knobs.' On this, the featherworkers glued a thin film of cotton gauze and set it in the sun to dry.
(c) They spread glue on the leaf's surface and covered it with a piece of cotton gauze, stretched gossamer-thin, and again set the leaf in the sun. When partly dry, 'once again they spread glue on the surface, thereby making the surface of the cotton glossy, shiny.'
Another early source affirms Sahagún's account of this use of maguey leaves:

> These pieces of maguey leaves are much used by the craftsmen, called *amentecatl*...who work in gold and feathers. On these leaves *they make a paper of cotton mixed with paste, as thin as a very thin veil,* and on the paper—on top of the maguey leaf—they work all their designs. It is one of the principal things that they use in their craft. (Motolinía 1950, 273)

Apparently no such use of maguey leaves is made today.
(d) This translucent sheet of glue-covered, cotton gauze was peeled off the maguey leaf, and placed over the original painted pattern. Then the featherworkers began 'tracing the painting which appeared from underneath.'
(e) When the design had been traced, in all detail, on to the translucent sheet, the sheet was glued to a piece of coarse paper to provide a firm backing, and laid on a cutting board. It was then trimmed to conform with the original painted design: 'They followed the black line to form the design with a flint knife' (Sahagún 1950-1982, 10:76).
(f) Another maguey leaf was then covered with glue to which the cut-out pattern was affixed and dried in the sun. This provided a somewhat flexible panel on which to mount feathers.
(g) Quite apart from the above procedure, common 'feathers had been cut one by one; glue-hardened...suspended, dipped, in glue.' After being stuck to the flexible panel, 'their surfaces were smoothed with the bone blade.'
(h) After installing the common feathers in this bedding, the artisans 'matched whatsoever kind would harmonize, would serve as the basis for the precious feathers. Blue cotinga were provided blue parrot feathers as a basis'; only 'heron feathers corresponded to those of the quetzal'; and so on (Sahagún 1950-82, 89-96; see also Sahagún 1956, 82-83). Thus, 'common' feathers formed the basis 'on which all the precious feathers were bedded.' Only common feathers were glue-hardened, and among these many were 'yellow dyed ones' (Sahagún 1950-1982, 10:91-96).
Despite this detail, it is unlikely that the whole process can now be duplicated accurately from Sahagún's description: the procedure was complicated and involved highly trained artisans.

[5] There is some, as yet unexplained, connection between the use by the Aztecs of maguey leaves in making sheets of paper and the sheets used as a foundation for mounting feathers.

also Sahagún 1950-82, footnote 94). Strictly speaking, however, the device is neither a template nor a stencil since it was used only once, not repeatedly. The same remarkable glue is thought to have been used in featherwork and bookbinding (Lenz 1973, 162; also Christensen 1963, 361). Glue for the translucent sheet was made from *Epidendrum pastoris* or, according to some, from *amatzautli, Cranichis speciosa*: Despite their importance in native crafts, the chemistry of pre-Columbian glues is still poorly known. Almost all were derived from plant sources, although animal glues had been used in parts of the Americas.

Feathers were mounted on a support of cotton, glue, and maguey fiber. The first layer, composed of 'common feathers,' was dipped in glue. Only common feathers were 'glue-hardened' feathers, and among these were some 'yellow-dyed ones' (Sahagún 1950-82, 10:94).

The Mexican featherworkers distinguished between 'common' and 'precious' feathers. Whereas the former were taken from herons, ducks, and blue and green parrots, the precious feathers included those taken from the quetzal, hummingbirds, and the blue cotinga. Common feathers were often dyed to various shades of yellow, since yellow feathers were apparently in short supply. Those feathers that were to be dyed yellow were treated with a solution made from a plant called *zacapalli*, likely a native dodder (*Cuscuta* sp.), a leafless, parasitic, twining plant of the morning-glory family. Alum and saltpeter were added to the solution, probably as mordants to brighten the colors: 'The yellow color [was] cooked on the fire; it boiled; *alum* was added; and then it was provided with *saltpeter*' (Sahagún 1950-82, 10:95). The precious feathers were mounted on this foundation of glue-hardened feathers.

At the time of the Conquest few native artifacts were more admired by the Spanish than Mexican feather mosaics:

> I don't admire so much the gold and the jewel stones. What amazes me is the cleverness and superior workmanship [of this featherwork]...it seems that I have never seen anything which for beauty could more delight the human eye. (Martír 1964, 1:430)[6]

Maya women also used feathers to embroider garments: 'they also raise birds for their own pleasure, *and for the feathers from which to make their fine clothes...*' (Landa 1941, 127).

In Inca Peru, as in Mexico, great supplies of feathers were stored (especially those of the hummingbird), and equally fine feather mosaics made: 'The gloss, splendor, and sheen of this feather cloth was of such exceptional beauty that it must be seen to be appreciated,' writes Cobo.

> The feather cloths were the most esteemed and valued, and this was quite reasonable because the ones that I have seen would be highly regarded anywhere. They were made...in such a way that the feathers stand out on the wool and cover it like velvet...incredible numbers and varieties are found in this land with such excellent colors that it is beyond belief. They used only very small, fine feathers. These they fastened on the cloth with a fine, wool thread... (Cobo 1990, 225-226; also Cobo, 1956, 2:239)

Archaeological specimens of featherwork from the deserts of coastal Peru are surprisingly well preserved, many showing little or no signs of age. (A large collection, made by artisans of the Chimú culture, has been recovered.)

In Peruvian featherwork, there was only occasional use of glue.[7] Furthermore, archaeological collections from Andean South America show no evidence of dyeing:

> A few words may be said regarding *artificially colored feathers* in the work of the South American Indians. I have never been able to detect a single instance of their use among the ancient Peruvians... (Mead 1907, 17)

Among living South American tribes, except those influenced by modern commerce, feathers are rarely dyed.

In Peru, as in Mexico, feathers were much sought after. Birds were caught by a method well known in the Old World:

> To catch birds, they used nets and *birdlime*, which they got from certain types of trees. *They hunted birds more for the feathers than for meat.* (Cobo 1990, 241)

Birdlime is a European name for various viscous substances smeared on twigs and branches in order to entangle birds when they alight. The Aztecs, too, used a sticky substance—smeared on maize plants or on trees—to catch birds. The substance was made from root of a plant, *Iostephane heterophylla*; from this root 'a fine birdlime is made for catching foolish birds' (Hernández 1888, 169). (Hernández called the plant *tecpátli*; Sahagún called it *tecpaolotl.*)

There was also a highly advanced featherwork in Polynesia (Hawaii, New Zealand and Tahiti). In discussing the featherwork of South America, many writers have noted

> the very striking parallel in Hawaii...On the other hand, the technique by which the feathers are made fast to the fabric

[6] Another early source praises the Mexican 'beautiful gold and feather work. This land has many excellent masters of this art, so much so that in Spain and Italy people would consider them as of the very first class and would look at them in open-mouthed astonishment...' (Motolinía 1950, 91).

[7] For instance, in the Chimú style, feathers were usually tied on.

> This is the most common method of creating feather cloth in the Chimú style. Tiny feathers attached to a non-textile surface, as in the tassels, appear to be glued on. (Rowe 1984, 154)

is fundamentally different from that used in Peru. (Wissler 1957, 60)

Tapirage

With regard to plumage, it has long been recognized that certain fatty foods can change, to some extent, the color of some birds' feathers. For example, 'it is well known that hemp-seed causes [European] bullfinches and certain other birds to become black' (Darwin 1897, 2:269).

Although he did not use the word 'tapirage,' Alfred Russel Wallace (originator, with Charles Darwin, of the theory of natural selection) wrote of a practice by native Amazonian peoples of the Rio Negro and Uaupes. The Indians wore a head-dress of red and yellow feathers, made

> entirely from the shoulders of the great red macaw, but they are not those that the bird naturally possesses, for these Indians have a curious art *by which they change the colours of many birds.*
>
> They pluck out those they wish to paint, and *in the fresh wound innoculate with the milky secretion from the skin of a small frog or toad. When the feathers grow again they are of a brilliant yellow or orange colour,* without any mixture of blue or green, as in the natural state of the bird; and on the new plumage being again plucked out, it is said always to come of the same colour without any fresh operation. The feathers are renewed but slowly... (Wallace 1890, 202)

Later, when he traveled to the Malay Archipelago, Wallace noted the occurrence of a custom that he had observed in South America. He sent the additional information to Darwin who wrote the following note on Wallace's communication:

> The natives of the Amazonian region feed the common green parrot (*Chrysotis festiva*, Linn.) *with the fat of large Siluroid fishes* [i.e., catfishes], *and the birds thus treated become beautifully variegated with red and yellow feathers.* In the Malayan archipelago, the natives of Gilolo *alter in an analogous manner the colours of another parrot,* namely, the *Lorius garrulus*, Linn., and thus produce the *Lori rajah* or King-Lory. These parrots in the Malay Islands and South America, when fed by the natives on natural vegetable food, such as rice and plantains, retain their proper colours. (Darwin 1897, 2:269-70)

Thus feather color can be changed, apparently, by either feeding some birds certain unusual foods or by applying to their plucked body-parts a secretion of frog poison. South American featherworkers (like those in Mexico) apparently found that natural supplies of yellow-colored feathers were too small. A yellow feather, touched with pink, found in Peruvian collections from Chimú, serves as another example. It resembles

> ...the outer tail feathers of the Yellow-headed parrot, *Amazona ochrocephala*, but is *not a feather that can be found in any living species of the genus*...[Likely such feathers] were produced by the method known as *tapirage.* (O'Neill 1984, 147)

In the 16th century,

> two chroniclers...Soares de Souza (1) and Magalhães de Gandavo (2), reported that the ancient Tupinamba Indians of the Brazilian coast knew how to change the color of the feathers on living birds. They took young common parrots, plucked their feathers, *and smeared the bald spots with frog blood* to which 'certain other substances were added.' The new feathers grew in *yellow.* (Métraux 1944, 252)

According to the following statement, the Paressí of the Matto Grosso practiced tapirage also:

> They raise ara, parrots, and other birds as we do chickens, and pluck them to apply on their skin pigments [?] which determine the color of the new feathers. They pluck these feathers for their fabrics and again apply pigments to create feathers of several colors...(Métraux 1942, 162, quoting from Pires de Campos in 1862)

A number of ethnographers have confirmed that tapirage was practiced by tribes from the northwest Amazon to the Gran Chaco in the south. Nevertheless, there is still some doubt that the practice actually occurred. This may be partly because of the seemingly confused statement of the respected observer, Theodor Koch-Grünberg, who reported that the Indians of the Rio Negro and Uaupes area

> pull from the tame red macaws the green feathers at the base of the wings and *smear the wounds with the fat of the pirarara fish* [a multi-colored catfish] or of *a certain toad.* The new feathers become beautifully *orange-yellow* and retain this color, even if several times changed, as they are pulled out from time to time... (Koch-Grünberg 1909-1910, 1:84)

Here, Koch-Grünberg claims that *the fat was smeared on the parrot's skin, rather than being fed to the bird.* It is more likely that the effective agent was the secretion of a 'certain toad.'

It should be noted that nowhere else in the Americas, and indeed in the world, was there a comparable knowledge of the physiological effects of frog or toad secretions. (In Amazonia, as we have seen, such secretions were also used as hallucinogens and for making poison darts.)

CHAPTER 11

EMBALMING AND MUMMIFICATION

Some of the earliest and most sophisticated of the chemical arts were devoted to preserving the human body. An artificial mummification, 'the oldest in the world, began about 5050 B.C. and was abandoned about 1720 B.C.' (Arriaza 1995, 35). It was practiced by a people known to archaeologists today as the 'Chinchorros,' ancient fishermen who lived on the coast of the Atacama Desert (in northern Chile and southern Peru), well before pottery and metallurgy were known there.

The most sophisticated of Chinchorro mummification techniques was devoted to producing 'black' and 'red' mummies. In both cases, body cavities were emptied and 'dried with hot ashes or glowing coals' (Arriaza 1995, 43 and 45), as evidenced by charred tissue.

In the case of black mummies, the body was completely disarticulated, and

> converted into elaborate statues with an inner structure of bones, unbaked clay, and sticks and ropes for reinforcement... (Arriaza 1995, 43)

The body was then

> completely modeled with a *white-ash* paste in which the morticians embedded the bones, sticks and cords in an attempt to give the body its original form...The final step in the process involved painting the body with *manganese*. On the face the *manganese paste* was thicker to permit the modeling of eyes, nose, and mouth. (Arriaza 1995, 45)

(The Chinchorro manganese source was probably a black oxide of manganese, MnO_2; this oxide is found in relatively-soft, earthy masses as pyrolusite, a widely-distributed mineral.)

In the case of red mummies, after emptying the body cavities, the 'morticians stuffed the body with a mixture of materials such as ashes, camelid hair, feathers, grass, animal and bird skin, and soil' (Arriaza 1995, 46).

> The head received a treatment similar to, but more elaborate than, that of a Black mummy...A *white ash paste* was placed over the defleshed facial bones and covered with the facial skin; the face was *then painted and modeled with either black manganese paste or dark red ocher* to produce the semblance of a mask. (Arriaza 1995, 45)

As is often the case, the organic binders in such 'pastes' are the most difficult constituents to identify. A not-altogether-successful early attempt to analyze the substances used by Chinchorro embalmers at Arica is an example of this difficulty. The body of a male individual was 'enclosed by concrete, or, more correctly, 'sand mixed with an unknown liquid'' ' (Linné 1957, 150). The so-called concrete

> emitted a peculiar strong and disagreeable odour. This emanated from some *sort of glue and one or more resins*...It was evident that the concrete had been made as a paste and applied in moist condition...His account, surely fascinating to a chemist, is incomprehensible to a layman. Succeeding in part, he was unable to disclose the full secret of the embalming mass. He could make out that *elastic resins had been added to the sand*, and stresses that if *natural elastic resins and balms are far from easy of determination*, this is so much the more difficult in the case of compounds such as those which occur here. His minute examinations result in the following statement: The Arica people made a *decoction of animal glue and added elastic resins or some sort of balm species*. Here we have *a clear case of embalming*. (Linné 1957, 150-151, summarizing the examination by Professor G. Bodman, 1924)

Similarly mummified human bodies have been found (dating from roughly the same period as those made by the Chinchorros) in highland Bolivia and Peru. These mummies, made by Andean hunters, are found in what is now Aymara territory. There, too, the procedure is surprisingly elaborate:

> After removing the internal organs, including the brain, the embalmers cut away the skin and the hair, and removed the muscles. The body was then *smoked* and filled with [camelid] wool, feathers and rubble, and *covered with clay*, molding carefully the limbs and face. Finally, the skin was put in place again and the embalmed body *painted either red or black*. (Schull and Rothhammer 1990, 45)

Whether such methods of mummification had their beginnings on the west coast or in the Andean highlands (or even in the Amazon) is debated.

At the beginning of the 16th century, several historical records of mummification come from the coast of Venezuela to the northern Isthmus of Panama. For example, the Zenu people (living around the Río Sinú, west of Cartagena) customarily embalmed their caciques as follows:

> When one of their principal men dies, they remove his viscera, wash the body, and anoint it with certain [unspecified, but presumably sticky] substances. Upon that, they put cotton wool, dyed in various colors, which clings to the body [*pone lana de algodó teñidos a diversas colores q se pega en el cuerpo*]. Covered with this material, they put him in a hammock, which is their kind of bed, and hang him in a house near where they make the fire. (Enciso, 1530. I was unable to reproduce the diacritical marks on some of the bracketed Spanish words. The book is

> unpaginated, but the passages translated here are on the 64th and 65th leaves.)[1] (Enciso found twenty dead bodies preserved in this manner in the houses of a nearby place called Catarapa.)

The natives around Chiriquí Lagoon (located farther to the northwest in what was, at one time, called 'Veragua') are said to have used *caraña* 'for painting the bodies of the dead so they do not decay' (León Fernández 1886, 5:158). *Caraña*, as noted above, is a resin obtained from a forest tree identified in this area as either *Protium Copal* (Pittier, 1908, 83) or *Trattininckia aspera*.

When a chief died in southern Panama, they hung up his body 'by cords, placing *many pans of charcoal round it*' (Andagoya 1967, 15, according to Markham's translation). In both Panama and Peru, gold masks were made for the mummies. For example, Balboa upon entering the great houses built by inhabitants of Comagre in Panama, saw the 'desiccated bodies of [their] ancestors *hung from the rafters with gold masks covering their faces...*' (Sauer 1966, 222, citing Pedro Mártir). Excavated masks for the dead at Moche on the north coast of Peru (Mochica culture flourished between 100 and 700 A.D.) were of hammered gold and silver with eyes of shell-inlay (Emmerich 1965, 15).

Several early Spanish accounts describe mummification in Peru:

> In the province of Jauja, which is a very important place in these kingdoms of Peru, they put them [their dead] in a fresh llama-hide, and sew them in it, shaping on the outside face, nose, mouth, and all the rest, and in this fashion they keep them in their houses... (Cieza de León 1959, 310)

Apparently mummification was practiced only if the dead person was a king, or another of great importance:

> ...only a few years ago I saw one of these...This body was so well preserved and adorned that it looked as if it was alive. The face was so full, with such a natural skin complexion that it did not seem to be dead, though it had been for many years. The face was preserved in that way because there was a calabash rind placed under the skin of each cheek. As the flesh dried out, the skin had remained tight and had taken on a nice gloss. The artificial eyes were open, and this gave the impression that it was looking at those who were present. (Cobo 1990, 42)

A historical reference to the mummification of pre-conquest Inca kings states that the people 'preserved the body, embalmed in *certain aromatic preparations*, which preserved it without spoiling' (Montesinos 1967, 65).

An example of something like 'mummification' is the well-known Jivaro custom of shrinking human heads. (Head-curing seems to be a different process from tawing or tanning.) The procedure involves removing the 'skull and shrinking the head to the size of a fist, whilst retaining the shape of the facial features' (Nordenskiöld 1931a, 496). In former times, 'Specimens of these heads...have been obtained and brought to Europe. Their exportation is now forbidden by the South American governments, as the supply not unnaturally was apt to coincide with the demand' (Whiffen 1915, 122).

---

[1] Another early account states the following:

> In some parts of Tierra Firme, when a cacique dies it is the custom to place his body on a stone or log, and around him and very near, without coal or flame touching the body of the dead man, a large fire is built and kept going until all the grease and fluid comes out through the fingernails and toenails, and in sweat, and the body becomes so dry that the skin fits tight to the bones, and all the tissue and flesh are consumed. When the body is thus dried out without opening it (which is unnecessary) it is placed in a secluded spot in the house prepared especially for it... (Oviedo 1959b, 37)

CHAPTER 12

SALT MAKING

There are several natural sources of common salt (NaCl), but these can be found only in certain locations. In arid regions beds of crystalline salt sometimes outcrop in sedimentary strata—for instance, there are 'great rocks' of salt in Peru (Cieza de León 1985, 174); this salt requires no treatment other than crushing. However, since rock salt is a readily soluble compound, it is rarely exposed at the earth's surface in humid areas.

In some places, especially in arid or seasonably-dry areas, salt crystals can be gathered on land surfaces that are subject to flooding by briny waters. Sea coasts are, of course, favorable for the natural formation of salt from sea water. For example, in the 16th century in the bay of Paria on the coast of Venezuela, there was a *salinas* (the Spanish name for saltworks).

> Because of high winds the sea becomes agitated and is driven on to a large plain adjacent to the shoreline; and *when the sun comes out* and the waves subside, an *excellent white salt crystallizes*; and it is of sufficient quantity that natives gathering quickly at the site would be able to load all the ships that sail. (Mártir 1964, 1:183)

Getting salt from such sources is scarcely a chemical art. Elsewhere, however, salt is less easily available and obtaining it requires some sort of chemical processing.

In the Americas, 'saltpans' (broad shallow depressions) were leveled in the earth's surface near the seashore or near a salt spring. The salty water was then led by ditches, or carried in jugs, from the shore or spring to these depressions. (The site should be underlain by a relatively impermeable layer, and not subject to flooding by freshwater.) The purpose of the saltpan is to store the brine until all the water is evaporated, and the surface is coated with salt crystals. Since this salt, whether formed naturally or in man-leveled depressions, is produced by the sun's heat, the Spanish called it '*sal de sol*' or '*sal solar.*' In this way, the Maya obtained much of their salt from arid northern Yucatán: 'the best salt which I have ever seen in my life, since it is very white when ground' (Landa 1941, 189).

An early sixteenth-century note from Coiba, on the Pacific side of western Panama, probably refers to a system of saltpans there:

> we encountered saltworks (*salinas*) of the *finest and best artifice and the best organized that have been seen.* In these, it is possible to make salt for a population like that of Sevilla, as *white as snow*, as salty and of as good grain as the best there is in Castile. (Sauer 1966, 274, quoting Gaspar de Espinosa, leader of the first Spanish expedition to the area)

Because of its many uses—and the prospects for commercial profit—the Spaniards were much interested in the availability of salt.

Another method of acquiring salt is to boil briny water over a fire. Solid salt obtained in this way is called '*sal cocida*' (cooked salt).

> Before the twentieth century, all of the salt produced along the Pacific coast of Guatemala was made by means of the *sal cocida* process, which has been in use for more than 2,000 years. (Andrews 1983, 68)

Cooked salt (*sal cocida*) on the Pacific coast of Guatemala is

> usually brownish or grayish and dirty-looking. The [Maya] Indians seem to prefer it, however, explaining that it has 'more flavor' than the sun-evaporated *sal de sol*. The report is widespread that some [of today's] makers of sun-evaporated salt sprinkle small quantities of *playa* dirt into their pure white product to make it simulate the cooked salt with its inevitable ingredient of silt. This was unverified, however. (McBryde 1945, 58)

The same method of cooking salt was used near Anserma, Colombia (Cieza de León 1985, 172).

Several early Spanish accounts speak of 'placing marsh soil in large canoes, filtering *estuary* water through the soil, and then cooking the resulting brine *in large ceramic ollas*' (Andrews 1983, 68).

To make salt for sale in the sixteenth-century markets of Tenochtitlán, the Aztecs heaped salt-impregnated earth in mounds. The mounds were then drenched, and the water percolating downward was filtered into an earthenware pot. Solid salt was precipitated by evaporating the briny filtrate over a fire.

> The salt producer gathers (salty) earth, *hills it up*, soaks it, wets it, distils [i.e., filters water through it], *makes brine*, makes ollas [pots] for salt, *cooks it.* (Sahagún 1950-82, 11:84)

Forms were then made to shape the salt cakes:

> Those who trade in salt...sell cakes that are *round or long*...plump and clean, without sand, very white, without any aftertaste; and at times they may also sell cakes which leave an aftertaste of crude lime. Occasionally, they also sell cakes with fine sand... (Sahagún 1956, 3:147)

A rather different method of salt making was described in 1580 from Colima, Mexico—a method in which special pottery was used:

They make salt in small quantities and with some difficulty, watering first the beach *from jugs containing sea water*. After two or three waterings they *heap this watered sand into piles*. The heaps made, they take jars (*ollas or tinajas*), and place one upon the other. *The top jar has some small holes in its base*, like the openings in flutes, *on which are placed straw mats*. Into the upper jar they then place such water-drenched sand until it is a little more than half filled, and then they turn (rotate?) it to expel the sea water and this water drips into the jar below and this filtered water comes out very wholesome and is drawn off into their jugs, which they take to their houses *to boil*, placing it on the fire until it is boiled down and converted into salt. (Sauer 1948, 69, translating an account, written in 1580, of the Motin region)

Sauer gives the following explanation:

The procedure seems somewhat complicated, but the beach sand probably absorbed certain salts, other than sodium chloride, and served as a partial means of purifying the table salt. Such base exchange would take place in particular if the sand carries, as it does locally, ferro-magnesian minerals. (Sauer 1948, 69)

One account, written about 1699, describes perhaps the most unusual and laborious method of obtaining salt. It was practiced by the Indians (probably the Sumu) of Honduras and Nicaragua:

They make a great fire close to the seaside, which *when it has well burned the sticks asunder, they take them singly and dip the brand in the sea*, snatching it out again, not too soon, nor too late; for, by the first, the drops of salt water which remain boiling on the coal would be quite consumed through too much heat, the coal not being sufficiently quenched, and, by the latter mismanagement, would be quite extinguished, and want heat to turn those drops of water into corns of salt, which, as fast as made, *they slightly wipe off with their hand into a leaf*, then put that brand's end back into the fire again, and take out the fresh ones successively, that *in half an hour's time a man can make about one pound of grey salt*. (Conzemius 1932, 93, quoting Awnsham Churchill [see References Cited]; the passage is also quoted by Joyce 1916, 39.)

The success of this procedure depends on the capacity of the ash and charcoal on the wetted brands to adsorb salt.

Today, or at least until recently, in the market at Huehuetenango in the Quiché Highlands of Guatemala, the Indians sell cakes of salt. These cakes are made in their saltworks, which are located on

a broad plain of sand, which lies between the river [i.e., the Río Negro] and the base of the mountain, covered with a thin, white salty crust. At the edge of the plain, near the mountain slope, flowing from the rubble are two small, *weakly-salty warm springs*. Their water, spreading over the plain, evaporates and leaves its salt content in and on the sand...

The plain is divided among several families occupied in salt making who carefully keep it clear of rubbish and plants. A *thin layer of fine, dry earth is spread over the sand, and kept continuously moist* by sprinkling it with water [in earthenware jugs] from the springs for about six hours. During this time, *part of the salt which has accumulated in the underlying sand*, is drawn up [by capillary action] into the layer of fine earth. *This layer is then swept together in piles and carried to the salt-kitchens*, located on the slope above the plain. Each salt-kitchen has a leaching device which consists of a rectangular box mounted on piles—the bottom of the box being made up of reeds and reed matting [which holds the earth itself in the box]. Into this, the earth is poured and washed out with water. Under the box is a stone-lined basin which receives the brine. This basin is emptied and its contents boiled in the salt-kitchen.

The salt-kitchens are buildings on whose floors are rows of stones, rising to a height of about 20 cm, between which burn wood fires. These stones serve as [fire-proof] *stands* for the earthenware bowls in which the brine is evaporated...but the salt is still quite impure...[While some is sold in this condition], part is purified by repeated dissolving and evaporating. This pure white salt is made into small round cakes. *The leached earth is dried over the fire and again strewn over the sandy plain.* (Schaub 1947, 155-157, translated from the German)

Unfortunately, the 'fine earth' (probably a clay) which is used repeatedly for adsorbing the salt is not further characterized.[1] This Guatemalan process is curiously similar to that used in ancient China—though there it was ash, rather than a 'fine earth,' that was used over and over again (Li 1948, 56-59; see Figure 29).

Stone cylinder *stands*, similar to those described in the passage above, 'used to prop up the brine-cooking vessels,' are reported from old salt-making sites on the Pacific coast of Guatemala and Costa Rica (Andrews 1983, 90 and 111).

Salt sources are especially meager in humid tropical forests distant from the ocean. Knowledgeable observers, such as Humboldt, have written that Amerindians ate

[1] Helfritz (1963) visited the salt spring on the Rio Negro later than Schaub, but was apparently unaware of Schaub's paper; at least, he makes no mention of it. The chemical identity of the 'fine earth' that adsorbs salt would have been particularly interesting. However, neither author is very specific about the subject, and neither seems to have had the earth *analyzed*. The following is Helfritz's account, translated from the German:

over the hard floor of the salt spring, the Indians spread a 5-10 cm thick layer of *a particularly fine earth* which, they claim, they can only collect *at special 'mines,'* as they call the deposits of this earth. Due to the incessant strong sunshine, the moisture which has seeped up into the loose earth is soon evaporated. But along with the moisture the salt has come to the surface and is now absorbed in the earth lying loosely there. After 3-4 days enough salt has already been soaked up that is worthwhile to scrape up the layer of earth with the salt, fill it in baskets and carry it to the filters situated on the steep banks of the river next to the houses of the Indians. These filters are large rectangular wooden boxes with sides 1½ m long. The bottom is funnel-shaped and ends in an opening closed off with a fine sieve. This box is filled about half full of the salty earth and the same amount of water is poured in. Under constant stirring, the solution slowly filters through and drips into a large earthenware dish placed under the filter. When the dish is full, it is placed on a wood fire and is boiled for a day until all moisture has evaporated. Normally they smash the hard crust of salt which remains in the dish. The earth which remains in the filter is shoveled out and dried in the sun. Later it is *carried back to the river bed and spread again over the salt springs*, upon which the same process is repeated anew. (Helfritz 1963, 7:305)

much less salt than the inhabitants of Europe. In general, this was probably true, and it was certainly the case in the rainforests of South America. On the other hand, large amounts of salt were produced and consumed in southern Mesoamerica, especially in the Maya area where salt was a major trade item. Tribes living far inland in humid areas were largely dependent on plant sources.

Salty-tasting compounds can be extracted from a variety of plants. In tropical areas, palms were often used. Hans Stade gives an early description of making salt, or a salt-substitute, in eastern Brazil:

> ...they prepare it in this manner; for I saw it, and helped to do it. Having felled a *thick palm tree*, they split it into small splinters; and they make a stand of dry wood, lay the splinters thereon, and *burn them to ashes, from which they make lye*, and this they boil till the salt separates from it. *I thought it would have been saltpetre, and tried it in the fire*, but it was not; it tasted like salt, and it was of gray colour. But *the majority of tribes eat no salt.* (Stade 1874, 133)

Hans Stade, like most European soldiers of his time, knew that the substance would have burned had it been pure saltpeter. (Long employed as a critical constituent in the manufacture of gunpowder, saltpeter is still used for its flammable qualities—for example, in the making of matches.) Stade's 'salt' probably contained sodium carbonate, Na_2CO_3, and potassium chloride, KCl, which has a salty bitter taste.

Others, too, remarked on the fact that the salt-substitute extracted from palms tasted like saltpeter

> In the province of Venezuela the Indians make salt from the ashes of the hearts of certain *palms*, which, though *quite white*, has a burned and bitter taste *like saltpeter.* (Cobo 1956, 1:113)

The Sumu of Honduras and Nicaragua burned the leaf-midrib of a spiny species of palm:

> The ashes are collected in a vessel with hot water, in order to dissolve their contained salts. After removing all impurities, the solution is evaporated in a large earthen vessel by boiling it down over a slow fire, whereby a whitish crystalline matter becomes deposited, which furnishes a good substitute for salt. (Conzemius 1932, 93)

(Thus, the Sumu had ways of making salt other than that mentioned earlier—that is, by extracting it from seawater with a firebrand.) In South America, the Tupi get a salt substitute also from the leaves of palms (Lévy-Strauss, 1963c, 301).

As mentioned earlier, the ashes of plants vary considerably in their chemical content, and a number of plants other than palms are used as sources of salt—or, more accurately, as salt-substitutes. For example, the ashes of several plants used by the Ayoreo, a tribe of hunter-gatherers in the Paraguayan Chaco, were analyzed. The ashes of their main plant source of salt, a tree (*Maytenus vitis-idaea*, in the staff-tree family, Celastraceae), contained 2.27% by weight of elemental sodium, more than five times the amount contained in the ash of a palm species which is also used by the Ayoreo as a source of salt (Schmeda-Hirschmann 1994, 161).

Chapter 13

Building Materials and Architectural Decoration

In arid regions the walls of buildings are commonly constructed of blocks of sun-dried adobe. These clay-rich blocks are usually mixed with grass, gravel or other materials, as a temper. In the Peruvian Andes, the Indians would 'mix mud with ichu grass to make their adobes so that they will not crack' (Cobo 1956, 1:198). (Adobe buildings and walls of Chan Chan—the capital city of the Chimú kingdom, now in ruins—cover more than ten square miles.) Among the Pueblo peoples, when the first Spanish arrived, 'ashes were mixed with adobe for building material' (Robbins et al. 1916, 29). Whereas the Pueblos and others used unprotected adobe, Olmec and Maya builders—coping with a more humid climate—often faced their adobe with stone.

Hernan Cortés 'refers to a walkway at Ixtapalapa [near Mexico City] which was 'ladrillado'; that is, 'paved with bricks' (Thomsen 1960, 428). It 'seems certain that fired brick was used in some parts of Mesoamerica, principally in the State of Tabasco' (Thomsen, 1960, 431). For some time, however, it was believed that fired bricks were not made 'in the Western Hemisphere prior to the introduction of the art from Spain' (Thomsen 1960, 439). This may be partly because clay-rich soil can be baked by accidentally-set fires into a brick-like consistency. Although bricks were rare in Mesoamerica, Thomsen's contention has been conclusively verified at Comalcalco (Tobasco) where rectangular fired-bricks with inscribed 'crude graffiti' have been found. These bricks are 'unique to the Maya area' (in Mesoamerica), and are dated as Late Classic—that is, between 500 and 900 A.D. (Coe and Kerr 1997, 38). Similar fired bricks were made in Peru, in roughly the same time period. These are '*...like flat, oversized fired adobe bricks*...with corners carefully rounded...'; 'this is the only reported case of fired tile flooring in Mochica construction [i.e., between 100 and 700 A.D.]...' (Shimada 1994, 270 and 160).

Mortar and Mortar-Additives

Mortar made with lime was a principal building material in Mesoamerica.

> The use of lime mortar in the New World appears to be confined to the area of Mesoamerica. It is, in fact, one of the most important of the several traits characteristic of this area which have given rise to the concept of Mesoamerica as a distinct cultural unit...such an invention plus the development of its varied uses *represents a complex technical achievement* about which very little is known. (Littmann 1957, 135)[1]

In one old study, it is claimed that the practice of using lime mortar as building material spread throughout Mesoamerica from a *single origin*:

> The area where mortar was used is continuous, and this invention *was never achieved but once*, all the tribes which use it having apparently borrowed it directly or indirectly from one source. (Watermann 1931, 519)

Basically, lime mortar is a paste-like mixture of lime, water, and sand or some similar aggregate. As the mortar mixture dries, it slowly hardens—often becoming harder than the original limestone itself. This process, which at ordinary temperatures proceeds very slowly, depends on lime's ability to react with carbon dioxide in the air to regenerate solid calcium carbonate:

$Ca(OH)_2 + CO_2 \longrightarrow CaCO_3 + H_2O$

Lime mortar is a white or gray bonding-material, usually found between building-stones 'or as a levelling coat on the outside of the mass where it is used as a base for decorative or protective plaster' (Littmann 1957, 136).

The ruins of Teotihuacán, once one of the largest cities in the western hemisphere, lie about 25 miles to the northeast of Mexico City. After being destroyed by fire in approximately 750 A.D., Teotihuacán was abandoned more than seven centuries before the arrival of Europeans in Mexico. Many walls, ceilings, and floors of the dead city's buildings are covered with well-polished mortar.

At Teotihuacán, limestone being scarce in that generally-volcanic area, mortar was commonly made of clay mixed with a minimum of calcined lime, and tempered with

---

[1] There seems to be no publication that traces the chronology of burning limestone and the making of lime-mortar. Lime is one of the oldest known artificial chemicals. In the Old World, lime manufacturing has gone on since the Neolithic; it even antedates the making of pottery (Gourdin and Kingery 1975, 133-134). 'Wood charcoal was used in the lime calcination process...' (Goffer 1980, 311). In the New World, whether or not lime production antedates the making of pottery, or charcoal was used in burning limestone is uncertain.

Somewhat contradictorily, a book on the processing of limestone makes the following statements on the subject of lime mortar: 'Its earliest documented use was about 4000 B.C., when it was used in Egypt for plastering in pyramids' (Boynton 1980, 441). That author believes that both independent invention and cultural diffusion account for the knowledge of lime mortar-making from place to place:

> Strangely, many of the ancient civilizations appear independently to have discovered lime and perfected their own uses. Some civilizations undoubtedly learned from each other, like the Greeks from the Egyptians and the Romans from the Greeks. But who educated (at that time) such isolated civilizations as the Incas, Mayas, Chinese, and the Mogul Indians ...Somehow they all 'stumbled' onto it by primitive ingenuity or happenstance. (Boynton 1980, 4)

In the 16th century, the methods of processing limestone and making mortar were substantially the same in the Old World and in Mesoamerica. For instance, in China,

> Lime was prepared by heating limestone or the shell of oysters, clams, etc., and mortar was also early prepared by mixing lime and sand. (Li 1948, 54)

volcanic gravel, *tezonte*. This no doubt, is the 'well-known earth called *tezontlali*...it is used to mix with lime and makes it very strong; it is much sold here in Mexico for buildings' (Sahagún 1956, 348). Limestone at Teotihuacán '...was probably obtained from Tula or other equally distant sources' (Littmann 1973, 176). At a much later date, the Aztecs at Tenochtitlán levied an annual tribute of lime from other parts of their empire.

While the building techniques at Teotihuacán were quite similar to those in the Maya area, the materials used in the two areas were somewhat different. Limestone being abundant in much of the Maya area, the mortar used was commonly made with calcined lime, with an aggregate of *sascab* or *zahcab*. *Sascab* is a friable form of calcium carbonate that occurs in pockets within the harder limestone.

In Peru lime mortar, so important to Mesoamerican architects, was unknown. Building stones were more precisely cut than in Mesoamerica, so that they would fit snugly together. Although no lime mortar was used, occasionally cracks between stones inside the walls were filled with a mortar of clay:

> In order to fit the stones together, it was necessary to put them in place and remove them many times to check them...*in some buildings not even a pin will fit into the joint*...We said that the Indians *did not use mortar in these buildings, that all of them were made of dry stone*; the first reason for this is that *they did not use lime and sand for construction*...and the second reason is because they set the stones together with nothing between them on the exterior face of the structure. But this does not mean that the stones were not joined together on the inside with some type of mortar; in fact it was used to fill up space and make the stones fit. *What they put in the empty space was a certain type of sticky, red clay*... (Cieza de León 1998, 328-29; also Cobo 1990, 229)

In Mexico, two early Spanish sources, Bernardino de Sahagún and Bernabé Cobo, disagree on whether limestone was fired in some sort of kiln or furnace. Perhaps more credence should be given to the words of Sahagún since he was a long-time resident of Mexico, whereas Cobo had only visited there. Probably they were describing procedures in different areas of the country.

Sahagún commented on the making of lime sold in Aztec markets:

> Those who deal in lime, crush the rock from which they make lime and heat it, and later slake it...the rocks good for making lime are gathered and *thrown into a furnace* where they are heated with firewood...The product is sometimes sold *as quicklime*, sometimes *as slaked lime*... (Sahagún 1956, 3: 143)

Cobo, on the other hand, claimed that the inhabitants of Mexico did not fire their limestone in kilns:

> Though the Indians of Mexico understand the use of lime [in making mortar], they do not mix it with sand, but with a ground rock called *tezonte*; *nor do they heat it in furnaces*, rather they make a heap of the rock, and apply fire. (Cobo 1956, 1: 122)

It should be noted here that while limestone frequently outcrops in highland Guatemala, it is the predominant surface-forming material on most of the Yucatan Peninsula, thus preventing the building there of a kiln-like surrounding structure. This fact may explain Cobo's statement.

Because of the high temperatures required (over 650°C, for a number of hours) to calcine a pile of limestone, it is best to stack the stones in a dome-like heap, and make in the center of it a thin, vertical cylindrical vent or flue by which an updraft is created. (This chimney-like air passage is, perhaps, the antecedent to the updraft kiln.)

Two comparatively recent descriptions of lime-making are of interest with respect to Sahagún's and Cobo's disagreement: One concerns descendants of the Maya on the Yucatan Peninsula, where a heap of lime is made using a chimney-like vent, but with no surrounding kiln-structure. The other, concerning the Mixtecs in the State of Chiapas, is an updraft kiln with baked-clay walls for firing limestone.[2]

In any case, lime-burning kilns were used elsewhere in the Mesoamerican area. In Honduras, a walled Maya kiln for burning limestone has been excavated at Copan; this kiln dates from the end of the Late Classic period, between 750-900 A.D. The calcined limestone was probably used for architectural purposes—that is, not for

---

[2] The first account describes a process apparently very like that discussed by Cobo:

> The Yucatecan [Maya] method of producing lime at the present time *is an ancestral heritage* that has come down through the centuries...*No sort of kiln or oven is used, perhaps because the only stone available is limestone, kiln walls of which would be destroyed by fire*. Nor is there earth suitable for the making of mud walls, *nor clay banks deep enough to permit the excavation of ovens in them*.
>
> The lime burner selects a spot where there is mature forest in close proximity...At the centre [of this space] a straight post about nine feet high and five to eight inches in diameter is set upright...The neighboring timber is felled and cut into proper lengths...[Limestone] is broken up into small pieces grading in size from one's fist downward. These are carefully piled on top of the wood until they constitute a layer about one foot eight inches thick at the margins and two feet six inches at the centre...The upright post is withdrawn from the centre, leaving in its stead a cylindrical [tube-like] hole. Into this, ignited material is dropped. *In the old days it consisted of a mass of burning charcoal*...The tube serves as a chimney [for the updraft] and the air readily sucks in through the crevices of the wood mass so that a strong draft is soon set up...dry chips and wood are thrown down [the cylindrical hole]...until it is evident that the green wood of the heap has ignited...*Preferably the lime is not touched until it has been thoroughly slaked by dew and rain*. Gradually what seemed but an insignificant little mound *swells to a dome of fluffy white powder of five or six times its original bulk*. The powder can then be used, but the masons who take most pride in their work do not like to prepare mortar from lime less than a year old. *This is especially true if plastering is to be done, for plaster made from fresh lime will check and crack, due probably to the fact that the last stages of slaking take place after it has been put on the wall*. (Morris 1980, 235-238)

In the other description, a simple updraft kiln or oven made by the Mixtec people accords with Sahagún's account:

> The ovens were *like chimneys* dug vertically into a clay bank, with lateral openings at the bottom for firewood. The choice of a *clay site was said to be important*, since otherwise an oven would cave in. (Pike 1980, 1)

making *nixtamal* (Abrams and Freter 1996, 424; see a reconstruction drawing of the kiln on p. 426).

Though Maya masons made much use of rough stone, their finest work was done with cut-stone—that is, dressed-stone laid in lime mortar ('*cal y canto,*' as the Spanish called it). For instance, the city of

> Chiurutecal is situated in a fertile plain, and within it there are *twenty-thousand buildings of dressed stone laid in mortar*, and there are many others in its suburbs. (Mártir 1964, 2:456)

Since limestone is almost omnipresent on the Yucatan Peninsula, a substitute there for common sand was usually needed for tempering mortar: 'there is an extraordinary abundance of stone, lime and a certain white earth [*sascab*] which is excellent for buildings' (Landa 1941, 18).

> [Sascab] served the ancient builders, as it does those of the present day, as a building material *to mix with lime in the place of silicious sand which is practically unknown in Yucatan.* (Thompson 1897, 77)

In northern Yucatan, the Maya center of Chichén Itzá, which florished between 692 and 1187 A.D., was largely abandoned several hundred years before the Spanish conquest. Chichén Itzá's architects spread lime mortar between their buildings, thus paving short stretches of roadway. The following is a sixteenth-century note describing a building there:

> This building had around it, and still has today, many other well built and large buildings and *the ground between it and them covered with cement* [i.e., lime mortar], *so that there are even traces of the cemented places, so hard is the mortar of which they make them there.* (Landa 1941, 179)

As indicated above, lime mortar gradually hardens for a long time—even for centuries—before it sets completely. More recently, others too have spoken of how strong Maya mortar and stucco are (Holmes 1931, 189). The traditional technology of living Maya builders suggests that vegetal additives were a fundamental ingredient of the lime mortars used in earlier times, and that these additives were an essential component in attaining the hardness and durability of mortar.[3]

Plaster and Stucco

Plaster and stucco have much the same composition as mortar, though usually less coarse. Plaster may be defined as a flat, external coat commonly used 'as a protective medium or as a surface for mural painting.' Stucco is a fine, high-quality plaster, usually used 'for decorative or symbolic purposes' (Littmann 1957, 136). Stucco surfaces, too, were often painted in Mesoamerica.

Although the plasters and stuccoes used in Mesoamerica were occasionally made of clay, they were more commonly made of lime. In contrast, the plaster applied to the buildings in Inca Peru was made of clays. Like the mortars in Mesoamerica, however, Peruvian plaster was sometimes mixed with organic adhesives, such as the mucilaginous sap of certain cacti, to make it cling better to the walls. For example, before the arrival of the Spanish, when parts of the city of Cuzco were being rebuilt and the new buildings were otherwise finished, the Inca ordered his workers to bring

> a large amount of cactus...so that the mortar that was used for plaster both inside and outside the [finished] houses would stick and not crack. They were *to spread the juice* [viscous sap] *of these cacti over these walls*. The mortar would be very well mixed with a large amount of wool [camelid hair] and put on the walls over the surface made wet by those cacti...If they did not want to use wool in the mortar, they should use straw which was very finely ground. This would give a glossy finish to the walls and structures. (Betanzos 1996, 70)

Washes

Washes are scarcely distinguishable from plasters except by their thinness. For instance, a wash (wash coat) may be defined as

> a coat of plaster which, because of its thinness, was probably applied by means *other than troweling.* Wash coats may have been applied by brushing on vertical or horizontal surfaces or by pouring a thin slurry of plaster over horizontal surfaces. Such coats are usually less than 1 mm. in thickness. (Littmann 1957, 136)

New World washes and colorants applied to building walls and other broad surfaces are usually water-based, whereas those colorants applied to such varied surfaces as pottery and the human body are often mixed with fixatives which are insoluble in water, such as fats or resins.

---

[3] In the old days when there was less hurry, the *maestros* took great pains in its preparation. It [the lime mortar] was *thoroughly stirred and remoistened once a day for two weeks, or longer if needed for floor or roof construction. These latter features, which astound one with their hardness considering that the cementing material is only lime*, owe their hardness to two things. They were tamped for hours on end with wooden mauls, until they were poreless and compact as stone. Moreover a special liquid was used for remoistening the surface paste as it was being tamped and finally trowelled. The bark of the *chocom* tree [*chucum* tree?] was *stripped off and put to soak in vats.* After standing for a number of days the water had drawn off enough of the soluble chemicals from the bark to fulfill the intended function. *Lime, moistened with it, takes a magnificent polish under the trowel and is practically impervious to water. It turns a bright red, and does not check upon exposure to the sun.* (Morris 1980, 240)
Littmann suggests that it was tannin in the tree bark that works 'to improve the workability or strength of the lime' plaster (Littmann 1960, 593-594). In an experiment meant to replicate and analyze this procedure, it was found that the chucum extract 'yielded a plaster with a reasonably dark color which changed from a purplish tan when added to the lime to a warm, red buff after a few minutes' (Littmann 1960, 596). The *chucum* tree is *Pithecolobium albicans*, a member of the legume family (Standley 1926, 397).

The Aztec, Maya, and others in Mesoamerica applied whitewash to their buildings, as did the inhabitants of Spain itself. For instance, the Spanish conquerors when approaching Texcoco (not far from Tenochtitlán) described their view as follows: 'Texcoco with twenty thousand houses, whiter than a swan, because all of the buildings were *whitewashed with lime and gypsum*' (Mártir 1964, 2:674). Whether or not the Aztecs actually used gypsum in making their whitewash is uncertain.

Calcined gypsum, nowadays known as plaster of Paris ($CaSO_4$+½H_2O), is gypsum ($CaSO_4 \cdot 2H_2O$) from which more than three-quarters of the water of hydration has been driven off by raising its temperature to between 130°C and 190°C; it is then ground to a powder and slaked. However, if the temperature is raised to over 204°C, the water of hydration is completely removed, and the product is useless as a plaster or a mortar (Eckel 1928, 36, 63, and 71). Gypsum mortars are quick-setting as compared to lime mortars—and, unlike lime mortars, they are rather soluble in water and are unsuitable for outside use in all but the driest of climates. One reason for using gypsum instead of lime for making plaster, at least in the Old World (Egypt), 'was the scarcity of fuel; lime requires a much higher temperature, and consequently more fuel, for calcination' (Goffer 1980, 104). In Mexico, calcined gypsum (*yeso cocido*) was sold in Aztec markets (Sahagún 1956, 3:157). The purposes for which it was used are unspecified, but the fact that calcined gypsum was a commercial item suggests that it was fairly common and important. Furthermore, there is some archaeological evidence that calcined gypsum was used, occasionally at least, in making stucco for application to pottery:

> One sample of gypsum stucco, which *corresponds to plaster of Paris*, has come to our attention. This was from a shell, stone, and bone plaque, purchased at Paso Real, near Toliman, Jalisco... (Shepard 1977, 276)

Because whitewash is not especially weather resistant, it must be reapplied from time to time—a fact illustrated by the events scheduled inside Tenochtitlán:

> The streets and highways of this great city were so well swept that there was nothing to stumble over...What can I say of the cleanliness of the temples of the devil, and of their steps and courtyards, and of the *houses* of Moteuczoma and of other lords, which were *not only white-washed, but polished, and repainted and repolished for every festival.* (Motolinía 1950, 211)

Among the Aztecs, a mason was referred to as 'one who makes mortar...who *smooths, polishes, burnishes the surface;* who *whitewashes*' (Sahagún 1950-82, 4:28).

The Maya of Yucatan, 'were accustomed to have in each town a large house, whitened with lime, open on all sides, where the young men came together for their amusement' (Landa 1941, 124). Whitewashing seems not to have been so common in Yucatan as in parts of Mexico, perhaps because many structures in Yucatan were built with blocks of limestone which are whitish already.

Buildings were not only whitewashed: other colored pigments were added to the wash coating. Arriving at one town in the neighborhood of Mexico City, Cortés spoke of 'the reddish-purple houses, of exquisite taste' (Mártir 1964, 2:461). The Aztecs also used a kind of yellow earth (most likely yellow ocher) as a wash for their walls (Sahagún 1956, 349).

In excavating Teotihuacán, archaeologists have been able to find and identify a number of mineral pigments used in painting the city's walls and murals. Washes were deliberately mixed so that the finished surfaces would sparkle; for example, finely ground mica (probably muscovite—a light-colored, glittering mica) was added to ordinary calcareous whitewash (Linné 1934, 160). Red wash and paint for walls and pottery was also mixed so that it would reflect light:

> Generally there are noticeable in the red paint, e.g. *in the case of red ware*, minute particles which are black and shiny. Magnification shows them to consist of hematite crystals *with reflecting surfaces* and irregular edges [apparently specular hematite, which is black to steely gray with a metallic luster]. This mixture gives good effect, and was very greatly in vogue, even the red decoration of the house walls showing the presence of *glossy grains of hematite.* (Linné 1934, 160)

In making washes for Maya buildings and murals (as well as paints for pottery), the harder minerals used as pigments were first calcined, then crushed and ground to powder. In addition to lime (or possibly gypsum) in *whitewash*, the following pigments were used both for wall wash and for painting pottery: hematite and cinnabar for red colors; celadonite (a hydrosilicate of iron) for green; and yellow ocher for yellow (Linné 1934, 160-161). (Since celadonite and cinnabar decompose at high temperatures, they can be applied on pottery only after firing.) Painters in both Mexico and Peru were still using red and yellow ocher as pigments when the Spanish arrived. In Mexico, 'The [Aztec] Indians mine a certain kind of ocher, called *tecoçahuitl*, or yellow earth [probably either yellow ocher or limonite] with which the painters applied a yellow color...' (Hernández 1888, 309).

The Incas of Peru 'were not accustomed to *whitewash* [their houses] as we [the Spanish] do, although the caciques usually had walls *painted in various colors and figures...*' (Cobo 1956, 2:242). Long before the rise of the Inca empire, the inhabitants of Moche also painted exteriors of their buildings with colored washes, as evidenced by illustrations on their pottery.

Some of the pigments used by painters are unknown. For example, in seventeenth-century Peru a copper mineral (called *coravari*) is simply described as giving painters 'a pleasing green color,' and used for 'many things': 'servir la coravari a los pintores por su gracioso verde, *aprovecha para muchas cosas...*' (Cobo 1956, 1:127-128).

Maya Blue

A truly unusual substance, a clay plaster or stucco combined with indigo, was used by the Maya to color their buildings, murals, pottery, and perhaps their books; this substance is now known as 'Maya Blue.' The colorant differs from

> any blue ever identified on ancient or medieval paintings from Europe or Asia. It is not based on copper or on ground lapis lazuli or lazurite, which are common in European and Asian paintings. (José-Yacamán et al. 1996, 223)

As noted earlier, blue was sacred for the Maya, and the color was frequently applied on murals and pottery. On ruins and potsherds, Maya Blue has retained its color for centuries, despite having been weathered in a humid tropical climate. At one time, Maya Blue was thought to be exclusively inorganic, since it cannot be destroyed by 'boiling nitric acid' (Gettens 1962, 557). More recently, however, it has determined that the colorant is largely an organic material made from the indigo plant; this colorant had been combined with a nearly colorless clay, identified as palygorskite or attapulgite. Moreover, has been found that Maya Blue is resistant not only to diluted acids, but to 'alkalis, solvents, oxidants, reducing agents, and even biocorrosion' (Jose-Yacamán et al. 1996, 223). Maya Blue 'is remarkably stable: the color is not destroyed by hot mineral acids or by heating to about 250°C' (Van Olphen 1966, 645). Clearly, this is an example of advanced decorative technology.

Whereas indigo was anciently known throughout much of Mesoamerica, the clay palygorskite or attapulgite is a fairly rare mineral. Although it has been found in several places in Central America, a principal source for the ancient Maya may well have been Sacalum in Yucatan, located not far from the archaeological site of Uxmal (Arnold and Bohor 1975, 25).[4]

In dyeing fabrics, the indigo plant is often used in its water-soluble condition—that is, in its unoxidized form, indoxyl. Whether the insoluble indigo (the blue pigment itself) was extracted before its use in making Maya Blue is not clear, though it seems 'unlikely that the Maya would first isolate solid indigo and then solubilize it [that is, reduce it] for application to a clay base' (Littmann 1982, 405).

> The procedure possibly used consists of impregnating attapulgite with an extract of indigo plant leaves and stems and subsequently heating the impregnated clay for an extended period of time at the temperature of boiling water. (Littmann 1982, 404)

'Palygorskite crystals in the paint form a superlattice that probably occurs as a result of mixing with indigo molecules'; however, impurities in the clay, such as calcium and magnesium oxides and carbonates, may partly account for the color (José-Yacamán et al. 1996, 223 and 225).

Though it had been made for more than a thousand years in prehispanic Mesoamerica, use of Maya Blue disappeared during the Colonial Period. Its manufacture is now a lost art. And although the colorant has been the subject of research for over half a century, its precise composition, and the procedure used in making it, remains uncertain. It is now possible, however, to prepare indigo-clay complexes 'that cannot be differentiated from authentic Maya Blue' (Littmann 1982, 407). In whatever way it may have been originally discovered, whether by serendipity or by reasoned experimentation, the art of making Maya Blue has been exceedingly difficult to rediscover—even using the techniques of modern science.

The marine blue employed at Teotihuacán for making murals was apparently similar to Maya Blue, but used other materials: an *unidentified organic colorant* over an inorganic layer consisting of 'pyrolusite [MnO_2] and gypsum' (Magaloni 1996, 21).

Mural-Painting Techniques

That colored dwellings are much older than colored man-made buildings can be seen in the European caves of Lascaux where the walls were elegantly painted some 30,000 years ago. Much simpler (but purportedly equally old) are the paintings found on the walls of an Amazonian cave, the Caverna da Pedra Pintada at Monte Alegre, between Manaus and Belem. The paintings were made with 'a fine red powder composed of quartz...and ferruginous laterite rich in hematite' (Roosevelt et al. 1996, 378). The cave was ostensibly inhabited in the late Pleistocene, well before Asian emigrants crossed the Bering Straits into North America.

The murals of the walls of man-made buildings in Mesoamerica are large multicolored illustrations, consisting of designs and other figures, painted on walls over stucco plaster. Some of the finest murals were made at Teotihuacán; many of these survive despite the city's age

> The murals of Teotihuacán are characterized by their *brilliant colors*, their permanence, the very large areas over which they are used, and, above all, the *exceedingly smooth surfaces* that were decorated. (Littmann 1973, 175)

The murals on old Maya structures, too, were described in the 16th century as 'painted with great elegance'

---

[4] In the archaeology of Ecuador, between 200 B.C. and 600 A.D., 'colors [of the ceramic figurines] include a local version of 'Maya Blue,' a turquoise color made by mixing indigo with white clay' (Bruhns 1994, 201).

(Landa 1941, 198). At Chichén Itzá (which, like Teotihuacán, had long been in ruins when the Spanish arrived) various wall surfaces 'have paintings upon them as clear-toned as if the pigment had been applied but yesterday' (Thompson 1932, 184).

In the pre-Columbian New World, but only in Mesoamerica, two mural painting techniques were possibly employed: *a fresco* (true fresco) and *a secco* (dry fresco or *fresco secco*). In *a fresco*, the pigments are mixed with either water or limewater and applied to a stucco surface while that surface is *moist*: if 'the surfaces were painted *while still damp we then have a true fresco painting technique* regardless of whether or not the painter [or anyone else] recognized the process' (Littmann 1973, 177). Calcite crystals form as the plaster dries, protecting the intermixed pigments from weathering, thus making their colors more-or-less permanent. (True fresco was much used by artists of classical Greece and Rome, and Renaissance Italy, as well as those of modern times.)

In contrast, in the *a secco* technique, the dry plaster is washed with a thin mixture of water, lime, and some organic materials as a binder; the colors are painted on this surface. This technique yields colors that are usually less durable.

At Teotihuacán, murals were made using the true fresco technique (Littmann 1973, 175 and 177); most specialists agree on that (Magaloni 1996, 17). It had been thought that most Maya murals were also painted 'in true fresco technique' (Covarrubias 1954, 116). However, there is some disagreement now among the authorities about the presence of gums in the mixture applied to the walls, or whether the latter were damp or dry. In analysis of 25 sites in the Maya area, it was shown 'for the *first* time, that Maya murals were made with a mixture of two or more unidentified vegetal gums...' (Magaloni 1996, 22); and so, Oaxacan 'murals are not frescoes *since they use gum as a vehicle*' (Falcón 1999, 38). Consequently, 'all Maya wall-painting was done *a secco* rather than *a fresco*: that is, applied in an aqueous or tempera-like medium on a *dry* surface, and not on a *damp* one...' (Coe and Kerr 1997, 134).

However, one of the most recent accounts states that most Mesoamerican murals were made in a slightly-modified true fresco method. The

> essential mural technique in *Mesoamerica* was affixing colored media to the surface of a building. *Stucco plaster* invariably served to support the application of pigments that were oxidized in diverse proportions. These pigments were of mineral origin, with the exception of *blues and greens*, which have organic components. In this the artists created a type of tempera, bound by *gummy plant fluids* such as the sap of the nopal cactus, and *a fresco* that was applied *over still-damp plaster*. (De la Fuente and Staines Cicero 2001, 493)

(Tempera is a painting medium in which pigments are mixed with water-soluble glutinous materials.)

The mineral pigments used for coloring the murals of Teotihuacán include malachite, for various greens; malachite and gypsum for 'clear green' (*verde clarito*); for dark and sky blue, a still unidentified *organic* colorant over an inorganic substrate of 'pyrolusite [MnO_2] and gypsum'; copper sulfate (bluestone) for various shades of blue; pyrolusite for bluish black; soot for black; and hematite and ochers for yellow, orange and red tones (Magaloni 1996, 21; see also Littmann 1973, 178).

Since most of the organic materials applied to Mesoamerican murals are still unknown, some hint of their identity may be based on the last-century, but traditional, practices of Maya workmen in the area. The following passage describes how various plant materials (probably interacting with lime or with other minerals in the mortar) may have been used to color buildings at the old Maya city of Chichén Itzá:

> The white was generally made up from a finely divided lime mortar mixed with the diluted juice of a plant called *chichebe*...The brown pigment used for the wide spaces on the carved stone surfaces and the common earthen vessels is made from burned *kancab* [i.e., the lateritic soil, rich in weathered iron oxides, so common in the tropics], ground into fine powder with a stone mill [metate] and applied first with a sop made of fibre, then rubbed with a fine fibre mat.
>
> The fine red and brown pigments, those used *in the mural paintings* and the figures outlined *on the finer earthen vessels*, are made from the red splinters and chips of the heartwood and the tinted sap of the tree called *chacte* by the Mayas [*Caesalpinia platyloba*, according to Standley 1926, 424]. Different shades and effects are produced when this *chacte* pigment is mixed with the sap and the latex of the *habeen* [i.e., *jabín, Piscidia communis*] and *chucum* tree [*Pithecolobium albicans*]. The sap and latex of the *chucum* were also used to change the tone and harden the white lime surfaces, the hard finish on wall surface and platform floors.
>
> The blue pigments [e.g., Maya blue] are obtained from several closely allied plants called *anil* [*Indigofera* spp.]...
>
> Yellow was produced from the fruit of the *achiote*—annoto—and from the boiled and strained chips of the fustics [for example, *Chlorophora tinctoria*, a mulberry-family tree]...
>
> The finest [black pigment] was made from the carbonized resins, the copal incense, and the resin of the *chacah* tree [probably *Bursera simaruba*], ground into fine powder. (Thompson 1932, 185-186; italics in the original text)

Although a number of minerals was used in Andean South America, paints there were colored mainly with red and yellow ocher (i.e., hematite and limonite). This is much as Father Cobo claimed when he reported that, in Peru, there is found

> a yellow earth much used by painters, called ocher [probably goethite] which in the Aymara language is named *quellu*; and another kind of orange earth [probably hematite ocher], called *pitu* in the same Aymara language, which is also *used by painters.* (Cobo 1956, 1:113)

Murals are much more numerous in Mesoamerica than in South America where, except for the tombs of Tierradentro in Colombia and a few other locations, they are mostly confined to the central Andean area.

Technically much simpler than those of Mesoamerica, Peruvian murals were *a secco*; that is, made on dry surfaces. The

> ...colors were mixed with some *glutinous substance to fix them.* The colors of these paintings were, in general, matte [i.e., relatively dull?] *and impermanent* (especially when exposed to weathering); they covered small areas that were almost always protected; and they were applied to *clay surfaces exclusively*, on walls that were not very smooth. Mineral pigments found in the natural state in the country were used. (Bonavia 1985, 185)

Several analyses of Peruvian colorants have been made, and these 'allow us to state firmly that *only mineral pigments* were employed in the preparation of mural paintings in ancient Peru' (Bonavia 1985, 181).

Since only mineral pigments 'in the natural state' were used in Peru, mural paintings there contrast strongly with those of Mesoamerica. At Teotihuacán, for example, lime was combined with a red pigment (hematite, Fe_2O_3) before the preparation was applied as a paint on murals; thus, it was 'a man-made and not a *naturally* found substance' (Littmann 1973, 178). The soot used for black at Teotihuacán is also a man-made pigment. Perhaps another man-made black pigment was copper oxide (CuO), having 'been made by heating one of the copper carbonates' (Littmann 1973, 177). Moreover, Maya Blue and other organic pigments were used in murals in Mesoamerica.

CHAPTER 14

POTTERY MAKING

Well-baked pottery—and even its polished surface—is remarkably resistant to weathering. Thus, pots and their sherds are among the primary evidences used by archaeologists in reconstructing the chronology of prehistoric events. And so pottery technology in the Americas is probably best known in the American Southwest, where intensive archaeological surveys have been made and where various native pottery industries survive. Consequently, in considering pre-Columbian pottery-technology, the Southwest (though by no means the oldest ceramic area in the hemisphere) serves as something of a reference when discussing materials and processes used by Native American potters.[1]

Studies undertaken within the last century make it possible to distinguish, in most cases, the practices which are indigenous from those adopted after the arrival of Europeans. Nevertheless, it is plain that present practices of potters in some areas are a complex mixture of indigenous and Old World elements which are not always easy to separate.

In the Americas, pottery 'began in Amazonia about 7500 years ago, more than 1500 years earlier than elsewhere in the hemisphere' (Roosevelt 1995, 115). The knowledge of making pottery may have diffused northward from Amazonia through cultural contacts (Clark and Gosser 1995, 209).[2] Later, there seems to have been a second introduction of pottery from Asia: at least some northern American pottery is almost indistinguishable from that of Kamchatka Peninsula (Linton 1944, 371-372).

'American potterymaking originally lay in the hands of the women, *the inventors of the ceramic art...*' (Linné 1934, 169). This speculation is, of course, debatable, but even today the manufacture of pottery 'at the nonspecialized household level is almost always a women's craft' (Skibo and Schiffer 1995, 86). Women, if not the inventors of the 'ceramic art,' were the probable inventors of the *potters art*.[3]

With regard to the role of women and female children in the development of pottery, among the Cayapa of Ecuador 'at an early age the girls are instructed in household duties of all kinds, and often begin to spin and weave, as well as to make pottery, basketry, and other articles, at the age of five or six years' (Barrett 1925, 41). In this, the Cayapa are probably pursuing a long-standing custom, one shared with other peoples.

Finding Pottery Clay

Most clays are the end-products of weathering processes. Largely colloidal, they are the finest of all geologic sediments.

Suitable pottery clays are available in many areas. They are sometimes obtained from ancient sedimentary beds, especially in arid regions where the lack of plant cover makes their deposits visible. In such outcropping beds, claystone—a compressed and hardened clay—is often interbedded with shale, a composite of clay minerals and larger particles.

A more common source of potter's clay is recently-formed sedimentary horizons, often cut and exposed by rivers. For example, in the Ecuadorian jungle, Quichua women 'find adequate clay deposits in river banks and brooks, or by penetrating an underground deposit in the forest' (Whitten 1976, 89). Tribes living along rivers in tropical forests obtain alluvial clay, and are able to transport it easily with sturdy dugout canoes. Among the Cayapa Indians, 'usually a considerable quantity [of clay] is gathered at a time and brought by canoe to the house...' (Barrett 1925, 174). The Shipibo of eastern Peru do the same (DeBoer and Lathrap 1979, 110). A Quechua-speaking potter, also in eastern Peru, provides another example: 'She digs her own clay...at a spot along the riverbank that is 2 hours away by canoe. The clay is used moist, as she finds it' (Litto 1976, 210). Thus weighty items such as clay and pottery are much less of an impediment for those living in humid lowlands like Amazonia than for those living in dry or more mountainous areas. Potters in the high Andes often rely on llamas to transport their clay (Sillar 2000, 92). Except for the llama and alpaca, no burden-carrying animals

---

[1] A thorough disussion of the subject is given in Anna O. Shepard's classic, *Ceramics for the Archaeologist*, upon which (as well as her other publications) the writer has depended heavily.

[2] Whereas in Amazonia pottery-making may have begun earlier than agriculture, in Mesoamerica agriculture began several thousand years earlier than pottery. Likewise, in 'the Southwestern United States, pottery began to be used shortly after the initial use of cultivated plants...' (Hoopes and Barnett 1995, 5).

[3] In the Old World, at least, crude baked-clay figurines long antedate pottery itself. Ceramic figures have been found in European sites which were occupied during the late Pleistocene, about 30,000 years ago: these artifacts are '*the first thoroughly artificial objects*' (Hoopes and Barnett 1995, 2). The gender of the figurine makers is, however, unknown. The use of pottery, itself, began much later in both the Old World and the New. Although the 'archaeological record makes it clear that pottery was most commonly produced by sedentary, agricultural societies' (Hoopes and Barnett 1995, 2), the first pottery was produced *not* by agriculturists but by fishermen, in the early Holocene. According to present knowledge, the earliest pottery was made at Jomon, on the Japanese island of Kyushu, around 12,700 years ago (Aikens 1995, 11)—probably before the first immigrants (who knew little or nothing about baked-clay objects) reached America.

were domesticated in the New World. (The dog and the travois were used to a limited extent for drayage in North America, but probably rarely for conveying clay and pottery.)

Some of the best clays, however, are not found in sedimentary beds but in rarer sites where their parent rocks have been altered in place, either by weathering or often hydrothermally.

Clays and claystones appear in various colors. Kaolin, whose chief constituent is the mineral kaolinite, $Al_2Si_2O_5(OH)_4$, is a white clay, relatively free of impurities. Among the Conebo (Conibo), who live along the upper Ucayali River in eastern Peru, '*white clay* [probably mostly kaolin] is collected from the river banks at *low water* and the pottery, on this account, is made in the dry season' (Farabee 1915, 94). But though essentially hydrated aluminum silicates (which tend to be whitish), clays and claystones often include iron oxides which give them a yellow, red, or brownish color; if organic materials are included, they may be gray, bluish, or almost black.

Sorting, Pulverizing, Tempering, and Kneading

Pottery clay usually contains undesirable coarse particles, most commonly extracted by handpicking. In the arid Southwest, Hopi women excavate lumps of hardened claystone from old sedimentary beds and 'carry it in shawls or cloths on their backs to their homes...' (Colton 1938, 5). It is then

> covered with water and left to soak. When sufficiently disintegrated, it [becomes plastic, and] is thoroughly kneaded until it becomes of a dough-like consistency, and *hard particles are removed with the fingers*. The more thorough the kneading the less likely are the vessels to crack and flaw [when fired]. (Colton 1938, 5)

(When mixed with enough *fresh water*, clay distributes itself fairly evenly in suspension throughout the fluid. In contrast, salty or strongly brackish water may cause clays to flocculate and settle.)

Although handpicking is a common and old way of sorting, other methods may also be pre-Columbian, especially in dry climates. In the American Southwest,

> San Ildefonso potters [along the Rio Grande] winnow the ground clay in a light breeze by letting it fall through the fingers...Yuman potters [along the Colorado River] used winnowing trays with a tossing, a to-and-fro, and a rotary motion to separate the coarse from the fine particles. (Shepard 1965a, 51-52)

Nowadays, native potters frequently sift clays with iron sieves to rid them of coarse particles, but in the village of San Bartolo Coyotepec in Oaxaca, Mexico, Zapotec potters ('the most truly indigenous of all Oaxacan ceramics') still extract foreign substances by hand: '...a sieve, common enough in other Mexican pottery centers,' is not used (Van de Velde and Van de Velde 1939, 10 and 28). Although sieves and strainers made of basketry, barkcloth, textiles, and perforated tree gourds (Sahagún 1956, 143) were common for other purposes, metal sieves were apparently not manufactured in pre-Columbian America, even where metalwork was highly developed. Quechua-speaking potters at Otavalo in Andean Ecuador sift their clay 'through a sieve of punctured sheepskin, *homemade if not aboriginal*' (Parsons 1945, 24).

As far as known, the process of decanting was never used to separate coarse materials from clay in pre-Columbian America (Velde and Druc 1999, 238), though the process had long been used to separate food products. Clay particles are minute and flake-like, because their surface is large compared to their weight. And so, when clay is mixed with water, the clay particles remain longer in suspension than the heavier and coarser materials which tend to sink; consequently, the two groups of materials become separated. The suspension containing the finer materials 'can be poured off and concentrated by longer settling with subsequent decantation of the water'; concerning her description, Shepard notes, 'this method is easily practiced, and it may yet be found' (Shepard 1965a, 52).

The mixture of clay and water from which the body of the pot is formed is called the 'paste.' Kneading makes this paste more plastic: the moisture is distributed evenly, most pockets of air are eliminated, and the texture becomes more homogeneous.

At Coyotepec the moist clay is first dried, then pounded 'until it is reduced to lumps of pebble size.' After foreign substances are removed, the clay is placed in a container, 'covered with water and left to stand until saturated.' It is then placed on a clean reed mat, and is subjected to a preliminary kneading: the mixture of clay and water is 'worked by stamping with the bare feet. This...is usually done by children of the family...When the clay has reached the proper consistency, it is divided into small portions and *kneaded by hand*' (Van de Velde and Van de Velde 1939, 28).

Tempering refers to the act of incorporating non-plastic materials in the paste by further kneading. Adding temper is done for several reasons. One is to make the clay of the vessel-to-be more rigid, thus preserving the shape that it is molded in: untempered clays usually crack when fired. Another reason for including temper is to make the vessel, after it is fired, either more permeable or less so, according to the kind of temper selected. (It should be noted that permeability in pottery may be an accidental fault or a purposely-made benefit.) Pottery with impermeable walls is useful for storing water, whereas pottery with porous walls is useful for keeping water cool. And so permeability is useful in hot climates.

Tempering with vegetable material, which is commonly destroyed during firing, may leave behind tiny pores in the vessel's walls through which small amounts of water can seep and evaporate into the surrounding air. During evaporation, heat is used up; this heat is withdrawn from the vessel's surface, cooling it and the water inside. The effect is especially strong if the vessel is set in a breeze or current of air. An example is the black pottery made in Coyotepec: 'All Coyotepec vessels, being non-glazed, will 'sweat' when containing liquids. This is desirable in a water-jar, the slow evaporation keeping the contents cool' (Van de Velde and Van de Velde 1939, 40). Although Coyotepec is in mesquite and cactus country, where the relative humidity is often low, even in areas of high average relative humidity the air is rarely saturated with water vapor. An example of this circumstance is found in the northwest Amazon, definitely a hot, humid climate.

> The clay is commonly to be found on the riverbanks, and with it the Indians mix *wood ashes*, either to stiffen it or...to render the finished article *more porous*, so that its contents are *kept cool* by evaporation. (Whiffen 1915, 96)

A great variety of materials is used in tempering: sand, the most commonly-used material; pulverized potsherds; volcanic ash; tree bark; chopped grass; and so on. Among the Aztecs, the clay used in making comales was tempered with cattail fluff (Sahagún 1956, 146) or with soft reed fibres (Sahagún 1950-82, 4:83).

Although the practical value of adding a tempering material has been recognized since the earliest days of pottery making, the particular material used is often a matter of tradition. Sometimes the choice is limited by the environment. For instance, on the Yucatan Peninsula, limestone is the common temper in Maya pottery because other suitable materials are scarce there. Using ground pure-limestone or mollusk shells as a temper can damage pottery. If a pot is fired at too a high temperature, its temper (calcium carbonate) may be decomposed to quicklime and carbon dioxide (as described above, in Chapter 2). Quicklime, exposed to water or moisture in the air, yields slaked lime, a powder, thus reducing the vessel, or a part of it, 'to a loose pile of grains...This destructive process can be prevented by keeping the firing temperature below that where calcium carbonate decomposes (i.e., 750-800°C)' (Rye 1981, 107; see also Velde and Druc 1999, 143-144, concerning the disadvantages of using limestone and calcium phosphate, crushed bones, as a temper).

In parts of the Amazon basin an unusual tempering material is used—namely, *fresh-water sponges*, found on the roots of riparian trees.

> In the realm of ceramics the application of sponge spicules in the clay, for cohesion and durability, is an invention made in Amazonas and, so far as I am aware, *not known from any other part of the world.* (Nordenskiöld 1931a, 496)

'All freshwater sponges have a skeleton of needle or sausage-shaped siliceous spicules...' (Linné 1965, 29-31). The spicules, which contain opaline silica, add greatly to the strength of paste material. 'Women potters close to Santarém on the lower course of the Amazon [as well as those farther upriver] used dried and crushed sponges.' Outside the drainage of the Amazon, tribes on the rivers Araguaya and Uruguay also used sponges in the production of potter's clay (Linné 1965, 29-31).

Tropical forest potters, especially in the Amazon drainage system, are notable for their use of *ashes made by burning siliceous tree bark*. The clay is mixed with 'burnt bark to prevent undue shrinkage and cracking of the ware. There is no sand available and to crush sherds in wooden pestles—rock does not exist—is difficult' (Linné 1965, 28).[4]

Ground potsherds are said to be ideal tempering materials because their chemical and physical properties are generally similar to those of the pottery being made. The Chiriguano women of Tarapayu, Bolivia, taking a

> practical view of antiquities, have found the rich archaeological ceramics of the district especially suitable as an ingredient, 'takupé', in their pottery-making...This ancient pottery, in its turn, also contains takupé. (Linné 1925, 36)

Much harder rocks than claystone are sometimes used as temper after being calcined to facilitate their crushing. For example, in a number of villages south of Cuzco, Peru, 'field blocks of basalt-andesite are collected, heated, and crushed to a powder,...and this is added as temper' (Ixer and Lunt 1991, 141 and 159). Likely the prehispanic Inca potters of that neighborhood made their temper similarly, since an analysis of archaeological and modern sherds shows that both were made in much the same way.

Composition of the processed clay varies according to the purpose of the finished vessel—for example, whether it is meant for cooking, storage, or simply for decoration. Paste and temper may even vary according to the part of the vessel being fashioned (base, body, or neck), as among Shipibo-Conibo potters (DeBoer and Lathrap 1979, 116).

---

[4] Trees of certain families—among them Chrysobalanaceae (closely related to the rose family)—have the ability to incorporate soluble silica (that is, amorphous silica) into their cell walls. This family includes the genera *Licania* and *Hirtella*. For instance, ceramics made by the Jamamadis of Amazonian Brazil are 'hardened with the bark ash of...*Licania octandra*' (Prance 1972, 221). The Shipibo-Conibo who live in eastern Peru where stone (as mentioned) is 'a rare commodity,' process the bark of *Licania* trees: 'The processing involves charring, pulverizing the charred bark beneath a *stone rocker pestle*, and sifting through a loose-weave cloth. The resultant temper is customarily stored in an old pot until needed' (DeBoer and Lathrap 1979, 111). Two tribes living in the Bolivian lowland also make use of siliceous bark as a tempering material: 'The bark used by the Cavina [one of the tribes] contained 48.63 per cent silica (SiO_2) which amounted to 95.97 per cent of the weight of the ash' (Linné 1965, 29). The fact that several genera of trees in the family Chrysobalanaceae contain silica makes their wood ideal for pile dwellings because it is resistant to termite attacks.

If hard claystone is the source of clay, the claystone is crushed and then ground to a powder. For example, the Indian potters at Otavalo in Ecuador use hardened claystone:

> The pebbly, sandy clay, placed on a mat in the yard, is beaten with a flat wooden flail, then ground on the mat with a *large unworked boulder*, naturally rounded or convex on one side so that it can be readily rolled from side to side. (Parsons 1945, 24)

(This 'unworked boulder' is but a simple variation of the rocker-pestle used, as described above, to grind dry maize kernels.)

Similarly in the Andes of northern Peru, the claystone used by native potters is a 'hard and well-compacted sediment...[and] must be crushed with a hammer' (Velde and Druc 1999, 239). Such dry and powdered clay must then be mixed with water. (It should be noted too that, while clays and claystones may appear completely dry before being mixed with water, they are *hydrous* minerals and have water in their molecular structure.) Although this is often done in shallow holes, simply scooped in the ground, the prehistoric Hohokam of the Southwest made special stirring and mixing-basins for the purpose. Better to hold the water, these mixing-basins had a 'caliche-clay lining, stabilized by burning' (Haury 1976, 196).

In the 16th century Aztec markets of Mexico City, potter's clay was sold, already kneaded (Sahagún 1956, 3:157). Widely-scattered reports suggest that for best results the clay, after being kneaded, should be kept moist and stored for some time in order to become 'ripened' or 'allowed to sour.' This may be in order to allow some of the clay's organic component to decompose. For instance, such reports come from the Navajo in Arizona (Hill, 1937); from the Zapotec in Mexico (Van de Velde and Van de Velde, 1939); as well as from the Andean Aymara (Tschopik 1963). 'But one cannot be certain in these cases that the custom was not learned from Europeans' (Shepard 1965a, 52).

After being tempered and kneaded, the clay is shaped into a vessel of the desired shape. Since the chemical changes undergone in shaping the vessel are minor, little is said here about the various methods used: direct molding, coiling, paddle and anvil, and so on. In this regard, it should be noted that the potter's wheel, a device with a rotating horizontal disk for molding clay into vessels, was unknown in pre-Columbian America.

Drying, Polishing, and Slipping

It is important that pottery, after being tempered, kneaded and shaped, but before being baked, be dried as gradually and thoroughly as possible—often a difficult task in the humid tropics. Usually the pottery is dried in the sun. Among the Zapotecs of Oaxaca this drying is accomplished by placing the vessels first

> in the shade of a tree in the yard to be warmed gradually by the filtering rays of the sun...They are felt and turned every so often until they are evenly warmed on all sides...When sufficiently warmed, the jars are placed in the bright sunshine, where they are allowed to stand until thoroughly heated...In the meanwhile the kiln has been fired and slightly heated. *When the temperature of the kiln becomes exactly the same as that of the jars that are being heated in the sun, the time for firing has arrived.* (Van de Velde and Van de Velde 1939, 32)

In some places, there is a preliminary heating ('firing') in which the still-unbaked pottery is dehydrated, with only moderate temperatures applied. For example, the Yuma potters 'place their jars around a brush fire for about 15 minutes, and many Pueblo potters follow a similar practice, taking care to turn the vessels in order to heat them evenly' (Shepard 1965a, 81). In Amazonia, Witoto potters accomplish much the same thing: they set their vessels out to dry in the sun and, afterwards, heap hot ashes over them (Whiffen 1915, 96).

After being dried, the vessels are given a smooth and shiny surface by polishing or burnishing with a hard, smooth object. The effect has been explained as follows:

> The clay particles tend to align themselves perpendicular to the pressure, and parallel to the surface of the pot. This clay orientation gives a *lustrous shiny effect...* (Velde and Druc 1999, 85-86)

During polishing, parts of the vessel are sometimes slightly moistened, thus temporarily interrupting the drying process. Referring again to Hopi practices,

> when the vessel has been successfully dried, it is rubbed with a piece of sandstone [or, elsewhere, often a piece of dried gourd] to remove all inequalities and to thin the wall of the vessel. *It is then ready to be polished or slipped.* For polishing, *a bowl of water is kept at hand* from which the potter wets the portion of the vessel upon which she is working and then polishes with *a smooth pebble.* (Colton 1938, 7)

Whereas most potters polish their vessels, many do not apply slips; for instance, the Cayapa Indians, who have a relatively simple ceramic technology, apply no 'slip or paint of any kind' (Barrett, 1925, 180).

A slip is a thin wash of fine clay with which a shaped pot is covered (often by wiping or painting, rarely by dipping) usually soon after it has been air-dried. Because they are more exposed to heat, or have a lower melting temperature, slips are melted more readily than the underlying paste. And so, when the slip melts during the firing process, part of it sinks into the surface pores of the pottery vessel and therefore adheres more strongly.
Among the Hopi, slips are (or were)

> applied with a rabbit tail or a cloth, used as a mop between the fingers. After *slipping*, they [i.e., the vessels] are again

> dried, and polished. (Colton 1938, 7)[5]

In Guatemala, in the Maya-speaking pueblo of San Sebastian, northwest of Huehuetenango, the potter 'simply wets her hand in the slip and spreads it over the vessel surface in long longitudinal strokes. Before the slip has dried completely, she burnishes the vessel well with a smooth river-worn pebble' (Reina and Hill 1978, 102).

Slips serve several purposes: Like polishing, slips have the effect of sealing the vessel's walls, thus making them somewhat less permeable. They also hide defects in the vessel's surface. And since slips usually differ in hue from the paste used in making the vessel, they provide a contrasting background for designs that may be painted on it. Mineral pigments are often added to the slip; these pigments may include—as they do among Shipibo-Conibo potters—white from kaolin, red from ocher, and black from 'manganese pigment' (DeBoer and Lathrap 1979, 111). Food is less likely to stick on slipped vessels and so they become easier to clean.

In the Maya area of Guatemala, roughly between A.D. 1300 to 1500, *comales* were made with an interior slip of talc-rich clay, '*to provide a nonstick cooking surface*' (Morley and Brainerd 1983, 389). Talc is a heat-resistant, hydrous magnesium silicate, $Mg_3Si_4O_{10}(OH)_2$. Among the softest of minerals, it is lustrous and scaly, with a greasy or soapy feel. (Talc is probably best known nowadays in its powdered form, as the chief ingredient of talcum powder.) Soapstone, which is composed mainly of talc, is quite soft and easily cut. Archaeological artifacts made of soapstone are quite common in the eastern United States.

In a number of modern Maya-speaking villages in Guatemala, the interior surface of *comales* and other cooking utensils is coated with talc. In Santa Apolonia,

> if the vessel is intended for cooking, the [woman] potter *coats its interior with a thin paste made from finely ground talc*...This should *prevent food from sticking*...The potter must do a thorough job of spreading and pressing the paste into the surface, however, because talc does not readily adhere to clay and may flake off as the clay dries... (Reina and Hill 1978, 63)

Firing (Baking)

Pottery clay contains two forms of water: liquid water, which has been mechanically mixed with the clay to make it plastic; and crystalline water, which has been chemically combined to become part of the clay's molecular structure. In firing, as the temperature climbs above 100°C, any remaining liquid water vaporizes. Thus, the thorough drying, discussed earlier, is necessary before firing, so that the excess water will not suddenly pass off as steam: otherwise, the vessel might crack. When heated to sufficiently high temperatures, crystalline water in the clay is also eliminated. The loss of this water takes place at different temperatures for different clay minerals. Depending on the heating rate, ceramic clays lose their crystalline water between 500°C and 1000°C, when 'the clay is transformed forever into another mineral or an amorphous form.' Kaolinite looses its crystalline water at about 550°C (Velde and Druc 1999, 53).

Thus, in the process of firing, the vessel looses weight, shrinks somewhat in volume, and becomes permanently altered and hardened.

Firing Sites and the Question of Early Kilns

There seem to have been two main methods of firing pottery in pre-Columbian times: (1) open firing—that is, firing on the surface of the ground or in shallow depressions; and (2) firing in pits; these features are commonly described as 'pit-kilns.' Both techniques are still found today.

(1) In many parts of the Americas, pottery was simply fired in open fires built more or less at ground level. For example, in a village near the coast of Venezuela (not far from Cumana), Humboldt describes the making of pottery at the end of the 18th century: The pottery of the village,

> celebrated from time immemorial...is exclusively in the hands of the Indian women. The fabrication is still carried on *according to the method used before the conquest*...As they are not acquainted with the use of ovens, they place twigs...around the pots and *bake them in the open air*. (Humboldt and Bonpland 1972, 2:280-281)

The Cayapa Indians also pile their unbaked pottery at ground-level and cover it with sticks. Of course, in open firing there can be little control of either draft or temperature. In this method of firing pottery, the heat is likely to be unevenly applied, particularly on large vessels. As a result, the Cayapa have great difficulty in firing the sizable containers used to make their fermented beverages:

> ...a considerable proportion of such large vessels fail in the early stages of their construction, *probably ninety percent of those successfully built up and cured, crack in the firing*. (Barrett 1925, 179)

Attempting to avoid this problem, the Shipibo-Conibo now practice two firings, and fire their large containers singly.

> The vessel, inverted and *supported on a tripod of old pots*, [or nowadays] metal cans, or bricks, is *gradually heated over a low fire placed in a small, shallow pit*. After this priming fire, the final firing involves completely covering

[5] On the other hand, some claim that the 'modern Hopi *do not slip* their vessels; the high polish produced by patient rubbing with a smooth pebble produces a background which *so simulates a slip that it is usually mistaken for one*' (Hawley 1929, 734).

> the vessel with a pyramid of bark strips or cane poles, the two preferred fuels. (DeBoer and Lathrap 1979, 120)

How potters who make the largest of ceramic containers solve this problem of cracking seems to be unrecorded. In the tropical forest of South America,

> Vessels of unusual size are seen in chicha jars; these range from 3 to 4 feet (1 to 1.3 m.) in diameter and height in the Montaña, to 3 feet (1 m.) high and 7 to 10 feet (2 to 3 m.) in diameter on the Rio Negro, where *manioc-pulp bowls* even attain a diameter of 10 to 14 feet (3 to 4 m.). (Lowie 1963, 27)

Only funerary urns, made more than five centuries ago, rival in size these chicha containers. In the Sinú area of northern Colombia, grave robbers (*guaqueros*) claim to have found huge funerary urns 'of baked clay, some having a *volume of four cubic meters*, each having *walls five inches thick*, and with caps of the same material and thickness' (Arango 1941, 316).

Although the temperature reached in an open fire probably averages between 500 and 700°C, the highest temperature that can be obtained depends on the fuel used and on the fire's exposure to wind; 'dung and desert woods give an even hotter fire.' By burning suitable woods in a good draft, temperatures of 940°C can be reached (Wertime 1973b, 675). In firing pottery, however, not only the temperature is important but duration as well.

(2) Vessels were often fired in pits made for that purpose. In Arizona, for example, Yuman potters

> dig a pit on the lee side of a hill for shelter from wind. The bottom is lined with small stones spaced so that twigs can be put between them. Vessels are inverted and surrounded by the fuel—slabs of oak bark gathered from dead trees. A second tier of pots is set on the first and the heap covered with large slabs of bark. (Shepard 1965a, 75)

Many pit-kilns have artificial walls, such as a lining of stone slabs or adobe blocks. In open firing and in pit-kilns, pottery is placed in with or covered by fuel.

The presence of early kilns in the Americas has been debated for some time, and authorities have differing opinions on the subject: '*as far as I know, no kilns of pre-Columbian origin have been found in the New World*' (Arnold 1975, 203; also Arnold 1985, 218). However, the question much depends on the definition of the term 'kiln.' Some define 'kiln' strictly as a structure with a closed top. In this oven-like kiln, made only in parts of the Old World, the hot gases are first directed upward, then deflected downwards by the closed top of the kiln, and pass 'through the pottery to exit near the bottom' (Rye 1981, 100).

In contrast, in an updraft kiln the heated gases sweep from the flame upward through the pottery and escape at the top. Thus, updraft kilns are chimney-like, and have open tops. In Mesoamerica (southern Veracruz, Mexico), simple updraft kilns for firing pottery have existed for over fifteen centuries (Pool and Britt 2000, 153 and 158).

Another updraft kiln, said to date from 'the last period of pre-Columbian inhabitants,' has been found in the Maya area, in a mound near the north coast of Honduras; it is described as formerly standing largely or entirely above ground. The kiln's clay walls were hardened by fire, and near the base of one wall was a small hole through which additional fuel could be fed. The kiln was, however, not closed—at least, no evidence of a top was found (Stone and Turnbull 1941-42, 40-41). The archaeologists who excavated it offer their own definition of the term 'kiln':

> *A kiln is built on the surface, not inside the earth*, and whatever dirt is to be found inside and over the kiln is there as a result of natural and not human forces. (Stone and Turnbull 1941-42, 40)

This definition of a kiln is somewhat unusual, since updraft kilns are not always above ground-surface, but may be carved in a clay bank.[6]

An archaeologist summarizes the firing of South American pottery in the following way:

> Firing was either in the open or in kilns or ovens...For the most part, open-firing was and is characteristic of the eastern lowlands and southern South America. *Kiln firing is typical of the Andean regions. Kilns with a strong draught, produced by openings at both bottom and top, fired pottery to a well-oxidized red.* (Willey 1963, 142)

More recently, however, archaeologists have found in pre-Columbian Peru 'relatively small, *double-chambered, semi-closed kilns* dug into the ground.' Made of baked clay, they were used by Mochica potters for firing relatively small-sized vessels: '...experiments reveal that these kilns are quite fuel efficient and easy to control' (Shimada 1994, 271; for a detailed description, see Wagner et al. 1994).[7]

Fuels and Firing Temperatures

Native potters choose their fuels carefully, knowing that woods burn differently and have dissimilar effects. For example, in Yucatan, Ticul potters

[6] A deep, updraft kiln is used by the Zapotec Indians at Coyotepec, Oaxaca. It seems to be dug in a clay bank:

> The vessels are placed one on top of the other until the kiln is filled *to the level of the ground* and the whole is covered with tepalcates (potsherds), which must be placed in such a way as to allow [chimney-like, a] free passage of smoke and flame from below. (Van de Velde and Van de Velde 1939, 32-34 and Plate VIIIa)

[7] Almost a century ago, the Conebo also used a ceramic 'kiln' in making their finest pottery: A

> *large pot with a hole in the bottom* is placed on three stones or more often three piles of inverted pots. The pots to be burned [baked] are inverted inside the large pot. The first one is placed over the hole and ashes poured around and over it, others are inverted over this until the pot is full or all are used. Then a slow fire is kept burning under the large pot until all are well cooked, when they are taken out one at a time... (Farabee 1915, 95)

> ...use undried (or 'green') wood and/or any variety of wood...that burns with considerable smoke for the initial warming stage of firing because smoke does not damage the pottery at this time. Conversely, only quick-burning, relatively smokeless wood...must be used during the last stage of the firing process. If potters use undried or smoke-producing wood during this latter stage, the pots will become black and fire-clouded and will have to be fired again. (Arnold 1985, 30-31)

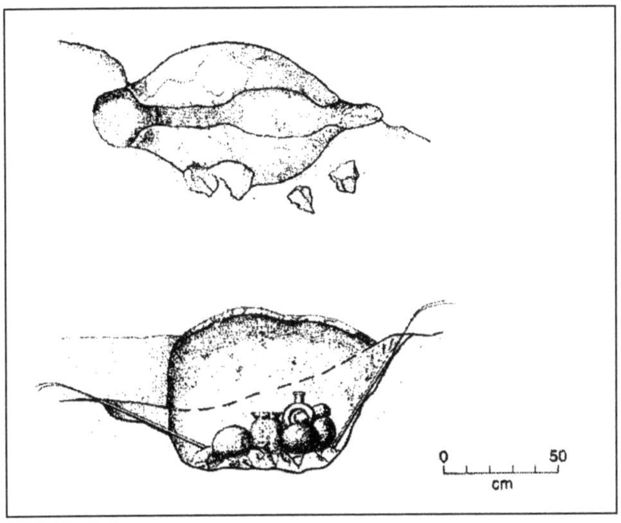

Above: Top view, Below: Section through the middle of the kiln. The Kiln was made of baked clay at Batán Grande on the North Coast of Peru.

'Nearly all are double-chamber kilns with flaring chimneys oriented north-south to north-south-west in accordance with the prevalent diurnal wind direction.' Typically the 57 kilns so far observed measure '60 cm in width, 70 cm in depth below ground surface, and 70-140 cm in length' (Wagner et al. 1994, 67-68).

Figure 42. Mochica Pottery Kiln (after Wagner et al. 1994, 68)

A wood 'may be chosen because it does not burn with a smoky flame; thus, the Pueblos choose juniper rather than piñon to kindle their fire' (Shepard 1965a, 77). If a quick intense flame is desired, cottonwood or other members of the poplar family may be used; oak or guava provides a longer burning fuel. In addition to wood, various other fuels—including leaves, bark, grass, and dung—are used to fire pottery.

Ruminant dung is recognized as a superior fuel by non-commercial potters in many parts of the world:

> it has a compact, porous structure that, when dry, will burn steadily, and completely, without too vigorous a flame, and it tends to retain its structure after combustion acting as an insulation to the pottery below. (Sillar 2000, 65)

Llama dung is a major fuel in the high Andes for the following reasons: It is

> easily gathered, because these animals [unlike bovine ruminants] defecate where they previously have; easily stored in dry huts through the rainy season; easily transported in sacks on llamas; and generally considered by the natives to be the most efficient producer of heat available, especially if some *bellows* or even the ubiquitous bamboo blowing-tube is used. (Webster 1975, 16:201)

(It should be noted, however, that inhabitants of the Americas *did not use bellows* before arrival of the Spanish, though they did use blowtubes.)

The use of llama dung (and later, cattle dung) for fuel seems to have met with Father Cobo's disapproval:

> A pit is dug in the ground, and in it *they bake the pottery not with wood but only with dung and grass*; and even today, pottery is baked in this manner; to be sure, they [the native potters] have been helped not a little in this by livestock brought from Spain, mainly by cows, which provide the sort of fuel they prefer. All of the pottery used by Spaniards in this kingdom, so far as it is *made by Indians*, passes through such flames. (Cobo 1956, 1:114)

Quechua and Aymara people of the Andean highlands still depend mainly on ruminant dung to fire their pottery (Parsons 1945, 25; Sillar 2000, 62-65). In the production of Moche pottery, llama dung may well have been 'employed to attain a lasting high heat and a *reducing atmosphere*' (Shimada 1994, 197).

Colors and Other Effects of Firing

Color in a ceramic artifact can be much influenced by controlling the firing program. Thus, color of the raw clay used in making pottery is a poor guide to its appearance after firing.

The free atmosphere contains roughly 21% oxygen. When baked in fires that are well supplied with atmospheric oxygen, the pottery is said to have been fired in an *oxidizing atmosphere*—a condition which is, of course, favored by a good draft.

Most pottery clays contain iron compounds. Such clays, baked with sufficient oxygen, will have their black ferrous oxide (FeO) changed to red ferric oxide (Fe_2O_3), also known as hematite. Colored by hematite, vessels' surfaces tend to be tan-to-red. However, pure kaolin clay, which lacks iron compounds, remains white even after firing. And so whiteware can be produced by selecting either a slip of low-iron clay or a paste of relatively pure kaolin-clay.

When a fuel is burned in an *oxidizing* environment (that is, with an ample supply of oxygen), most of its carbon is converted to carbon dioxide. The pottery paste itself usually contains organic carbon compounds, as noted above; nearly all of these, too, will be burned to carbon dioxide and driven off.

In contrast, if the fuel is burned in a poor draft (that is, with a deficient supply of oxygen), the pottery is said to have been baked in a *reducing* atmosphere. Cutting off the supply of oxygen-bearing air to the firing fuel is most often accomplished by limiting the firing time, smothering the pottery in fuel, or using wet vegetable material or green wood as fuel. As a result, variable amounts of carbon monoxide, CO, rather than carbon dioxide will be produced. Such an oxygen-starved fire will reduce the red ferric oxide to ferrous oxide, a black or dark-gray compound.

In addition, the incomplete combustion of fuel gives smoky flames. This black smoke is made up largely of minute, unburned carbon particles (soot), which adhere readily to the surface of the pottery during firing. By penetrating its pores, the carbon deposit turns the surface permanently black and reduces the pottery's permeability, a useful quality in cooking ware. Most blackware is made by this sooting technique, also known as smudging.

Judging from the scarcity of conquest-time reports on New World pottery, the subject was of no great interest to the Spanish. If this is so, a blackware which they found off the west coast of Costa Rica must have been unusually attractive. On the island of Chira, the Indians

> make very beautiful ware—plates, bowls, pitchers, jars and other containers—and as *black* as fine velvet and as shiny as much-polished jet. I took several pieces of this ware to the city of Santo Domingo on the Island of Hispañola. (Oviedo 1959a, 4:424)

(This blackware was so handsome that Oviedo recommended that efforts be made to find how the Indians manufactured it.)

In pre-conquest South America (as with Oviedo's example from conquest-time Costa Rica) 'black pottery was common in Perú, particularly the North Coast, where it is most characteristic of the Late Chimú Period' (Willey 1963, 142)—that is, around the first half of the 15th century.

Blackware is made by present-day potters of largely Indian descent in both North America and South America.

The lustrous blackware made by Zapotec potters at Coyotepec in Oaxaca is an example: Vessels are stacked in a deep pit kiln until it is filled to ground level. The open top of the updraft kiln is then covered with a thick, rounded heap of potsherds (*tepalcates*), appropriately packed to allow free passage of flame and smoke, and the fuel lighted.

> The pottery is *sufficiently fired* when flames emerge evenly all over the top of the kiln...the *tepalcates* that for hours have been just a smoldering black mass suddenly burst into fire. It is then necessary to *remove the fuel at once* and allow the kiln and its contents to cool slowly. Soot will come off on the fingers when handling freshly fired pieces; however, this soon rubs off, leaving the finished ware a bright shiny black. This color is due to the fact that clay contains no oxide of iron and also to the careful control of the temperature during the process of manufacture. (Van de Velde and Van de Velde 1939, 34)

This is a good illustration of a reducing atmosphere during firing: incomplete combustion, abundant smoke, and smudging of the pottery. Because soot would be burned off, proper blackening is prevented by excessively high temperatures. Successful smudging at Coyotepec is probably also due to the use of green wood as fuel: 'the writers have often wondered how the potters managed to keep their fires *burning with the fresh green sticks* we have seen them use' (Van de Velde and Van de Velde 1939, 24).

In South America, there are also examples of the latter method of creating a reducing atmosphere: 'Indians of the upper Rio Xingu, and on Mt. Roraima [in northernmost Brazil] put *green leaves on the fire* so that a thick smoke was produced' (Linné 1925, 122). In making blackware, a contemporary potter at Santander, Colombia, uses '*green branches* for the reduction process...Sometimes an unexpected patch of red clay remains where one piece [of pottery] prevents the smoke from coming in contact with the surface of another piece' (Litto 1976, 141).

Painting

Often paint is applied directly to the unfired vessel's surface, usually with a brush. On the other hand, the slip itself may include colorants. In either case, mineral pigments that can resist high temperatures must be used.

Only a limited number of mineral paints keep their colors (or produce desirable new ones) when subjected to the high temperatures required for baking pottery. Of the heat-resistant mineral pigments, iron oxides (the several ochers mentioned above) are the most widely used; they produce a broad array of colors, including yellow, buff, red and brown. Manganese oxides, as mentioned earlier, give darker colors; for example, prehispanic Costa Rican potters used a manganese pigment to produce a lustrous brownish-purple, and Pueblo potters use manganese oxides to paint their pottery dark-brown or black.

Slip Painting

Polychrome ceramics 'came to be a hallmark of southern Peru and the adjacent [Andes]' (Shimada 1996-2000, 396). For example, decoration of fine ceramics of the Nasca civilization (200 B.C. to 600 A.D.) was carried out chiefly by polychrome slip painting.

> In preparation for painted decoration, a background slip, usually light beige, was first applied and the design area demarcated...Fine-line outlines of the designs were drawn prior to or after application of slip paints, which were prepared by mixing locally available *ground manganese or iron oxides* (limonite, hematite, and magnetite) with fine

clay in water...Painting [with a brush] was followed by thorough burnishing to assure that paints adhered to the vessel surface. (Shimada 1996-2000, 400-401)

Iridescent Painting

This rare method of painting, limited to the Americas, is mainly confined to southern Ecuador.[8] The technique

> involves making a superfine slip by suspension of the ground clay particles in water. The fine particles remaining after the grosser ones have sunk to the bottom are dispersed *by adding an electrolytic agent such as tannic aci*d, so that the microscopic platelets of clay float in a parallel alignment. When a vessel which has been painted with such a solution is fired the painted areas acquire a hard glossy finish which will not absorb carbon as easily as the unpainted (or painted with an ordinary slip) parts of the vessel. If the vessel is then subjected to post-fire smudging, the fused microslipped areas will acquire an iridescent quality, particularly notable when the vessel is wet. (Bruhns 1994, 123)

(Electrolytes are substances that dissociate into ions in solution—namely, acids, alkalis, and salts. Thus, tannic acid is a common electrolyte.)

There may be some similar knowledge among present-day Andean Indians: Chiriguano women in Bolivia occasionally take turbid water from the Pilcomayo River, make little puddles of it, and purify the puddled water by pouring vegetable juices into it (Schmieder 1926, 163 and Pl. 25b). These juices, with their precipitating and clarifying properties, may contain tannins.

Glazes in the New World

A glaze is 'a coating of glass fused to the surface of a ceramic vessel' (Rye 1981, 146). The technique of glazing now widely used by most native American potters was introduced by the Spanish. That glaze contains a lead flux.[9] (In ceramics, a flux is a material that decreases the melting temperature and promotes vitrification.) At present, these glazing materials are common commercial items.

The following statement contrasts Asian and European ceramics with those of the Americas:

> ...the American archeological experience has never gone as far as the *porcelain stage* in technological prowess, though results are often stunning. Most archaeological wares are *rarely slip covere*d and *more rarely glazed.* In the European archaeological experience the full range of ceramic evolution has been produced... (Velde and Druc 1999, 130)

It is true that making porcelain requires both higher and better-controlled temperatures than were attainable in pre-Columbian America, but the knowledge of making porcelain came to Europe from China.[10] Regarding slips (in the passage quoted above), no numbers are available for how many places they were applied in fifteenth-century America, but slips were frequent on pre-Conquest pottery in Mesoamerica (for instance, at Kaminaljuyu and Chipoc in Guatemala) and elsewhere (for instance, in the Pueblo area).

In pre-Columbian Peru, an area where ceramics were well developed, the potters 'had not made the discovery of *glazing...*' (Cobo 1956, 1:114).[11] While strong in the fifteenth century, this contrast between Eurasia and the Americas was not always so marked. Granted, glazing was rare in the New World. Nevertheless, though the art was eventually lost, glazed or glaze-painted wares were made and traded by Native American potters in several archaeological periods.

Plumbate Ware

An lustrous glassy-surfaced pottery was made in early Mesoamerica, the so-called 'plumbate ware.' It is an attractive, thin-walled ware, and (while there are other hues) the characteristic color is light gray. The hardness of the surface is extraordinary: 'the gray parts are harder than steel, and the hardest areas cannot be scratched by quartz [7 in Mohs' scale]' (Shepard 1948, 93). Plumbate 'may properly be classed as a *glazed ware* although only the hardest parts of vessels are vitrified and those not completely' (Shepard 1948, 96).

Plumbate pottery's 'distinctive qualities are due to properties of the slip and the method of firing.' The ware has a thick slip, or often two slips—and analysis shows that no fluxing material was deliberately added to its clays. Because most of the vessels have slipped interiors,

[8] In addition to Ecuador, iridescent painting has also been found in Guatemala, perhaps indicating direct cultural contact between the two areas. In Ecuador, this method of 'painting, first developed during the Early Formative...and became a distinctive decorative characteristic of fancy pottery during the Late Formative' (Marcos 2003, 22). After that time (i.e., several centuries B.C.) iridescent painting was replaced by resist painting.

[9] 'The early glazes used in the Middle Ages in Europe were a *lead-based* silica composition charged with several percent copper, giving a *green* colour...' (Velde and Druc 1999, 93-94).

[10] Chinese porcelain ware was baked for several days at between 1350°C and 1450°C (Li 1948, 98). In eastern Asia porcelain was produced in a early period: the Han Dynasty (206 B.C. to 220 A.D.) can 'certainly be regarded as the time of the birth of porcelain' (Li 1948, 69).

[11] Archaeological research has since shown that, while Cobo's statement on the lack of glazing in Peru is true, his generally negative appraisal of Peruvian pottery was excessive.
Despite the lack of glazing, the ancient Peruvian potters have a long history of excellence in ceramic skills. For example, the wares of the Moche (100 A.D. to 700 A.D.) in northern Peru depict well the Andean and coastal culture: 'The modeled life forms and meticulously painted pictures on pots are veritable source books of contemporary ethnology...' (Steward and Faron 1959, 87).
In contrast to Cobo's assessment, Francisco Hernández described the pottery vessels sold in Aztec markets as 'lacking nothing in elegance' when compared to those sold in Spain (Hernández 1986, 103).

probably the ware was slipped by dipping (Shepard 1948, 94 and 97).

Iron is remarkably high in the slip, and ferruginous matter has an important influence on the color and hardness of pottery. During firing in a *reducing* atmosphere, red ferric oxide—which does not act as a flux—is changed into the less-oxidized form, black ferrous oxide—which 'acts as a strong flux and so promotes vitrification' (Shepard 1948, 98 and 96). That is, despite its name 'plumbate,' the glaze is *not due to a lead flux*.

First manufactured around 900 A.D., plumbate pottery had its origin near the coast of southwest Guatemala (Neff and Bishop 1988, 505). The ware was widely traded in Mesoamerica until around 1000 A.D., after which the art of making it was gradually lost. Plumbate ware is found in various sites as a trade item from Tepic, Mexico (near the northern border of Mesoamerica), as far south as Chiriquí in Panama. Because it was made only in a localized area, and traded in a restricted period of time, the ware is a prehistoric marker of great chronological value.

'The extremely hard, semivitreous slip of Plumbate is *perhaps the most interesting technical development of New World pottery*' (Anna O. Shepard, author of several paragraphs included in Linné 1965, 24).

Glaze-Paints in the Pueblo Area

Examples of simple glazing are also found in the prehistoric Pueblo area in the American Southwest. 'Rio Grande Glaze Paint is one of the few true glazes produced by the American Indian...' (Shepard 1965b, 66). However, this glazing did not cover the whole surface of the vessel, as is the case with plumbate ware; that is, 'glaze-paints' were used for making designs (with brushes) on only parts of the vessel. Moreover, in contrast to plumbate ware, the glaze-paints were fluxed by lead: 'an iron-manganese ore which *carried enough lead* to form a glass...Some of these glazes have a high proportion of copper, which also promoted fusion' (Shepard 1965a, 47).

In the late thirteenth-century, glaze paints began to be used in the Zuni area along the present boundary of New Mexico and Arizona. 'Zuni Glaze Ware is characterized by red and white slipped surfaces, with white and glaze black paint'; the vessels 'were tempered with crushed sherds, and were fired in an oxidizing atmosphere' (Duff 2002, 88-89). Shortly thereafter, such glaze-paints began to be used farther east along the Rio Grande, around the present site of Albuquerque.

Three elements 'are of great importance in their effect upon the properties of the [Rio Grande] glaze: lead, the principal flux; copper, which may either color the glaze or, under certain conditions, also act as a flux; and manganese, a pigment' (Shepard 1942, 220). The Zuni and Rio Grande glazing technique was practiced for over three hundred years, ending about the time of the Spanish conquest.

Applying Paints and Other Materials after the Main Firing

After it has been fired, materials are frequently applied to pottery for decorative or utilitarian purposes. For example, resist painting is an 'unusual form of acquiring a red and black vessel *without double firing*' (Bruhns 1994, 123). This method of painting (also known as 'negative painting,' 'organic resist painting,' or 'batik painting') is 'a variant of the more general practice of intentional smudging of pottery' (Lathrap 1975, 34). Once widely practiced in pre-Columbian times, resist painting had a curiously discontinuous distribution in the Americas, and 'seems to have been invented many times' (Bruhns 1994, 393).[12]

No American 'Indians have adorned pottery in this fashion recently and we have no written description' (Lothrop 1963b, 70). Although this decorative technique has puzzled archaeologists for some time, it may have been clarified by a gourd-decorating procedure. According to Linné, C. V. Hartman is the 'complete solver of this problem.' At the end of the 19th century, Hartman described tree gourds (i.e., calabashes) at the Pipil (Nahuatl-speaking) village of Izalco in El Salvador:

> The calabash vessels were manufactured exclusively by women, and were decorated in the following way: the outer skin of the calabash was rubbed off, whereupon the surface was finely polished. The ornaments were then painted on with a small hair brush *dipped in melted wax...When the wax painting had dried, the whole surface was coated with black paint consisting of powdered charcoal and the pods of a certain tree*, mixed with sugar. *The calabash* was then again left to dry. After that it *was dipped in boiling water. The wax then melted, and along with it came away the black paint that had covered the figures painted in wax. The pattern then showed up in the light colouring of the rind itself, against the black coating.* (Linné 1934, 162-163, citing and translating Hartman 1910, 137)

Following the example of the resist painted calabashes, some archaeologists put forward a proposal for creating the same effect on pottery. First the vessel is fired in an oxidizing atmosphere so that its surface is of a light color—usually buff-to-red. Next, the potter paints a figure in hot wax on the vessel's light-colored body. Then the vessel is covered with paint of a dark color (typically, made of a black carbonaceous pigment). When the pottery is heated, the dark paint is permanently fixed to the vessel's surface, whereas the wax is burned off,

[12] In pre-Columbian Peru, Amazonia, and Mesoamerica, resist (negative) painting is well-known archaeologically.
It was also present in the southeastern USA. Since the technique appears there 'with explosive suddenness without any apparent earlier stages' (Willey and Phillips 1944, 183), it may have been introduced, probably from the Antilles.

exposing the lighter design. 'Archaeologists familiar with negative [resist] painted pottery in Middle and South America are in agreement that this, or something very like it, was the technique employed' (Willey and Phillips 1944, 174; and Lothrop 1963, 70).

On the other hand, more recent and rather different explanations have been given of resist painting.

> The cooled, fired vessel was then coated with an organic material, probably of vegetable origin, which was liquid enough to seep into the porous fabric of the vessel. Next those areas which the potter wished to remain the base red or buff color were painted out with slip, leaving the design areas uncovered. Finally the vessel was heated until the exposed organic-material-coated areas charred. When cool the now dry slip was brushed off, leaving a vessel with black decoration...Although some have suggested that wax was used, in a batik-like technique, experiments have shown that wax does not char, but merely runs and burns, leaving a mess. (Bruhns 1994, 123 and 393; see also Trachsler 1965, 159-160; and Lathrap 1975, 34)

In any case with regard to Hartman's theory, gourds have played a significant part in the development of New World pottery. For example, vine gourds and pottery share two general qualities historically desirable in a water container: (1) the ability to store water without leakage and (2) the ability to keep water cool. (As noted earlier, the tree gourd has the first quality and the vine gourd, with its more porous walls, the second.)

The custom of applying stucco with paint to pottery and gourds is ancient in the New World. For example, at Tikal, the old Maya center in Petén, a number of vessels were found, 'one a finely *stuccoed and painted urn*, and the remains of two, *red-painted, stuccoed gourds*.' These vessels are 'estimated to date back to about 25 B.C.' (Coe 1965, 1412-1413).

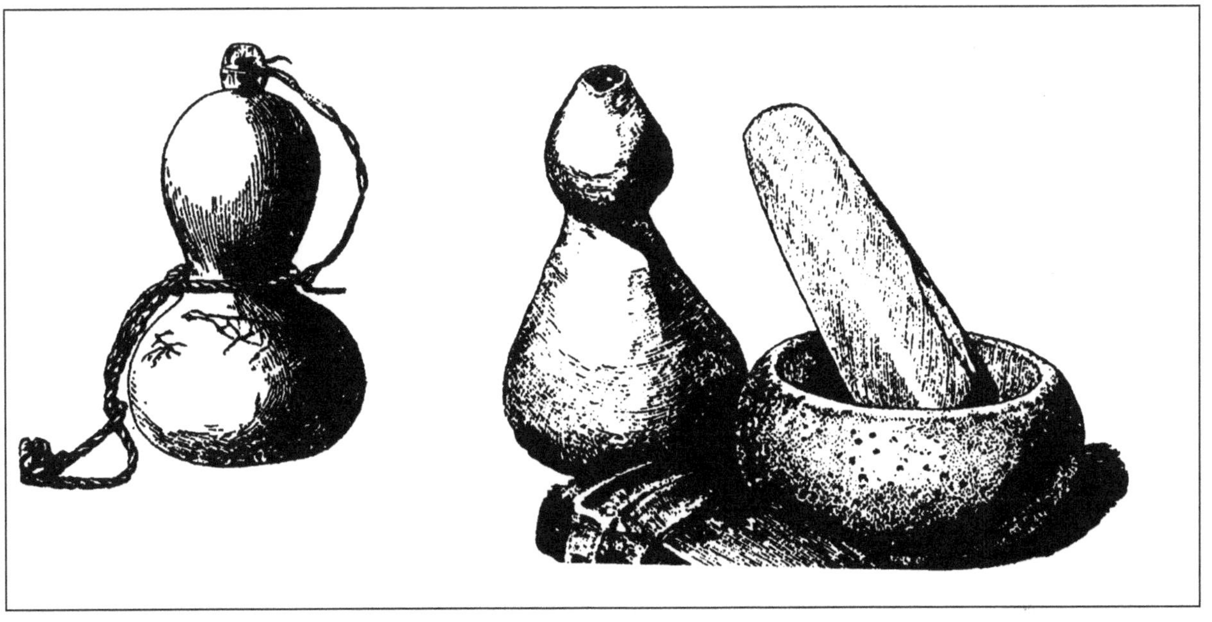

Left: Huichol vine gourd; height, 22.5 cm (Lumholtz 1973, 2:220). Like many Indian field workers from northern Mexico to Peru, the Huichol use hourglass-shaped vine gourds to carry cool drinking water; the gourd is carried by string tied around its middle.

Right: An earthenware vessel in the shape of a vine gourd, alongside a mano and metate, found in mounds near the Upper Piedras Verdes River in northern Mexico (after Lumholtz 1973, 1:81).

In Mesoamerica, the vine gourd, originally probably from Africa, may have been the earliest cultivated plant. Antedating pottery by millennia, it has been found 'in the lower levels of cave deposits in Tamaulipas, Mexico, where it is dated to between 7000 and 5000 B.C.' (Willey 1966, 1:22).

Vine gourds served as models in the shaping of earthenware vessels. In Ecuador, the ceramics of 'Valdivia phases 3-5 [a Formative culture, 2300-1900 B.C.] show a wide range of bowls, many of which are clearly imitations of halved, carved bottle [vine] gourds' (Lathrap 1975, 29). Occasionally old ceramic containers have the gourd-like knobs on their surfaces: in the Valley of Mexico, excavated 'bottles with gourdlike protuberances' are found (Piña Chan 1971, 165 and 162). Even in modern Guatemala jars for carrying water are sometimes modeled after an hourglass-shaped vine gourd. (See Figure 3 in Reina and Hill 1978, 25).

Figure 43. Vine Gourds and Pottery

> Pottery that was painted *after firing* was generally coated first with a lime plaster or, less frequently, with white clay or gypsum plaster. This technique, referred to as *stucco painting*, is comparable to that of mural painting; similar pigments were employed in ceramics and architecture...When a plaster coat is painted after it is *dry* (*fresco secco*), a medium must be used *to bind* the pigments...but *true fresco painting* has not been reported. (Shepard 1965a, 178)

(However, the discussion about mural techniques in Chapter 13 may lead to some uncertainty about the lack of true fresco on baked pottery or gourds.)

Among Zuñi potters, for instance, the 'pigments are ground in stone mortars and made into a paste with water to which a *syrup of yucca fruit* is added' (Robbins et al. 1916, 51). In 16th century Mexico, Aztec potters used 'an ocher-like earth which is colored red, and they call it *tlachichilli*; with it, they varnish plates, jars, etc., because it gives an excellent red luster' (Sahagún 1956, 3:349). This was probably hematite, though in this case it was likely applied with some organic fixative.[13]

Sometimes the pottery was fired a second time, at a much lower temperature, because most organic materials oxidize at temperatures between 200° and 300°C. For example, resins begin to decompose at around 200°C; and, since Maya Blue (a mixture of indigo and white clay minerals) becomes grayish at about 250°C (Van Olphen 1966, 645), it was necessarily applied to pottery after the main firing. The same is true of some mineral pigments. Following are several mineral pigments that decompose or change color if the temperature is too high: The green and blue colors of malachite and azurite are lost during the main firing process. Nor can the yellow of orpiment and the red of realgar be applied before pottery is fired. The vermilion provided by cinnabar will also disappear. Brown ocher (limonite), too, changes color on firing.

A late 18th century report from the Orinoco Basin gives a possible example of such refiring:

> The colours used by the Maypures are the oxyds of iron and manganese, and particularly the yellow and red ochres... Sometimes...*bignonia chica*...[i.e., carajura, is] employed, *after the* [baked] *pottery has been exposed to a feeble fire*. This painting is *covered with a varnish* of algarobo, which is the transparent resin of the *hymenaea courbaril*...The Otomacs, and even the Guamoes, are known at the Oroonoko for the fabrication of *painted pottery*, which extended formerly toward the banks of the Amazon. Orellana [a 16th century Spanish soldier who explored the area] was struck with *the painted ornaments on the ware of the Omaguas*, who in his time were a numerous and commercial nation. (Humboldt and Bonpland 1972, 5:157-158)

Many of the organic substances applied to baked pottery have not been identified as to source; nevertheless, they include the following categories: (1) resins, (2) waxes and fats, and (3) starches. Vessels, after they have been baked, are coated with these organic materials for both practical and decorative purposes.

(1) Resins:
Resinous substances were often applied in the Andes more than a thousand years ago by the potters of Paracas and Nasca cultures (Sawyer 1961, 274, 277, 279, 281, 289 and 498).

In a part of eastern Ecuador, Quechua-speaking potters buy their resins and slips from a neighboring tribe, the Canelos (Litto 1976, 210). Immediately after firing, while the vessels are still quite warm,

> the inside and then the outside are rubbed with a *copal-like resin*...The resin, gathered from a jungle tree, will melt and flow evenly over the surface of the hot bowls. In a few minutes it will cool, forming a hard surface...Without the protective coating, the painted designs would remain fragile and easily rub off. (Litto 1976, 211)

Two trees are the source of resins for the Shipibo-Conibo of eastern Peru. The first is from a tree of the *Protium* genus, 'for lending a glaze-like finish to white-slipped surfaces'; the second from *Hymenaea courbaril* (the tree mentioned above by Humboldt), is used 'for waterproofing the interior of liquid-containing vessels' (DeBoer and Lathrap 1979, 115). According to an earlier report, Shipibo potters of today use a black resin '*for repairs*' on pottery (Litto 1976, 215).

Resins were also used for repairing broken pottery near Lake Nicaragua. Oviedo mentions a fragrant 'myrrh-like' exudation (likely a resin from either *Bursera*, *Protium* or *Hymenaea* trees) which, when heated, makes a fine glue, useful 'for mending broken things, such as dishes and bowls' (Oviedo 1959a, 1:177).

The Tucano of northwest Amazon

> coated [pottery] with clear or black resins after firing. We saw some pieces with designs very freely daubed with fingers using red, yellow, or white clay slip. However, this was done after the piece has been fired and coated with resin, so that the decoration rubs off at a mere touch. (Litto 1976, 216)

Among the Kuikuru on the Upper Xingú, a (resinous?) tree bark was used. There, toxic-manioc pulp has to be rinsed with clean water.

> For this purpose *an enormous flat-bottomed pot is used.* This type of cauldron, sometimes measuring over two and one-half feet in diameter, is especially prized and ordinarily is *not put over a fire*. As with common pots, *the interior surface is sealed with several coats of rubbed soot from charred tree bark.* The exterior is *slipped* with an emulsion of *urucú* (*Bixa orellana*) and red clay, and later painted red

[13] In parts of the American Southwest 'hematite was occasionally rubbed on the exterior of....[the] vessels after they had been fired' (Wendorf 1953, 72).

> with a mixture of *urucú*...and oil of the *piqui* fruit (*imbene, Caryocar butyrosum*), forming a background on which designs in red, black and white are painted, especially on the bottom. (Dole 1978, 22)

(Like *Bixa orellana*, the *Caryocar* genus is reputed to contain an abundance of carotene.)

(2) Wax and fats:
Axin, a fat, was used in painting Maya pottery:

> There is a red worm [sic] from which a good yellow ointment is made...by only crushing them or kneading them together; and it serves as oil *for painting the vessels and makes the color fast.* (Landa 1941, 193)

In coastal Ecuador, vessels designed by the Cayapa Indians for making fermented beverages or for water storage

> are given a coating of *brea* [not tar, but a black beeswax], usually inside and out, which renders them impervious. It would be useless to apply the brea to pots that are to be placed over the fire, since it melts at a low temperature and readily ignites. (Barrett 1925, 181)

Nevertheless, apparently in conquest-time South America 'wax coating of the vessel is *solely an Amazon trait.* Resin varnishing, for appearance and impermeability, is more widespread...' (Willey 1963, 143).

(3) Starches and gums:
At Coyotepec in Oaxaca, when they wish a vessel to be non-porous, the Indians treat it with *atole*, the gruel-like substance serving the purpose of varnish... (Van de Velde and Van de Velde 1939, 40).

In parts of Guatemala, after being removed from the baking fire, vessels are commonly treated with nixtamal water, probably a weak solution of starch in alkaline water. For instance, at the Cakchiquel-speaking village of Santa Apolonia, the woman potter

> gives them a coating of *agua de masa*, a thin liquid drained from the mixture of lime and ground corn used to prepare tortilla dough. The water quickly evaporates and *the residue, rubbed to a high polish, enhances the appearance of the vesse.l* (Reina and Hill 1978, 63)[14]

If the vessels sold in markets do not receive this treatment by the potter, it is applied by the purchasers themselves.

Such examples of pottery making (mostly drawn from isolated areas of tropical forest where native American languages are still spoken) probably represent a continuation of pre-Columbian practices.

[14] On the other hand, lime water is often used alone. Although the Chorti Indians of Guatemala sometimes use the same starchy maize-material, the

> usual method is to boil lime water in the container for about an hour, so that the lime penetrates the walls and seals them...A second boiling with pure water is usually done to remove the taste of the curing agent. (Wisdom 1940, 170; also Reina and Hill 1978, 245)

CHAPTER 15

LACQUERS AND VARNISHES

The terms lacquer and varnish are sometimes used interchangeably. Both refer to the glossy protective or decorative coating obtained when a drying-oil, the essential constituent of both varnishes and lacquers, is spread—commonly in successive layers—on wooden or other surfaces. Drying-oils are organic liquids which—when exposed to the air in thin layers—take up oxygen and form resinous films that are tough, dry, and elastic; these oils contain double bonds which account for their 'drying' properties. In making a lacquer, various powdered, mineral fillers are usually added to the drying-oil to give it body and color. And so, varnishes are generally transparent, whereas lacquers are opaque.

A variety of methods is used in decorating an object with lacquer. For instance, some artifacts are coated with lacquer of only one color. If the successive layers are of different colors, however, a design may be scratched in the uppermost layer thus exposing the color of an underlying layer. On the other hand, some lacquered objects are inlaid—that is, a design is cut in the lacquer, and the cut-out spaces filled with lacquer of a contrasting color.

Although lacquering is best known from East Asia, it was also practiced in pre-Columbian times in parts of both North and South America.[1] In comparison to Asia, the lacquered objects which have found in the New World are of no great age.[2]

[1] Opinions vary as to where lacquering began in the western hemisphere, though the archaeological evidence gathered so far indicates that is older in North than in South America. As to the beginning of the craft in Mesoamerica, this is the judgment of one authority: 'The art of lacquering was practiced in Middle America from early times, having originated probably in western Mexico, spreading in Classic times to Teotihuacán and as far south as Kaminaljuyú in Guatemala' (Covarrubias 1954, 118).
Whether lacquering developed independently in the New World or was influenced by bits of knowledge introduced from the Old World is still, despite much discussion, an unsettled question, since the prehistorical evidence is fragmentary. Although suggestive similarities exist between lacquering in the New World and eastern Asia, the techniques and materials are by no means identical—in America, they are thoroughly intermixed with those used in decorating gourds and murals, in a way that seems to be absent in Asia.

[2] In China, 'as we know from archeological evidence,' lacquering started before 1100 B.C. (Li 1948, 104). Yet in Japan, lacquer was made in Jomon at a much earlier date—between 4000 and 2500 B.C.: a variety of coarse wooden specimens has been recovered recently, and 'more refined items that had been *painted with red or black lacquer* included a number of bowls and a wooden comb. Some fragments of *lacquered pottery* were recovered as well' (Aikens 1995, 14).
Not only is east-Asiatic lacquering older than that of the Americas, it is more elaborate. For instance, gold dust was incorporated in Chinese lacquer; lacquer-ware there was often incrusted with mother-of-pearl; and the layers of lacquer were sometimes separated by 'a layer of hempen cloth, linen, paper or...silk' (Li 1948, 111).

Lacquering in North America

Lacquering on vegetal materials is still common in Mexico. Nowadays, tree gourds (*jícaras*) and various woods are the favorite surfaces. (The wood of some trees is deemed unsuitable for lacquering, perhaps because of reactions between the wood and constituents of the lacquer.) Although most of the lacquering techniques now used in the area of Mesoamerica are indigenous, a few of the materials used (and many of the decorative designs) are obviously not.

In much of Mexico a superior lacquer is made using an unusual ingredient—namely, the animal product, *axin*.[3] (In contrast, in South America and in eastern Asia, as well as in parts of Mexico itself, the drying-oil was of plant origin.) A 16th century document on the Maya states that axin—called *ni-in* in the Maya language—'is very useful in this land for *lacquered trays, jicaras*, writing desks [probably a Spanish introduction] and other things painted with oil. It is very cheap because of the *great quantity* which exists' (Roys 1931, 294).

Axin, though it reportedly contains no resin (Jenkins 1966, 130), acts both as a drying-oil and a plasticizer. (A plasticizer is a substance used to impart viscosity and flexibility.) Other native drying oils are obtained from the seeds of chia (*Salvia* sp.) and the prickly poppy (*Argemone Mexicana*). Chia oil and axin are often mixed together. According to several historical reports, chia rather than axin was used in the west—for instance, at Olinala, Guerrero. To the east and south, lacquer is made with axin, but with no chia oil. (Flax is an Old World plant, and the custom of employing linseed oil, now frequent, is a Spanish contribution.)

Although historical and ethnographic literature is replete with references to decorated gourds, it is sometimes difficult in old written accounts to distinguish between those referring to varnishing and lacquering. For instance, chia oil and axin, though important constituents of lacquer itself, were often applied simply as a varnish or as a polish. Thus, in 16th century Mexico,

[3] Nowadays, the best-known source of all varnishing materials—i.e., shellac—is produced by the scale-insect, *Lacifer* sp. Grown in India and Ceylon, it is related to those scale insects in the Americas that produce the waxy fat, *axin*, and the red dye, cochineal. The earliest known use of the lac insect was as a red dye; its employment for a varnish was not known (at least to English literature) until the late 16th century. Nevertheless, shellac's use as a varnish was probably well established in the regional folk crafts of India at a much earlier date. It seems, however, that no true lacquer was made from shellac.

> ...the painters apply *chia* oil. With it they varnish things, make them glossy...And this *chia*, when ground, is *moistened with hot water. Then it is wrapped in a rag; with this the beautiful thing is washed.* (Sahagún 1950-82, 12:181)

In an example of this use, chia oil was sold in Aztec markets to 'preserve from injury from the rain or weather the statues of the gods' (Hernández 1986, 106).

Similarly, Guatemalan Indians nowadays blacken gourds with the smoke of a resinous wood, then varnish them with axin (McBryde 1945, 57). Nevertheless, the following comment on a tradesman who sold gourds in 16th century Aztec markets likely refers to a lacquered, not simply to a varnished, product: 'He sells gourds with raised (designs), with stripes, with lines, scraped, rubbed with axin...polished gourd bowls, burnished, varnished; painted gourd bowls' (Sahagún 1950-82, 4:78).

Tree gourds (*jícaras*) were a major trade item in 16th century Mesoamerica. They were burnished, 'painted,' and polished with *axin* (Sahagún 1956, 3:143). When the colored decoration on these gourds is referred to as a paint, it was probably a lacquer; an example is the following note on Maya custom:

> There is a tree [*Crescentia cujete*] from whose fruit which is very like round gourds, the Indians make their vessels and they are very good and they make them *finely painted* and handsome. (Landa 1941, 196-97)

The Tarascan Indians of Michoacán have long been known for their lacquer work, and the Tarascan town of Uruapan is one of the several surviving centers of Mexican lacquering. Inhabitants of Uruapan use two light-colored earths to make lacquer: dolomite, $CaMg(CO_3)_2$, and lithomarge (an impure form of kaolin) are mixed with a drying oil to make the 'stucco' applied on the object's surface. (Since drying oils, including axin, act as binders when mixed with these minerals, a lacquer can be considered a kind of stucco.) The following is an account of Tarascan lacquering at Uruapan as observed at the end of the 19th century:

> The vessel [a gourd bowl] to be lacquered is first covered with a coating of *lithomarge* (a clay). On this the men trace the designs, which are then cut out with a knife, and the women *fill in the incisions with various colours*, smoothing them over with their thumbs...Details are added by means of a finely pointed instrument. Then the varnish is put on, and the beautiful polish produced by patiently rubbing the surface with a bit of cotton. The lacquer thereby becomes so hard that *it will even resist for a time the action of water*. Gourds are lacquered only on the outside. *The varnish is produced from a plant louse* called in Spanish *aje* [axin], which is gathered during the wet season by the Indians of Huetamo, six days' journey southeast of Uruapan. (Lumholtz 1973, 2:444-445)

At another Michoacán lacquering center, Quiroga, there is no use of axin. Chia oil was used at one time, but was replaced in the late 19th century or earlier by oil from the seeds of prickly poppy. In more recent times, the poppy oil has, in turn, been almost entirely replaced by linseed oil (Brand 1951, 168). At Quiroga the minerals used to give body to lacquer cannot be classified simply as calcareous or argillaceous. Wooden bowls and trays are 'slipped'—that is, covered with a layer of stucco. The slip is made of two 'earths,' mixed with a drying oil:

> One of the earths was referred to as *tierra colorado*, and seemed to be simply *very hard clods of reddish earth* [likely colored by oxides of iron, as at Jomon]. All that is done to prepare the *tierra colorada* is to grind it to a very fine powder on a metate. The other blackish earth is called *tierra negra...*, and is country soil *mixed with humus* which is obtained beneath oak trees. This *blackish earth is burned or roasted*, ground to powder on the metate, *and mixed with lampblack*. (Brand 1951, 168)

The lacquering at Quiroga, however, has probably been more influenced by the Spanish than that of Uruapan.

There may be a close historical connection between the techniques and materials used in making murals in Mesoamerica and those used in lacquering and stuccoing pottery. As in the fresco secco technique, stuccoes (usually composed of calcareous and argillaceous earths)[4] were mixed with some binding agent and applied to fired pottery. For example, in making pottery at Teotihuacán 'the outside of the vessel was covered with a *fine coating of plaster upon which the design was painted exactly as in frescoes*' (Spinden 1928, 178).

At Teotihuacán the same artisans may have decorated both murals and pottery: 'similar pigments were employed in ceramics and architecture...; furthermore, style *suggests that in some instances the artist was a mural painter*' (Shepard 1965a, 178).

---

[4] Several such lacquers have been described as they appear on excavated pottery, dated between 1000 B.C. and 600 B.C., from the valley of Mexico.

A yellowish lacquer:

> This type...is characterized by a white wash and over this a yellow wash, in the style of lacquer wares; or perhaps *over a layer of lime or kaolin, with resin as an adhesive*, there is yellow painting, fresco or polished. (Piña Chan 1971, 162)

An orange lacquer:

> This is similar to yellowish lacquer, except that the second wash is orange, and *it may have incised decoration*...bowls and tecomates. (Piña Chan 1971, 164)

'Pseudo-Fresco [fresco secco] or Monochrome Lacquer':

> An infrequent type, it is linked...to the yellowish and orange *lacquer* types. Apparently here was begun the technique of placing on the brilliant surface of a vessel *a wash of lime, stucco, or kaolin, mixed with an adhesive resin*, or a wash of blue or turquoise-green al fresco, which is rubbed or polished, the same technique as with the lacquered ware. Later more colors were added and parts of the wash were scratched so that further layers could be applied in decorative patterns. (Piña Chan 1971, 165)

Calcareous and argillaceous minerals seem also to have been used in ancient China: the object to be lacquered was first coated with the sap of *Rhus vernicifera*, in order 'to fill up the pores of the wood...On this is applied a priming consisting of one or two coats of composition, made by mixing lacquer with finely ground burnt *clay* or similar materials...Some of the old writers have said that in this composition, the Chinese occasionally used pig's blood and powdered quicklime' (Li 1948, 111).

Moreover, the techniques used in making Mesoamerican murals and the stuccoing and lacquering of pottery may even have extended to the painting on books. For example, at the ancient Maya center of Kaminaljuyu (located in the outskirts of the present site of Guatemala City),

> stucco decoration [on pottery sherds] was not necessarily the work of the potter; *the brush technique of some specimens suggests that it may have been done by codex painters.* (Shepard 1977, 274)

Also in Guatemala,

> Codex-style ceramics employ a fine, *almost calligraphic*, black-brown line on white or yellowish-cream background and constitute a highly distinctive style of pottery in a manner *reminiscent of Maya folding-screen books.* (Bishop 1994, 41)

In China, too, lacquer 'was early used as a writing material' (Li 1948, 104).

Paint Cloisonné

Often paint cloisonné is called 'pseudo-cloisonné' or 'stucco-cloisonné.' Such terms are meant to distinguish it from *true* cloisonné, an Old World technique of enamel working in which colored designs are similarly produced by inlaying. However, in true cloisonné the lacquered areas are separated by thin metal bands—and the whole artifact *is baked at high temperatures.*[5] Paint cloisonné '...gives the effect of [true] cloisonné, but *there is no indication that the paint was hardened* [as in the Old World] *by baking*' (Lothrop 1956, 233).

In parts of Mesoamerica and northward into the American Southwest, paint cloisonné was applied to the surface of gourds, wooden objects, pottery, and plaques. It was done in the following manner:

> A layer of paint was first applied over the entire surface of the vessel [or an artifact]. The design was then cut into this, all of the paint within the design area scraped away, and paint [lacquer] of another color put into the hollows and smoothed off to the level of the original layer. Other cuts were then made and other colors inserted, either in the basic layer or within the areas of color applied in the first operation...*the cuts were always made at an angle*, that is, a narrow cut would be *v-shaped*, and a broader channel would have sloping sides in the same manner. (Ekholm 1942, 93)

Many gourd fragments decorated with paint cloisonné have been found in northwestern Mesoamerica at Guasave in Sinaloa, Mexico. Only a partial analysis of the lacquer's constituents was possible because of the material's weathered and fragmented condition. However, what remained of the lacquer appeared to be entirely inorganic; that is, no trace of the drying oil could be found. 'Only six colors were noted in all of the examples discovered: gray, orange, red, turquoise-blue, yellow, and white': The white is unpigmented calcareous material; 'the orange and yellow pigments...may possibly be iron derivatives.' It was concluded that colored pigments of the lacquer were mainly inorganic, since they did 'not burn out of the paint on heating to red heat in a crucible.'

The method 'is identical with what has been called pseudo-cloisonné technique and the modern 'lacquer' technique of the Tarascan Indians of Michoacán' (Ekholm 1942, 93). Around 1902, lacquered pottery was excavated in Jalisco:

> As far as my knowledge goes, this was the first ware of this kind that had been met with in Mexico. *The method of decoration was the same as that employed in making lacquer-ware among the present Tarasco Indians.* The accessible surface of each piece was evidently first covered with a thick bluish-grey coating of a kind of clay, *into which the patterns were cut, the incisions being filled with the different colours*, and the piece then was fired. (Lumholtz 1973, 2:461)

Or, perhaps, the piece was then heated at a much lower temperature, since most organic binders in the lacquer will not withstand the high temperatures of firing.

At Apatzingán in Michoacán, there is

> evidence of at least one occurrence of paint cloisonné, and scraps of *unfired pigment* suggest further instances. The pigments presumably were affixed to some perishable object, perhaps a gourd or a wooden tray. Colors include a cream, a light rose, and a pale green...The archaeological occurrence of paint cloisonné in this area is of interest because it lends credence to the belief that modern Michoacán 'lacquer' has a pre-Conquest background. (Kelly 1947, 145)

Examples of paint cloisonné have been found on severely rotted pyrite-incrusted plaques[6] excavated in Guatemala, at Kaminaljuyu, dated 'from the earlier part of the Classic Period' (Kidder et al. 1977, 3), a few centuries A.D. Several plaques had their backs covered with stucco; the

[5] Although cloisonné is generally thought to have been absent in the pre-Columbian New World, 'true cloisonné process may have been known in Southern Peru; however, few 'metal specimens exist with colors separated by soldered wires...' (Lothrop 1956, 233). Of several small metal statuettes excavated in Peru, one was 'a *silver* llama with a caparison [i.e., an ornamental saddle] of *inlaid lacquer* which has for the most part crumbled away. The edges of the caparison and the toes [of the llama] are of gold' (Lothrop 1937, 323; see Plate 38c).
In ancient China, the surface to which lacquer was applied was '*almost always wood*'—and in South America as well, it was most often affixed to wood. However, Chinese lacquer was sometimes applied to metal: 'porcelain, *brass*, and white *metal alloys* were also occasionally used...' (Li 1948, 110-111). This sparse evidence indicates that in Peru, lacquer was applied to metal also.

[6] The use of these artifacts is unknown. Although they are usually identified as mirrors, we think it is improbable that the pyrite-incrusted plaques served as such. The many small plates of the mosaic, no matter how accurately fitted, would have formed a poor reflecting surface, yielding a much broken-up image...[However,] nothing produced in aboriginal America seems to us to rival these plaques in the matter of skilled and meticulous craftsmanship. (Kidder et al. 1946, 130 and 131)

cut-away parts on one of these were 'filled with brilliant red, apparently cinnabar mixed with some adhesive substance to give it body' (Kidder et al. 1977, 130).

> The entire back [of another plaque] had been *coated with stucco* upon which the *decoration was drawn in thin black lines*, but the paint had so much 'body' that the *lines seemed to have stood out in bold relief.* Thick red, black, and perhaps green paint was then apparently filled in between the lines, which, if we observed correctly, *functioned as do the metal strips in true cloisonné.* (Kidder et al. 1977, 130)

The appearance of paint cloisonné not in Mesoamerica, but in Arizona and New Mexico was unexpected. At Snaketown, Arizona, paint cloisonné is inlaid 'on the beveled rims or on the backs of sandstone plaques which bear pyrites mosaic' (Ekholm 1942, 94-95).

Stripped Painting

Yet another method of producing designs on lacquered pottery was practiced—a procedure that has been termed 'stripped painting.' If the successive layers of lacquer are applied in different colors, an outlined portion of the topmost layer can be scraped away, thus exposing the delineated design in a contrasting color on an underlying background. Technically, 'this is not paint cloisonné, *since the cut out area is not filled with paint*' (Kelly 1947, 145). The cutting-out procedure is 'the essential operation in paint cloisonné' (Ekholm 1942, 95).

Lacquering in South America

In South America, lacquer was applied almost exclusively to artifacts made of vegetable material. In 'western South America the practice of lacquering wooden objects extended all the way from Ecuador to northern Argentina' (Nordenskiöld 1931b, 96). Most of the surviving lacquered artifacts are wooden cups, called *keros* or *queros*—a word of Quechua origin. Although this lacquer has a '*resinous vegetal base*' (Lothrop 1956, 233), generally the plants from which such drying-oils came have not, as yet, been scientifically identified.[7]

In Peru, the lacquering of cups seems to have begun only a century or so before the Spanish conquest. However this may be, the craft reached its highpoint under Inca artisans in Andean Peru, and designs on keros there '...are among the finest products of Inca art, and are worthy of comparison with the best work of the Mexican codices' (Rowe 1963, 245; some are illustrated in Linné 1949a). In 17th century lowland Peru, lacquered drinking cups continued to be made, sometimes from dried gourds, but more commonly of wood:

> On the outside, they are covered with a *certain shiny varnish and decorated in a variety of colors with various figures and paintings*; and these vessels of wood are called *queros.* (Cobo 1956, 2:242-243)

Some keros are decorated with paint cloisonné, but it is uncertain whether such cups antedate the arrival of the Spanish: 'One pre-conquest specimen from Pachacamac has *incised* representational designs *filled* with [white] pigment...' (Rowe 1961, 339). In any case, historical native peoples in South America have made paint cloisonné artifacts, and sources of their color have been identified.

> The wood is *engraved with figures subsequently filled with lacquer* which is...coloured with lead oxide [i.e., litharge, a yellow-orange pigment], antimony oxide [of a whitish color], ferrous oxide [black], finely ground turquoise [greenish-blue] and malachite [green], finely washed kaulin [i.e., kaolin, a white clay mineral], etc. (Nordenskiöld 1931b, 96)

An almost complete set of a lacquerer's equipment has been found in a grave at Ica in lowland Peru (Nordenskiöld 1931b, 98-99; the set is illustrated on p. 97). The set contained gourds; lumps of ready-made lacquer, wrapped in leaves; hide bags, partly filled with mineral pigments; and mollusk shells used as mixing saucers. Following is the result of an analysis made of minerals found in the set and the colors derived from them: red, from cinnabar; brick-red, from a mixture of cinnabar, gypsum, 'and some glutinous substance'; dark-red, from gypsum mixed 'with a very dark-coloured sand'; pale rose, from powdered limestone; orange, from realgar; yellow, from orpiment; grayish-green, from finely powdered obsidian; and dark green, from malachite. This represents a surprisingly wide range of inorganic pigments—perhaps surpassing those used in Mesoamerica where lacquering was more elaborate and apparently older.

In addition to cups, other wooden articles were lacquered; for example, a number of lacquered wooden artifacts (probably either paddles or spades) were found in the desert sand at Ica. One such artifact, decorated with bird figures, was 'both coated and *inlaid* with lacquer'; upon chemical analysis of the lacquer, cinnabar was identified as the source of red in the lacquer—and orpiment, as the source of yellow (Nordenskiöld 1931b, 95). (Cinnabar-red, vermillion, was also a favorite in ancient Chinese lacquering.)

The following passage deals with an inlaid lacquered cup made in more recent times at Tisaleo in the Ecuadorean Andes:

> It is made from the wood of the guayacán tree (*Guaiacum officinale*). The inlays consist of lacquer in six different colours: yellow, light-green, dark-green, light-brown, red and black. The resin from which the lacquer was prepared had been obtained by the Indians *from a plant* known as *mopa-mop.* (Nordenskiöld 1931b, 96-97, citing Jijón y

[7] The lacquerers of ancient Eastern Asia—for instance, those of Jomon—also used a resinous vegetal base for their lacquer. The Chinese, too, have long used the resinous sap of the cultivated sumac tree, *Rhus vernicifera*.

Caamaño and Larrea; except for its native name, the last-named plant is unidentified)

In the 16th century, the lacquering of gourds was probably practiced more widely than is generally assumed in the New World, since there may have been no interruption of the custom of lacquering between the two continents: In the late 17th century, it was noted that the 'Darien gourd is painted and much esteemed by the Spaniards' (Wafer 1934, 59 and 103). In the same century, the following comment about tribes on the north coast of Colombia may also be a reference to lacquering: 'with pitch [i.e., *betún*] they paint beautiful tree gourds' (Simón 1882-1892, 5:172).

In inland Colombia at Pasto, Moguex and Páez, 'gourds were ornamented with a special 'Pasto varnish'' (Steward 1963, 5:727). This may well be the 'Pasto lacquer' described in the 1880s by the great archaeologist, Max Uhle, as follows:

> The most important material is the resin of an opposite-leaved plant which apparently grows mainly in the *tierra caliente* of Mocoá. The resin is secreted in drops in the leaf axils of branches where a new leaf is beginning to form, disappearing with the growth of the young leaf if not harvested.
>
> The Indians collect this resin and bring it in small balls to the *tierra fria* of Pasto [2522 m. above sea level] where its processing takes place. In the *tierra caliente*, the resin is supposedly too soft for processing. Yet another product for the lacquer industry comes from *tierra caliente*: the latex *vija,* supplied by a species of rubber tree.
>
> The lacquer is made in Pasto as follows: The resin is divided into small lumps, chewed in order to extract impurities, and heated in warm water. [Resins are insoluble in water.] Then, with the help of two fingers of each hand and the teeth, it is swiftly stretched over a warming charcoal fire to form a membrane which is applied to the warm vessel to be decorated, to which it quickly sticks.
>
> This first lacquer cover is placed on all vessels as a foundation. But on top of this a patterned decoration is added, for which, these days, the following colors are available: golden-yellow, orange, light gray, copper-red, various other reds, blue, green. Red is produced by adding achiote (*Bixa orellana*), and the latex *vija* in order to fix the colorant during chewing, or by the addition of *pacau* (a colorant scraped off the backside of certain leaves). Blue is displayed by adding indigo, green by adding indigo and saffron [the latter, an introduced plant], light gray by adding white lead. Golden-yellow is produced by laying a piece of silver leaf between a simple lacquer membrane and one colored with saffron, this three-layered membrane being applied like a single layer and adhering just as well...
>
> The papery substances are laid on the rewarmed vessel according to the colors desired, and the required outlines are carved out on the bowl. (Uhle 1889-1990, 2-3; translated from the German)

(Since the membrane clearly consisted of the resin, the 'papery substances' may be still-unused pieces of the membrane, colored as described.)

Lacquering, or a procedure very like it, was described in 1849 by the naturalist, Henry W. Bates, at Brevis on Marajó Island, near the mouth of the Amazon River:

> A few Indian families reside here, who occupy themselves with the manufacture of ornamental pottery and painted cuyas [tree gourds], which they sell to traders or passing travellers. The cuyas—drinking-cups made from gourds—are sometimes *very tastefully painted.* The *rich black ground-colour is produced by a dye made from the bark of a tree* called Comateii, *the gummy nature of which imparts a fine polish.* The yellow tints are made with the Tabatinga *clay*; the red with the seeds of the Urucú, or anatto plant; and the blue with indigo, which is planted round the huts. The art is indigenous with the Amazonian Indians, but it is only the settled agricultural tribes belonging to the Tupí stock who practise it. (Bates 1962, 138)

(The clay called 'Tabatinga,' according to Bates, is 'in places stratified alternately pink and yellow' and occurs frequently throughout the Amazonian region.)

Another site of probable lacquering in the 19th century was Montealegre on the Tapajos River, affluent of the Amazon. Decorated calabashes were made there in great quantities.

> They are very neatly finished, scraped thin, and either stained of a shining black or painted in brilliant colours and gilt. The designs are fanciful, with sometimes figures of birds and animals and are filled up with much taste and regularity. The Indian women make the colours themselves from various *vegetable juices or from the yellow earth*, and they are so permanent that the vessels may be constantly *wetted for* a *long time without injury*. (Wallace 1890, 99)

Sometimes a simpler kind of lacquer is made—especially in South America—which has no mineral constituents in it: tree gourds and other woody surfaces are processed with only organic materials. Several examples of this practice are described below.

Among tribes on the upper Rio Negro in Guiana, the

> ...lacquering is done in an unusual way. The calabash after being well smoothed on the inner surface and washed down with a decoction of *carayuru* (*Bignonia*) leaves [i.e., carajura leaves] is turned upside down over some *cassava* [i.e., manioc] *leaves* sprinkled with human urine, where it remains until such time as the inside becomes *black* and *shiny*. (Roth 1929, 302)

As Salser reported more recently, women of the Cubeo tribe in northwest Amazonia 'blacken *lacquered* gourds' very similarly. The tree gourds are halved, dried, and their insides thoroughly smoothed. Manioc leaves 'are plucked, sprinkled with water and placed in a small basket' to decompose. A woman chooses an (unidentified) species of tree, growing 'in a very old second growth forest...' When, with her machete,

> she had scraped off a double handful of *the resinous bark*, she took *some leaves from a nearby vine* [also unidentified], *wrapped the sticky resin in them* and returned to the village...

To make the lacquer, water was added and the mixture was squeezed through a 'cloth to eliminate all particles of bark that mar the surface of the painting.' The reddish lacquer was painted on the inside of several gourd halves; then

> the gourds were laid upside down on the manioc leaves...When examined the following day, it was found that...[the] *lacquer had turned a deep purple, almost black color*...As an experiment to determine what causes the gourds to darken...a gourd was dried without placing it over any [manioc] leaves...[The gourd] has retained the *reddish color* of the lacquer. (Salser 1975, 243)

Likely, the 'vine' in which the resinous tree sap was wrapped was the carajura liana. As mentioned above, carajura grows spontaneously (or is planted) next to a tree, often in second growth forest. No other constituent of the mentioned lacquer gives a reddish color. Probably both the carajura leaves and the decomposing manioc leaves caused the gourds to blacken.

The Chivaro, a tribe of the Peruvian Amazon, decorate their wooden spearshafts with a gleaming lacquer of red, black, dark-brown and white colors. The basis of this lacquer is also vegetal, being derived from

> *a tree whose identity is unknown to me*...Were it not for the simplicity of the design (stripes, wavy lines, diamond-shaped patterns), a spearshaft segment could well be taken for a product of Japanese art. (Tessmann 1930, 738)

A viscous oleoresin such as copaiba, which is obtained from tropical South American trees of the leguminous genus *Copaifera*, may have been used to make such lacquers.

CHAPTER 16

METAL–WORKING AND METALLURGY

Metallurgy, according to common usage of the term, includes such processes as melting a metal or smelting its ore. Defined in this way, metallurgy is distinguished from the mere shaping of native metals by such stone-age skills as hammering, cutting, and abrading. Athough theoretically smelting can be carried out in a hole or trench in the ground, baked-clay containers or implements are usually needed. So far, there is no archaeological evidence that metallurgy was practiced in the New World in pre-ceramic times.

In contrast to several other chemical arts, metallurgy seems to have been almost exclusively the province of males. Although pottery was chiefly made (and perhaps invented) by women, metallurgy 'lent itself directly to war [largely a male enterprise], in a way that ceramics could not' (Wertime 1973a, 876).

Because native ceramic industries still survive and the practices of historical native potters have been recorded, it is possible to analyze the making of prehistoric pottery. This, however, is not the case with native metallurgical skills and industries which collapsed shortly after the Conquest. Unfortunately for for these crafts' survival, native artisans did much of their best work in gold, silver, and copper—and the two former metals were especially valuable to Europeans. With remarkable zeal and thoroughness, the conquistadors sought out, confiscated, and melted-down artifacts made of these precious metals. Most of the surviving metal artifacts have been found by *guaqueros* or by archaeologists.

Until the Spanish conquest, metallurgical knowledge and skills were spread widely over a number of more-or-less discontinuous regional centers, each of which contributed technical or other innovations. The primary center was located in coastal Peru, Andean Bolivia, and northwestern Argentina; a second center was in Ecuador; Colombia and Panama was a third regional center; farther north, there were two other centers: western Mexico and the Huastec area of eastern Mexico.

Coastal Peru and the Neighboring Andes

The following are the main metal-working and metallurgical processes known from the primary center:

Hammering (Repoussé) and Annealing

While hammering and annealing metals do not fit the strict definition of 'metallurgy,' these methods were much used by Peruvian metallurgists (as they are by the most skilled of modern metal workers). In Inca times, there were several professional occupations, among them

> ...weavers, stonemasons, and *silversmiths*, which were learned and practiced by persons who dedicated their entire lives to them and, as has been said, practiced them in the service of the nobility. (Cobo 1990, 240)

Although working with hammered metal sheets was a specialty of Peruvian artisans, their tools were surprisingly simple; for instance, the hammers—made usually of stone, but sometimes of metal—were without handles.

> I have seen dinner services *made with the use of pieces of copper and two or three stones*, so finely worked, and the goblets, platters, and candelabra all embossed with leaves and designs, that master workmen would have their work cut out for them to do as well with all the instruments and tools they have. (Cieza de León 1959, 176)

To pound silver into thin sheets, the Inca workers

> used hard stones, of a color between green and yellow as anvils. They planed and smoothed them against one another; and esteemed them highly since they were very rare...Despite these handicaps they executed marvellous works... (Garcilaso de la Vega 1966, 130-131)

Such stones were of varying hardness and weight, probably because of their use for different hammering tasks.

Among the hammer stones unearthed by archaeologists, two made of rhyolite (a fine-grained igneous rock) are

> smooth as ivory and are lovely things to handle. One can well understand why their owners 'held them in high esteem'. They exhibit nine and twelve flat working surfaces of varying shapes and sizes. It is clear that control of these hammers could have been mastered only after a long apprenticeship. Placed in inexpert hands, it would not be possible to fabricate the paper-thin metal artifacts found... (Lothrop 1950a, 161)

In Peru, however, the shapes of hammers and anvils used—as well as the materials from which they were made—varied from place to place. Another archaeologist found what

> appears to be a complete gold worker's toolkit... [consisting of] three small cylindrical hammers [cut from granular basalt, fine-grained sandstone, and garnet-quartz hornfels] and a larger, almost mushroom-shaped, evenly worked anvil [made from a fine-grained greenish porphyry]

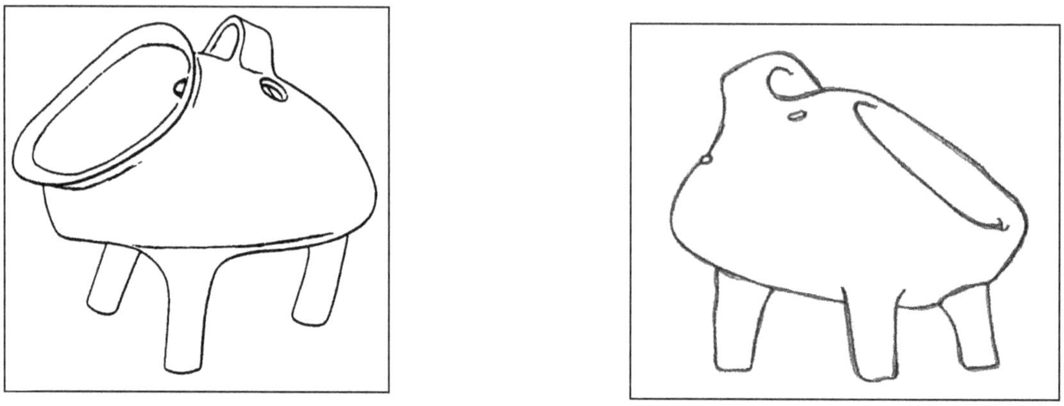

Left: Three-legged ceramic 'brazier' from Machu Picchu:Handle band-shaped, attached vertically to top...Three openings or vent-holes in top...Interior and exterior much fire-blackened. Height: 17½ cm; width: 15 cm; depth: 18 cm...These braziers were probably used for re-heating or annealing small pieces of metal, presumably [because of their small size] by means of a charcoal fire (Bingham 1962, 268-269). This artifact is also said to have been used for heating metal (Rutledge and Gordon 1987, 593).

Right: A comparable three-legged heating container from Sacshuaman, north of Cuzco (Julien 2004, 11) which is said to be a maize toaster. A very similar ceramic utensil (but without three legs) is shown from Ollantaytambo; this utensil was only 10 cm in height (Gibaja O. 2004, 187).

Figure 44. Peruvian Ceramic Heating Containers

> ...the anvil had a broad slightly curved platform...All three hammers contrasted with the anvil in that their striking surfaces were flat, while in the other the platform was slightly curved...Although not strikingly different in size, their [i.e., the hammers'] differences in weight suggest that they may have been used in sequence as the foil became progressively thinner. (Grossman 1972, 274)

The anvil was probably held vertically between the knees. The presence of the foil 'tells us that these pottery-producing people possessed gold foil as early as 1500 radiocarbon years B.C.' (Grossman 1972, 275).

Some of the finest pieces of Peruvian and Ecuadorian metal work are ornamented with repoussé, designs in raised relief on a metal sheet. Repoussé work is produced by hammering on the metal's reverse side; chisels and punches—made either of stone or of gold and copper *and their alloys*—were also used by the Andean artisans. (One might question whether tools made of metals would be sufficiently hard for this task, but as noted below alloys may be much harder than their constituent metals.) Repoussé work was also common in coastal northern Ecuador.

When some metals are hammered, they harden, become brittle, and crack. Annealing is a process by which they are made malleable and ductile again. This is done by first heating the metal, then allowing it to cool slowly. After annealing, the metal can be successfully hammered again.

The process of annealing metals may well have grown from the ancient practice of heat-treating stones to be worked for arrow points: 'Heat improves the chipping quality of certain kinds of stone, notably chert, which was commonly used [for example] by the Hohokam' (Haury 1976, 273). Some 'American Indians, and possibly other prehistoric peoples, annealed their flints and other stone materials to as high as 300°F [148°C], again paving the way for annealing of metals' (Wertime 1973a, 886, ftn. 48). In the Great Lakes region of North America—outside the New World's area of true metallurgy—'spearheads and knives [made from native copper] evidence an advanced annealing and reworking technique' (Wertime 1973a, 880). Ordinary wood fires provide sufficient heat for annealing; charcoal is not required.

Pure gold, the most malleable of metals, 'is the easiest metal to work as it remains soft when hammered and thus does not need to be annealed' (Root 1963, 211). However, most native gold contains up to ten percent silver, and so annealing is necessary.

Silver, though next to gold the most malleable of metals, hardens when worked; this is true of copper as well, which is also quite malleable. In annealing, silver and copper are heated to above 300°C. On the other hand, such heating 'has the opposite effect on [such alloys as] tumbaga [copper + gold] and bronze [copper + tin]. Annealing hardens them' (Root 1963, 217).

Although annealing native metals may have been only a preliminary step in the development of true metallurgy, the procedure requires special care: while temperatures must be raised high enough to reverse the changes in internal crystal structure made by hammering, they must be well below the metal's melting point. As smiths have long experienced,

> ...if certain alloys were overheated, they got 'burned', and no amount of heat treatment would restore their malleability. All the smith could do [in that case] was to abandon the work and melt it down. (Easby 1956, 406)

Melting and Smelting

As mentioned earlier, it is often difficult to separate physical from chemical processes. For example, melting (making a solid into a liquid by applying heat) is mainly a physical process; metals are made liquid in order to alloy them or to cast them with molds. Smelting, on the other hand, is more definitely chemical: by the application of heat and the use of fluxes, ore-minerals are converted to metals. (Fluxes are substances added to ores in the smelting furnace to promote the fusion of minerals, to help separate metal from slag, or to prevent the formation of undesirable oxides: 'in general deliberate fluxing was at the heart of metallurgy' [Wertime 1980, 16].)

Apparently metallurgy and mining had special religious significance for the Incas: 'Those who went to the mines worshipped the hills where they worked and the mines themselves' (Cobo 1990, 45). A symbolic smelting of ores was part of their religious ritual:

> They took two braziers made of silver, copper, or clay, which were about the size of a large still without a beak that had many openings all around it and another larger opening in the top from where the flame came out. The braziers were placed facing each other, full of slivers of wood soaked in fat [a mixture probably easier to ignite than charcoal]. The braziers were lit up, and the attendants blew into them with tubes.... Around these braziers many vessels made of gold, silver, wood and clay were set with various kinds of food and drink. Later the principal attendant, with the others, chewing coca, first chanting then weeping, started with words that they knew for this purpose...The spirits were invited to come to the banquet that had been prepared for them... (Cobo 1990, 169)

Smelters

According to historical and archaeological sources, two general kinds of forced-air smelters were used by Andean smiths: natural-draft and blowpipe. (Prior to the Conquest, bellows, used by many Old World smiths, were unknown in the Americas.) Without a source of forced air (either natural or artificial), metals such as copper and gold could not be liquefied: the average temperature of a charcoal fire is well below the melting point of copper (1083°C) or gold (1063°C), but with the application of forced air charcoal fires can attain a temperature of well over 1000°C.

Natural-Draft Smelters

One kind of natural-draft smelter was a 'wind furnace', called a *guayra* or *guaira*. This ingenious, but rather complicated device was unknown outside the Andean area. Made of baked clay and portable, the guayra was cylindrical, stood upright (about a meter tall), and was perforated with many holes. The diameter at the base was about a third of its height, but increased toward the top; the sides were about a centimeter thick (Boman 1908, 2:546-547). The charge (ore, charcoal, and flux) was placed in this furnace through the open top. Then the guayra, dependent on strong up-drafts of air, was carried by Inca smiths to windy slopes, and the charcoal was ignited.

The Spanish used Indian laborers and methods when they began working the famous silver mines around Potosí, located in the Bolivian Andes at more than 3940m, almost 13,000 feet above sea level. Cieza de León describes the guayra and its use at Potosí in the mid-16th century. The Indians

> made *certain molds of clay*, something like flowerpots in Spain, having many air holes. In these molds they put *charcoal, with the metal on top*, and they are then placed on peaks or slopes where the wind is strongest, and the silver is extracted...they call these molds *guaires,* and in the night there are so many of them on the fields and slopes, they look like festival lights. (Cieza de León 1985, 375; the guayra is also described by Garcilaso de la Vega 1966, 538)

The following description, written about a century later (in 1640 A.D.), is more complete and accompanied with an illustration:

> The natives of this Country, who have not yet gotten to the point of using our Bellows, employ, for smelting, furnaces called Guayras (Wind Furnaces); the same are still used in this Imperial Village [i.e., Potosí], and in many other Parts. They...[have] the walls perforated with many holes, through which Air enters when the wind blows, at which time alone they can smelt. *From the lower side of each one of these holes there projects a piece like a small Ear, which holds lighted Charcoal, so that the entering air will be heated.* They are placed on high locations where wind is usually blowing (Barba 1923, 197-199.)

According to one chronicler, a cruder version of the guayra which did not require charcoal was also used around Potosí in the 16th century. It was made without using baked clay or adobe blocks and consisted simply of a small stack of rocks. At the highest altitudes,

> ...they make certain small furnaces *by stacking loose rocks one on top of the other without clay;* these are *hollow like little towers* and as much as *two spans high.* In them they place the metal ore with the *dung of their cattle* [llamas or alpacas] and a little kindling because *they lack charcoal.* The wind, blowing through holes between the rocks, smelts the metal (Capoche 1959, 109-110).

(When he says that the Indians 'lack charcoal,' Capoche means that they lacked suitable fuel for making it, not that they didn't know how.)

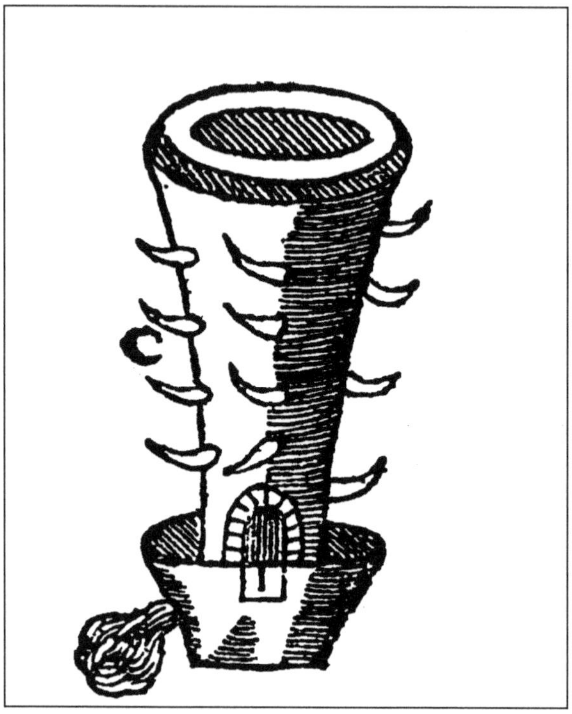

Figure 45. An Inca Guayra (Barba 1923, 199)

At high altitudes in the Andes certain plants grow that contain compounds with higher calorific values (terpenes, resins, etc.) than common tree wood:

> Nature has provided in great abundance what we call Yareta [*Azorella* sp.], more suitable for burning than ordinary Fire-wood because it is so greasy, and *full of resin.* (Barba 1923, 173)

In addition to the guayra, some claim that a different type of natural-draft smelter, using tunnels or shafts to channel the wind, existed on crests of the Andes. The

> pre-Inca ruins of Curamba...stand on an elevation of approximately 12,500 feet [3788m]...*it was one of the Inca centers of wind ovens for smelting gold, silver, and copper ore.* At the highest part of the crest, overlooking Curamba, are numerous oval-shaped wind ovens eight feet [2.42m] in diameter with walls two feet [.6m] thick; the mouths of the tunnels face northeast in the direction of the winds from the Amazon. (Cieza de León 1959, 133, according to notes written by V. W. von Hagen)

In the Atacama region of northern Chile, ruins of copper smelters have been found. Dated 'to the first centuries B.C.,' these smelters are 'the earliest furnace ruins' reported so far in the Americas. These 'furnaces were small, in the vicinity of 50 cm in diamter'; they may be 'a type of *natural draft furnace*, and perhaps ancestral to the well-known *huayra* [guayra] furnace of historic times.' 'Laboratory tests are positive in identifying the...*slag* [found with the smelters] as a copper smelting by-product.' The 'composition of the slags shows that a temperature of 1100° to 1250°C was attained' (Graffam et al. 1996, 101 and 111).

Blowpipe-Smelters

In one kind of smelter, the air was artificially supplied by blowpipes. (The blowpipe is a tube through which intense jets of air are forced into a flame in order to concentrate and increase its heating action.) Metalsmiths frequently placed flame-resistant ceramic tips, nowadays called tuyeres, on the distal ends of cane tubes.

Long before the Inca empire arose, probably Mochica smiths (around 550 to 700 A.D.) placed ceramic tips on tubes when smelting (Shimada 1994, 96, 203 and 272). Sicán smiths (900-1100 A.D.) in the Sechura Desert of northern Peru definitely placed ceramic tips, tuyeres, on perishable (probably cane) tubes when smelting copper ores.[1] The air stream from the blowpipe shaft was compressed by a small hole in the tuyere:

> Although the overall size and shape of the tuyeres changed over time, *the diameter of the hole at the distal end remained constant,* at about eight millimeters. (Shimada and Merkel 1991, 84).

Among some Inca metallurgists, the stream of air in the tube was bent, apparently at nearly a right angle: the

> pipes were *blocked at one end,* but *had a small hole through which the air came out compressed and with greater force. It might be necessary to use eight, ten, or twelve* [canes?] *at once.* They walked round the fire blowing, and still do today, for they do not like to change their habits. (Garcilaso de la Vega 1966, 131)

Because the operators were on their feet, shafts of these blowpipes were long.

In the blowpipes used by Colombian smiths, the air stream was compressed in a different manner: metals were melted 'with the aid of ceramic blowpipes; *the interior of the pipe narrowed toward one end to increase the force of the air as it left the tube*' (Plazas and Falchetti de Sáenz 1979, 41).

According to Sir Walter Raleigh, in northwestern Venezuela at the end of the 16th century, the natives used *canes* as blowpipes—though Raleigh doesn't say whether the tubes were tapered:

> they vsed a great earthen potte with holes round about it, and when they had mingled the gold and copper together, they fastned canes to the holes [fastened with tuyeres?], and so with the breath of men they increased the fire till the mettell ran, and then they cast it into moulds of stone and clay, and so make those plates and Images. (Raleigh 1848, 96)

[1] Similar tuyeres were used on *metal tubes* in Egypt as early as 2500 B.C. Illustrations there show smiths with clay–tipped blowpipes blowing air into a flame for melting metals (Forbes 1967, 578–579). Blowing on flames through canes or reeds may have been practiced much earlier (even in the Paleolithic), but such tuyeres were non–existent in pre–ceramic times.

Figure 46. Sicán Tuyeres (Shimada and Merkel 1991, 84)

Some of the blowpipe-tubes used by the Incas were made of metal. The tubes were 'about the size and shape of an harquebus [a small caliber long-gun, dating from about 1400]. These tubes were made of copper from the middle upward, and the other half was silver' (Cobo 1990, 169); they were 'three or four spans long' (Cobo 1990, 239). Since silver melts at about 960.5°C, obviously such tubes could not be used for smelting most copper sulfide ores or to melt pure gold or silver. One would think that such metal tubes would have some sort of non-metal mouthpieces. Otherwise, the heat conducted through the tubes might have burned those using them, but that metal tubes (especially if long) would become too hot to handle may be an unwarranted assumption (Donnan 1973, 296).

Although some blowpipe-smelters were portable,[2] others were immovable. An example of the latter was made in north Peru by Moche (as well as Sicán and Chimú) smiths, who dug trenches in workshop floors to make subterranean furnaces (for illustrations see Shimada et al. 1983, and Shimada and Merkel 1991). In the furnace, they put a charge of copper-oxide ore, first-rate algarrobo charcoal, and iron oxides as a flux. They heated the charge by blowing on the flame with canes. Despite all this, the furnace

> did not produce enough heat to liquefy the charge completely. The charge was instead transformed into a thick, viscous slag containing *prills*, droplets of nearly pure [copper] metal several millimeters in diameter. (Shimada and Merkel 1991, 83)[3]

After the slag had cooled, it was crushed with large rocker-pestles[4] and hand-sorted to extract the embedded prills; these were then remelted into ingots.

Inca smiths at Quito smelted gold and silver; the following early account describes one of the ways that this was done:

> In the first place, when they smelt gold and silver they put the metal in a long or round crucible, made of a piece of cloth smeared with a mixture of earth and *powdered charcoal*; when the crucible is dry, they put it in the fire, filled with the amount of metal that will fit into it; *then with five or six tubes of cane* (more or less), they blow until the metal is smelted and rid of impurities; later, the goldsmiths sitting on the ground, with specially-formed black stones, help each other make (or to speak more correctly, *used to make* in the period of their prosperity) whatever they were commanded to produce—that is, hollow figurines, vases, sheep [llamas], jewels... (Benzoni 1985, 118-119)

Identification of ores by color and with blowpipes:

Ancient Peruvian miners recognized and named numerous minerals; no doubt this identification was aided by handsorting and observing the color of each one. More important for identification, however, is the color of the mineral's powder: when the surface of a mineral-making ore is bruised by another rock or a hammer (or reduced to a powdered state with a mortar and pestle), the resulting powder's color is generally lighter than, and sometimes quite different from, the unground ore itself. (For determining color, the modern mineralogist scratches the mineral on a streak plate of unglazed porcelain.) This color is helpful in distinguishing dark-colored minerals—especially, the metallic oxides and sulfides of the heavy metals mined as ores.

Clearly, ancient South American smiths made various kinds of blowpipes. For instance, along the north coast, they must have had 'a fine blowpipe [for soldering], whose jet would have allowed' only a tiny area on a small artifact to be heated (Lechtman 1988, 368). Although

[2] A remarkable example of the former has come to light, modelled apparently by Moche potters around 600 A.D.: a number of seated human figures surround a hollow, dome-shaped furnace (clearly, some sort of a capped melting or smelting furnace) blowing through pipes. (See Donnan, 1973 for a photo and a description—as well as a dicussion by Shimada 1994, 272). In the 17th century, a similar capped-furnace was still used by Peruvian Indians: such furnaces were called '*Tocochimbos* in this Province, and rich ore, in small quantity, is smelted in them. The Indians used them for *refining only*' (Barba 1923, 198–99). Apparently this furnace had been considerably modified by the Spanish, since it was sometimes used with a bellows.

[3] If, on the other hand, the smelting furnace can attain a higher temperature, the metal and slag will separate into separate layers, the slag floating on top. In this case, before it cools, the slag can be drawn off, leaving only metal. After cooling, the slag solidifies into a glassy waste–product.

[4] As mentioned previously, smaller rocker–pestles were used in South America and Panama for grinding maize, but in Peru and Bolivia larger rocking stones were also used as a mill for crushing ores:

> ...a boulder with a curved underside and a flat top—a half–moon shaped piece of stone to whose upper surface a beam was lashed, extending outwards from each side so as to allow two men, alternately pushing down on the ends of the beam, to rock the boulder to and fro, crushing any material placed under it. (Bakewell 1984, 15, citing Petersen 1970)

used for heating smelters by ancient metalworkers, such blowpipes are a rudimentary form of the mineralogist's mouth-blowpipe—that is, a tube narrowing toward the distal end (or having a small orifice at that end) to concentrate a stream of air into a flame. Since flame color is an identifying test for minerals, such blowpipes may have helped early metallurgists to recognize their ores.

Smelting Copper Ores

There are numerous copper ores available in the Andes, and New World metallurgy was principally based on copper alloys. The difficulty encountered in smelting a copper ore depends on its composition. Probably, among the copper ores smelted in pre-Columbian times are the following: (1) the oxides and carbonates (cuprite, Cu_2O, and malachite, $Cu_2(OH)_2CO_3$); (2) a simple sulfide (chalcocite, Cu_2S); and (3) a mixed sulfide (chalcopyrite, $CuFeS_2$).

(1) If the copper ore is malachite, a hydrous carbonate, it is first calcined (heated enough to drive off the water and carbon dioxide):

$Cu_2(OH)_2CO_3$ =	2CuO	+	H_2O	+	CO_2
Malachite	copper oxide		water		carbon dioxide

The resulting copper oxide can then be reduced by further heating with carbon: 'the temperature at which copper oxide can be reduced by carbon is well below 600°C' (Thompson 1958, 3); however, 'to obtain the reducing conditions necessary to divorce a metal from the oxygen with which it is combined in a simple oxide, it is necessary to have close and intimate contact *between the ore and the reductant, charcoal*' (Tylecote 1980, 183). (Ores were reduced by heating them with charcoal in both in the Old World and the New.)

(2) Copper sulfide ores were smelted, not only in the Old World but also, many believe, in ancient Peru. Fifty years ago, opinions were quite different. Sulfur had already been detected in a copper ingot from the Province of Ica, Peru—indicating that it 'was not produced by melting native copper nor by smelting an oxidized ore' (Caley and Easby 1959, 63). This sulfur could conceivably have been a contaminant, but more recent analyses have shown that various mainly-copper artifacts contain not only sulfur but arsenic and antimony, common constituents of sulfide ores.

> That sulfide ores were smelted is indicated both directly and indirectly by the analytical data. The presence of small proportions of residual sulfur in four of the samples is a direct indication, and the presence of considerable arsenic in all the samples, along with antimony, bismuth, or both, in most of them is an indirect indication. (Caley 1973, 57)

Copper sulfides (for example, chalcocite, Cu_2S, and chalcopyrite, $CuFeS_2$), unlike the oxides, carbonates, and sulfates, cannot be reduced to the metal by simply heating with carbon. A preliminary 'roasting,' with limited access of air (oxygen), is necessary for sulfides. Consequently, smelting copper sulfide ores is a two-step procedure in which the sulfide is first converted to an oxide and then to metallic copper:

$Cu_2S + 1\frac{1}{2}\,O_2 = Cu_2O + SO_2$

$Cu_2S + 2Cu_2O = 6Cu + SO_2$

(This formulation of the process was kindly suggested to me by Dr. Hanns Ahrens.) Since step 1 produces more heat than is needed in step 2, the production of heat must be controlled carefully and the procedure needs much skill on the part of the metalworker.

After the ore is crushed and mixed with charcoal, the charcoal is lighted, and all that is needed

> ...in the way of temperature is sufficient heat to ignite the sulfur and burn it off, 350°C. or more depending on the composition of the ore and how finely it is crushed. *Too much heat is fatal, for the metal in the ore will begin to fuse, the roasting action will come to a stop, and the elimination of the sulfur will be prevented.* (Caley and Easby 1959, 61)

(3) The operation of smelting chalcopyrite ($CuFeS_2$) is yet more complicated, for 'the amount of iron normally far exceeds that of copper, and it is the removal of this iron which is the real problem' (Thompson 1958, 5).

Cu_2S and FeO are formed by roasting the ore.

> Fortunately for the early workers, iron has a greater affinity for oxygen while copper has a great affinity for sulfur...Consequently where, as here, insufficient oxygen is present, it is attracted to and unites first with the iron, while the insufficient amount of sulfur present unites by preference with the copper. (Caley and Easby 1959, 61)

The FeO is removed as slag when siliceous rock or sand is added to the molten charge. Then the Cu_2S is converted into metallic copper by oxidation: $Cu_2S + O_2 = 2Cu + SO_2$ In order, however, to achieve the metal purity which was attained by ancient Andean metalworkers, it is probable that 'a series of repeated smelting operations were carried out to extract the copper' (Caley and Easby 1959, 62).

The 'roasting-reduction' process (essential for the smelting of sulfide ores) was probably employed in both in the Old World and the New.

> It cannot be sufficiently emphasized—and this is a fact that non-metallurgists are prone to overlook—*that there is an enormous difference between the simple melting of metal and the winning of metal by roasting or reduction.* (Bergsøe 1938, 23-24; italics in the original)

In whatever way these processes may have been discovered (or knowledge of them may have been diffused), archaeological evidence shows that both were practiced more than three thousand years earlier in the Old World than in the New: 'If the [pre-Columbian] Indians really understood how to extract metals from their *oxygen- or sulphur-containing ores*, their period of apprenticeship must have been a *fraction* of that passed through by [the Old World] Eastern peoples' (Bergsøe 1938, 24).

> A fundamental step in the whole history of civilization was taken when some early genius [in the Old World, and probably another in the New] realized that copper also occurred in the unweathered sulphide ores...[Thereafter, and before the 16th century,] processes involving *relatively very advanced metallurgical knowledge* were gradually devised. (Thompson 1958, 3-4)

'That sulfide ores were smelted in the Andes in Pre-Columbian times may come as a surprise to some' (Lechtman 1975, 10 and 11; Lechtman 1976, 1; Bruhns 1994, 177 and 393). And, in fact, an opinion persists that South American metallurgists did not smelt copper sulfide ores, 'because there was an abundance of oxidized ores available' (Patterson 1971, 316). For instance, at Batán Grande, a Middle Sicán site (900-1100 A.D., one of the most thoroughly convassed metallurgical sites in South America), it is contended that malachite was the principal copper ore smelted; there is said to be no evidence at the site to suggest roasting and smelting of copper sulfide ores (Merkel et al. 1994, 214 and 221; see also Shimada 1985, 372 and Shimada et al. 1983, 39).

The fuel used for guayras in the neighborhood of Cobres on the altiplano of northwestern Argentina was *llareta* or perhaps *taquia*, llama dung (see Lechtman 1976, 38-40). Such combustibles give sufficient heat to smelt the copper ore, chrysocolla, a hydrous copper silicate, $CuSiO_3{\cdot}2H_2O$ (Boman 1908, 2:540). There seems to be no record of either *llareta's* or *taquia's* being made into charcoal. Nevertheless, the latter was used for smelting silver during the Spanish colonial period: 'large mining centers such as Potosí [Bolivia] *annually consumed more than 800,000 loads of llama dung in the smelting of ore*' (Browman 1974, 194, citing Troll 1958, 30).

In the Bolivian and Peruvian Andes, silver occurs in numerous ores, including a silver-rich galena, PbS. Silver can also be found in 'the mineral cerargyrite (AgCl), which can be smelted to silver merely by heating to about 800°C when the chloride oxidizes and the oxide dissociates' (Tylecote 1980, 205).

In the Old World,

> a good deal of early silver was derived from the *cupellation* of lead. This process takes advantage of the fact that when silver-containing lead is heated (preferably above the melting point of silver, 960°C), *the lead oxidizes and in turn vaporizes, leaving silver behind.* Normally this would be done in shallow clay plates, or cupels. (Tylecote 1980, 206)

According to some authorities, 'silver was *not cupelled* in South America' before arrival of the Spanish (Patterson 1971, 361).

> Although it is certain that native silver and/or other silver minerals, such as cerargyrite (AgCl), were accessible to Pre-Hispanic miners, it is not generally understood that these minerals, although abundant, were disseminated in the iron-rich *pacos* [as they are called in the Mantaro Valley] and seldom found as solid masses. Thus it would generally not have been possible to extract the silver from these ores without the use of lead to collect and *dissolve* the silver during smelting. Lead, if not already present within the ore, would have been added in the form of lead, lead metal, litharge, or lead-containing slag. Subsequent refining, the separation of the silver from the lead, would take place [as in the Old World] with *cupellation.* Evidence of cupellation is always visible in the presence of residual lead in the silver. (Howe and Petersen 1994, 190-191; see also Lechtman 1976, 37)

Silver is usually found with lead and copper ores. Several early Spanish sources maintain that silver was extracted from lead ores in pre-Columbian times:

> The way the Indians refined and extracted the silver was by smelting, which means melting the mass of ore in a fire, a process that leaves the dross [waste products] on one side and separates the silver from the lead and tin and copper and other mixtures that it has.
>
> The richest ore is refined in those little wind ovens called *guairas*; this is the ore that contains the most lead, and lead [serving as a flux] causes it [the silver] to melt. To make it better the Indians throw in what they call *soroche* [or *suruchec*, galena?], which is an ore resembling lead. When subjected to fire the dross falls below, the lead and silver melt, and the silver [specific gravity 10.5] floats on top of the lead [sp gr 11.3] until it is skimmed off; then they refine the silver many more times. (Acosta 2002, 171 and 182).

In the central highlands of Peru, between 1000 and 1533 A.D., 'the chemical evidence strongly suggests that silver-lead ores were smelted and the silver then refined through the use of cupellation' (Howe and Petersen 1994, 194).

Alloying

In the New World, *tumbaga,* a mixture of copper and gold (and sometimes silver) was one of the favorite alloys used in making ornamental artifacts. Tumbaga (sometimes spelled *tumbago*) is a Malayan word, but the Spanish often called the alloy by the (probably Arawak) name, *guanín.*

Tumbaga alloys 'swept through the Americas from Peru to Mexico and were in common use in the entire [metallurgical] region when the Spanish invaded...' (Lechtman 1980, 293). Though the alloy was probably first made in Peru, tumbaga technology was even better developed in Colombia and Panama.

Alloys often have desirable characteristics that are lacking in their constituent metals; for instance, lower melting points and increased hardness. While the melting point of gold is 1064°C, and that of copper is 1083°C, the melting point of tumbaga is much lower. (The proportions of the alloy's constituents are important: a mixture with a melting point lower than that of any other mixture of the same constituents is said to be eutectic.) In addition to its low melting point, tumbaga has other advantages: 'pieces cast in tumbaga reproduce decorative details from the mold more finely than those made in almost pure copper or gold' (Plazas and Falchetti de Sáenz 1979, 42).

The Indians valued the alloy guanín (tumbaga) especially for the reddish color given it by copper:

> The gold-copper alloy was not to economize the gold, but for the sake of hardness, better and easier casting, and especially for the red-gold tones. (Sauer 1966, 246)

The Spanish (whose interests were concentrated on the alloy's other component) melted quantities of guanín artifacts in order to extract the pure gold. The copper content of the alloy varied, but was usually estimated as two thirds or more.

In both the Old World and the New, bronze was made, and in both hemispheres two different alloys were used: copper-tin and copper-arsenic. At 'about A.D. 900, a new alloy was developed by metalworkers on the north Peruvian coast—copper-arsenic bronze' (Lechtman 1988, 356). Traditionally only the copper-tin alloy is called 'bronze,' but the two have similar physical characteristics and many authorities call both alloys 'true bronzes.'

In both the Old world and the New, use of arsenic in the alloy was unquestionably intentional: 'This deliberate attempt to reproduce copper-arsenic alloy...represents a significant development in the ancient history of metallurgy' (Merkel et al. 1994, 203). Although copper-arsenic bronze was long used in northern Peru, copper-tin bronze was more common in southern Peru and Bolivia. (Since copper-tin bronze was preferred by the Inca, it eventually spread throughout their empire.) The mineral cassiterite, SnO_2, can be reduced to metallic tin by smelting. Then, the tin (which fuses at only 231.9°C) and the metallic copper can be melted together to form bronze. At Machu Picchu, pure tin was kept as a stock metal for making bronze (Rutledge and Gordon 1987, 582).

In the making of copper-arsenic bronze, such toxic compounds as arsenic oxide may have been partly controlled by Inca metalworkers (and probably by those of much earlier times), because they had discovered

> that smoke from any metal was bad for health, and thus made their foundries, large or small, in the open air, in yards or spaces, and never under roof. (Garcilaso de la Vega 1966, 131)

Inca metallurgists also made an alloy by adding native bismuth to their copper-tin bronze. Since there have been no previous reports of the intentional addition of bismuth to bronze, this alloy is apparently unique to Peru (Gordon and Rutledge 1984). A small knife, or *tumi*, with a cast handle was found at Machu Picchu: 'The average composition of the handle is 73 percent copper, *18 percent bismuth*, and 9 percent tin...The introduction of bismuth...[into the alloy] was used to facilitate the composite casting procedure by which the tumi was made.' The alloy used for casting the handle 'was deliberately made by the addition of *native* bismuth to molten tin bronze.' The addition 'of small amounts of bismuth to bronze helps make sounder castings'; it may also produce better adhesion to other objects because 'of reduced shrinking during solidification' (Gordon and Rutledge 1984, 585-586).

Along the west coast of South America, weights and measures, two important features in any chemical procedure, were used. For weighing, both the balance-beam scale and the steelyard were employed (Nordenskiöld 1931b, 54); rubbed stone and pieces of lead were used for the balances (Nordenskiöld 1930c, 218). Such scales were used for weighing silver and gold (for historical documentation, see Saville 1925, 273-274); whether they were also employed in metallurgic operations is unknown, but likely—for instance, in alloying.[5]

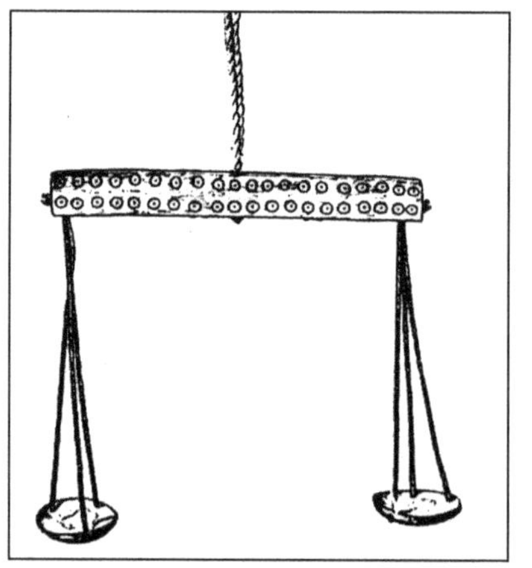

Figure 47. Peruvian Scales (after Baessler 1902-03, pl. 161, fig. 434)

Whether the *quipu*, a Peruvian calculating and recording device, consisting of a series of variously colored and knotted strings, was also used in metallurgical or other chemical procedures is uncertain—but not entirely

[5] With regard to the use of the balance–beam scale, there are

> exceedingly remarkable parallels between the higher civilizations of the Old World and that of western South America...[Among these are the] occurrence of the beam scale...and weights that are multiples of a unit. (Nordenskiöld 1931b, 46)

Despite this parallel, Nordenskiöld believed that the balance–beam scale was independently invented in both hemispheres, and that this discovery was 'suggested by some previous device. This, to me seems to have been the balanced double–load pole' (Nordenskiöld 1931a, 502). This explanation might apply to the balance–beam scale, but less so to the steelyard.

improbable.[6] At least, quipus are known to have been used for tallying *precious metals* and other commercial items:

> ...by *these knots* they [the Incas] kept the account of the tribute to be paid by the natives of that district in *silver*, *gold*, clothing, flocks, down to wood and other more insignificant items... (Cieza de León 1959, 174)

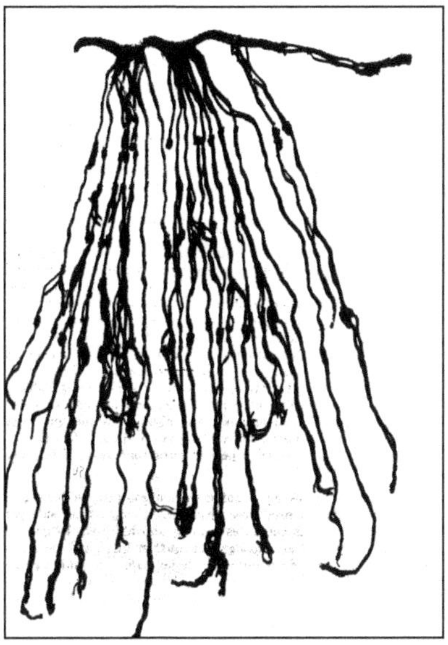

Figure 48. Peruvian Quipu (Museum of Archaeology, University of Cambridge, after Radicati di Primeglio 1980, 92)

Casting and the Lost-wax Process

In casting, an artifact is formed by pouring molten metal into a prepared mold and letting it harden. In the central Andes, most castings were in open molds, often simple two-piece molds made of fired clay. But closed molds were also made through the lost-wax (*cire perdue*) process, a method of casting metals by first making a model of the desired figure in wax and enclosing it in a clay shell. Then the wax is melted and drained out of the clay shell; the latter is used as a closed mold for casting the molten metal.

In Peru the lost-wax method was known, but rarely used, perhaps because stingless bees of the family Meliponidae are nearly absent 'from the highlands and west coast valleys of Peru...' (Bird 1979, 500). Nevertheless, the best examples of lost-wax artifacts have been found on Peru's north coast; while the age of these examples is uncertain, they appear to have been made after 1000 A.D. (Lechtman 1988, 348).[7] The lost-wax casting method was far better developed in Colombia and Mexico, as discussed below.

Soldering and Welding

Andean goldsmiths mastered both soldering and welding by 500 B.C. Because in coastal Peru and the central Andes, soldering and welding are of less chemical interest than other practices, they are not discussed here. Nevertheless, several are distinctive—for instance, the sweat-welding carried out on Peru's north coast (Lechtman 1988, 365 and 368).

Gilding

Several methods of gilding were practiced in Peru. Gilding is the art of applying a thin layer of gold over some baser metal, such as copper.

Depletion-Gilding

Depletion gilding, also known as *mise en couleur* gilding, is probably the best known of the gilding methods practiced in pre-Columbian America. When heated, a tumbaga alloy oxidizes, forming a dark layer of copper oxide at the surface. Much of this oxide can be removed (i.e., 'depleted') by an acid solution, leaving behind a surface layer of apparently pure gold content. The acid used on tumbaga in ancient Peru is unknown, but metalsmiths there were acquainted with acid solvents.

At a Sicán site, at around 1000 A.D., a gold-silver-copper (tumbaga) artifact was found that is markedly depleted in copper (with the enrichment of gold and silver) on the surface. The Sicáns would have had acids, including 'aging fermented beverages,' that 'would permit dissolving and removing the oxidized copper layer' (Gordus et al. 1996, 83 and 90). During the later part of the Chimú period, depletion gilding was practiced 'with consummate skill on the north coast of Peru' (Lechtman, Erlij, and Barry 1982, 9). Nevertheless, the process was probably most common in Colombia and Panama.

Gilding by Electrochemical Replacement

On the north coast of Peru, ancient Moche metalsmiths gilded (gold-plated) copper objects. These exceedingly thin-and-even coatings were not made by a depletion technique, nor were they 'by mechanically applying thin foils of metal to their surface'; moreover, the copper objects were not 'gilded by immersion into a bath of molten gold' (Lechtman, Erlij, and Barry 1982, 10-11 and 13).

The following electrochemical process would explain the gilding on a number of Moche copper artifacts: The metalworkers made use of '*naturally occurring* corrosive

[6] 'In Oceania, too, quipus are known, as e.g. from the Marquesas Islands, and it is usual to count quipus among such culture elements as the Indians have in common with Oceania' (Nordenskiöld 1930a, 211). Nordenskiöld, however, believed that independent invention explains the occurrence of the trait on both sides of the Pacific Ocean.

[7] In Mesopotamia and the Levant, the lost–wax method of casting was already well developed before 3000 B.C.—that is to say, several thousand years before it appeared in the New World. In China, this method of casting was not used until around the sixth century B.C.

minerals common to the dry environments' of Peru. The gold was gently heated for several days 'in an aqueous solution of equal parts' of common salt, NaCl; saltpeter, KNO_3; and potash alum, $K_2SO_4{\cdot}Al_2(SO_4)_3{\cdot}12H_2O$ (Lechtman, Erlij and Barry 1982, 26 and 15). However,

> the resulting gold solution is highly acidic and immediately attacks copper dipped into it...It was thus necessary to neutralize the gold solution before introducing the copper. Among various salts that can be used for this purpose, we have found that sodium bicarbonate [baking soda, $NaHCO_3$] is most effective. (Lechtman, Erlij, and Barry 1982, 15)

Reaction between ions in the solution and the copper artifact produces electrochemical plating 'without the use of an external battery or other source of electric current' (Lechtman, Erlij and Barry 1982, 24). (Silvering can be done in the same way as gilding.) Such 'plating through electrochemical deposition has never been reported for Andean artifacts (*nor for other ancient materials*, as far as we are aware)' (Lechtman, Erlij, and Barry 1982, 13).

Both depletion gilding and electrochemical replacement 'involve sophisticated chemistry, and Pre-Columbian surface metallurgy is surely as much chemistry as it is metallurgy' (Lechtman 1988, 373). In ancient Peru, this complex and technical information was transmitted verbally from generation to generation without the aid of written records.

From Peru, metallurgical knowledge seems to have spread northward to other geographical areas, in much the same way that similar knowledge was spread in the Old World (for example, from the Middle East to Northern Europe)—that is, by diffusion. (Here, diffusion is an anthropological term, meaning the transmission of learned traits from one culture to another.) In both cases, innovations were added along the way.

Ecuador

Ancient Ecuador seems to have been a pivotal area for the communication of metallurgical knowledge in the Americas. Apparently through this area, Peruvian metallurgical information reached highland Colombia. And through a maritime connection between Ecuador and Mesoamerica, evidenced by a number of shared cultural traits, South American metallurgy reached western Mexico.

The artifacts from southern Ecuador

> come mainly from tombs and burials...Several of them indicate some relationship to the later coastal cultures of northern Peru...Gold was beaten, hammered, welded, granulated; the smiths were aware of alloying and gilding, and the art of soldering was highly developed. During the Integration period, circa 800 C.E. [800 A.D.] to the Inca conquest, the lost-wax process appeared. (Rehren and Temme 1994, 267)

Lead Smelting

Lead ore is another sulfide that requires roasting before the smelting can be completed. Lead is never found in the native state, but almost exclusively as the sulfide ore galena. In northern coastal Ecuador, the

> Esmeraldas Indians employed charcoal and blowpipes. If one places a fragment of galena or pulverized galena onto a piece of charcoal and tries to melt it, the attempt will be unsuccessful. But if one continues patiently, the sulphur will gradually burn away and the lead oxide formed will then be *reduced* to metallic lead...With the apparatus at their disposal it would have been an easy matter for the Indians to procure *small* quantities of lead by *a combined roasting and reducing performed in the above way.* But procurring lead by the pound [without a furnace] is quite another story. (Bergsøe 1938, 24)

Although lead was widely known in pre-Columbian America, whole lead artifacts were rarely made (unlike in the Old World). Among the few examples are a lip plug from western Mexico (Caley and Easby, 1964) and a pair of nose rings from the Esmeraldas coast of Ecuador (Bergsøe 1938, 22).

Platinum Working

Platinum is among the heaviest of elements, being even heavier than gold and almost twice as heavy as lead. Thus, like gold dust, small grains of platinum were collected by panning stream-sands—that is, by a process of sedimentation. Although platinum is found scattered sparingly throughout the world, it was the Ecuadorian Indians of Esmeraldas who, in the early centuries A.D., used it in metallurgy. This was a remarkable technical accomplishment, because pure platinum has a very high melting point, about 1772°C; pre-Columbian metalworkers had no way of attaining such heat.

> Of course, the native gold will contain some silver and the native platinum some iron, both of which will lower the melting point...With a charcoal fire and blowpipe it is not normally possible to attain temperatures higher than 1400°C, while a more usual temperature would be expected to be in the range from 1000 to 1200°C [almost certainly over the melting point of native gold]. (Scott and Bray 1994, 304)

The Esmeraldas artisans had apparently learned to make platinum and gold into a homogeneous mass, through the sintering of refractory metals. (Sintering, a modern metallurgical term, means to break solid metals into very tiny pieces—that is, almost to powder them—and heat them together, upon which strong bonding takes place between the metal particles.)

Professor Paul Bergsøe was the first to describe the sintered gold-platinum alloys in the New World. He has reconstructed the process from the native artisans' ornaments:

> The small grains of platinum were mixed with a little gold dust and small portions placed upon a piece of wood-charcoal. When the gold runs it will coat the grains of platinum with gold...If the piece is now further heated by means of the blowpipe, let us say, the following will place: a portion of the fused gold permeates the platinum and simultaneously a little of the latter is dissolved in the molten gold...a pasty, sintered mass is formed. This mixture of gold and platinum can now withstand a light blow of the hammer, especially when hot. By alternately forging and heating it is possible gradually to build up an homogeneous mixture...(Bergsøe 1937, 24)

This practice was:

> independently reinvented [in Europe] only during the nineteenth century. This process also forms the basis of modern powder metallurgy, which has made possible the utilization of metals with extremely high melting points such as tungsten, carbide and titanium, essential to the production of advanced machinery such as jet engines which have to resist enormously high temperatures. (Emmerich 1965, 55)

Fusion (Wash) Gilding

Fusion gilding can be counted among the native arts that were lost well before the Spanish conquest; it is known only from archaeological objects. Although this process, which 'has not been identified anywhere else in the world' (La Niece and Meeks 2000, 233), was once used by Moche metallurgists in northern Peru, fusion-gilded artifacts come mainly from coastal Ecuador. Fusion gilding was first described from an analysis of these objects:

> The copper object was heated with the blowpipe on charcoal to a temperature of at least 850°C and then brought into contact with a melted alloy of gold and copper (preferably 20% copper). This alloy [a mixture of gold and copper at close to their eutectic proportions] fuses at a temperature 200° lower than the copper. This alloy will run rapidly over the copper just as tin runs over a clean copper surface. [Such 'tinning' is a common Old World process.] The use of the blow-pipe would make the use of flux unnecessary, as the reducing flame creates a surface *free from copper oxide*. After the copper had thus received an overlay of molten alloy and cooling had taken place, the rough surface was burnished or the gilded object hammered out into a sheet, if such was desired (Bergsøe 1938, 29).

Despite Bergsøe's analysis, there is still some doubt about how the fusion-gilding process was carried out.

Northward from Peru and Ecuador, metallurgy appears at progressively more recent dates: in Colombia, it appears 'around or slightly before 200 B.C.'; in West Mexico, 'after 650 A.D. by a system of Pacific maritime trade' (Hosler and Stresser-Pean 1992, 1215-16); and in eastern Mexico probably only a century or two before the Spanish conquest.

Colombia and Panama

In the Colombian Andes, most gold was obtained from placer mining, but there was hard rock mining as well: 'In the northwest of Antioquia are the remains of *shafts dug by the Indians to work quartz veins*...The ore was ground in stone mortars and panned for the gold' (Plazas and Falchetti de Sáenz 1979, 38). Mined-gold often contains various metallic impurities—for example, lead and zinc. If there are large quantities of such impurities, the gold is

> almost impossible to work. The pre-Hispanic Indians probably refined impure gold in the following manner: the gold dust was heated to a dull red with common salt and clay, and the chlorine in the salt then combined with the impurities to form chlorides, which were driven off as gases. In this way the grains were covered with a layer of fine gold, which became thicker as the procedure was continued and the temperature increased. The process, *known as cementation*, was used in the mint of Santa Fe de Bogotá from 1627 until 1838, supposedly following Indian tradition. (Plazas and Falchetti de Sáenz 1979, 41)

As in Peru, in Colombia and Panama metal artifacts were made by master craftsmen, using similar implements. When the Spanish arrived, the inhabitants of a village in northwestern Venezuela (a part of the Colombian gold-working center) had

> their forges and anvils and little hammers made of hard stone. Some say they are of a black metal [mineral] like emery. The hammers are of the size of eggs or smaller, and the anvils are as big as a cheese of Majorcas, made of other stones of the hardest nature. The bellows *are a reed about as big as three fingers or more*, and as long as two hand-breadths. (Oviedo 1959a, 3:11; translated by Saville 1925, 271)

Sir Walter Raleigh describes how the Indians of northeastern Colombia—much like the metalworkers of Peru—used blowpipes in melting and alloying metals:

> I after asked [of an old Indian] the manner howe the Epuremei wrought those plates of golde, and howe they coulde melt it out of the stone; he tolde me that most of the gold which they made in plates and images was not seuered [severed] from the stone, but that on the lake of Manoa, and in a multitude of other riuers [rivers] they gathered it in graines of perfect golde and in peeces as bigg as small stones, and that they put to it a part of copper, otherwise they coulde not worke it, and that they vsed a great earthen potte with holes round about it, and when they had mingled the gold and copper together, they fastned canes to the holes, and so with the breath of men they increased the fire till the mettell ran, and then they cast it into moulds of stone and clay, and so make those plates and Images. (Raleigh 1848, 96)

In contrast to their counterparts in Peru who were specialists in shaping solid sheet-metal, the metallurgists of Colombia and Panama, although skilled with hammers, usually 'treated metal as a liquid' (Hosler and Stresser-Pean 1992, 1216). Moreover, in contrast to other regional centers, there was 'almost no use made of [relatively pure] silver or copper for the manufacture of tools or ornaments' (Root 1961, 242): Artisans of this area strongly preferred working with tumbaga, the gold-copper alloy.

The lost-wax process is one of the best illustrations of using molten metal to fashion artifacts. For example, the Quimbaya along the Cauca River and the Zenú near the coast of the Caribbean Sea were masters of the method. The most famous exhibit of their work is in the *Museo de Oro* at Bogotá.

Use of the lost-wax process in the New World apparently started in northern South America: 'The first lost-wax cast artifacts made in the Americas are found in this region [Colombia] and date to around A.D. 100..., predating the appearance of the technique in West Mexico by at least 500 years' (Hosler 1994, 99). Significantly, it is in this region where lost-wax casting was so commonly practiced that one of the most intensive centers of native apiculture is reported: In the 16th century, a Spanish soldier claimed to have seen 80,000 hives in the neighborhood of Valle de Santa Marta, Colombia (Nordenskiöld 1930b, 202 and 207). Even allowing for considerable exaggeration,[8] wax and honey must have been produced in the area in great quantities.

Using the lost-wax process, the Quimbaya covered many of their tumbaga figures

> with bands of fine, wirelike decorations known as 'false filigree', a technical term referring to the fact that these decorations *are cast* as part of the figure they are found on, in the same flow of metal, rather than made of *separate of gold wire*...[These] filigree ornamentations were built up with fine wax threads on the original wax model... (Emmerich 1965, 72)

The following has been written about ornaments made by the Zenú: 'So intricate are their lacelike patterns that it has been suggested that they may have been cast in molds shaped originally by woven fabric or netting' (Emmerich 1965, 79). This cast-gold wire, or false filigree, is so fine and uniform that it is often mistaken for drawn wire.[9]

As noted before, natives of this area commonly applied depletion gilding. The practice may have appeared in Colombia between 400-700 A.D., possibly coming from Peru; however, the earliest account from the New World of depletion gilding was written in 1519. It describes an Indian custom along the north coast of Colombia near present-day Santa Marta.

> The Indians possess much gold and copper. They also possess much gilded copper. The Indians say that they treat the copper [that is, the guanín] *with an herb* that is in this country—which is crushed, *the juice extracted*, and the copper washed with it. Placed in the fire, it turns the color of very fine gold; the color increases in intensity according to the amount of herb [juice] given it. (Enciso 1530)

A short time later a similar account was given:

> The Indians know very well how to gild the pieces and things they make from copper and low-grade gold...They give such a bright luster to what they gild, that it looks like fine gold of 22 carats or more. They produce this with a certain herb, and it is a great secret that any goldsmith of Europe, or any part of Christendom, would be very happy and very rich if they had this manner of gilding. (Oviedo 1959a, 1:165)

The identity of the herb used as a source of acid is uncertain. An assortment of plants has been proposed—especially various species of oxalis. (Oxalic acid, mentioned above, is so-named because it was first discovered in a species of oxalis.) A more likely source than oxalis, at least in such tropical lowland areas as the Sinú (a major Colombian goldworking center) would be *caña agria* (*Costus* sp.); the plant grows abundantly in swampy places, along stream courses and in regrowth vegetation, and is sometimes cultivated by descendants of the Zenú in northern Colombia.[10] Another possibility

---

[8] The Spanish soldier in question based his number of 80,000 hives on the fact that there were some 10,000 houses in the Valle de Caldera near Santa Marta, each keeping upward of ten hives. The bees were kept in 'big pots or pitchers,' and made their honey from flowers of the guamo tree (probably a species of *Inga*). There is little doubt that the soldier observed the bees closely; for instance, he noted that they were small, did not make combs, and deposited their honey in 'large pockets of wax.' (Although this number of hives seems excessive, Oviedo also saw apiaries of a 'thousand or two hives' in Yucatan.)
Oviedo and Landa were the first to report that stingless bees do not, like honeybees of the Old World, deposit their honey in hexagonal cells. Rather the honey is put in more or less globular or oval containers—'four times as large as those made by Spanish bees' (Oviedo 1959b, 72).

[9] Probably wire was not drawn in the New World; for example, it 'was never drawn in the Andes' (Lechtman 1988, 362). Instead, wire was often made by rolling thin–cut slices of sheet–metal between two polished stones. Bits of ancient wire found in Ecuador (Bergsøe 1937, 43), however—as well as some found in western Mexican—are so fine and uniform that some believe they were drawn. Drawing wire requires a draw plate made of very hard material, with a small hole in it; the metal is forced through the hole which is tapered toward its distal end. The question is whether Mexican artisans were capable of working material hard enough to make a draw plate:

> it must be borne in mind that the Tarascan lapidaries made fine perforations in such hard stones as jadeite, rock crystal, and amethyst, and, moreover, that such perforations were tapered as required in a draw plate...[Moreover, Mixtec] master craftsmen appear to have discovered the principle of extruding through a narrow orifice the wax threads for their cire perdue models. (Easby 1962, 23)

[10] Upon reading Root's translation of Enciso's comments in 16th century Spanish, I was hopeful that the mysterious herb could more certainly identified as a *Costus* species. Root, an outstanding authority on South American metallurgy, translates the critical part of Enciso's description as follows: 'The Indians say that they gild the copper with an herb..., crushed, and with the top taken off; and they wash the copper with it' (Root 1961, 251). This would favor *Costus* over *Oxalis*, because the former has an often–dry flowering top which would probably have been taken off before the plant was used. (The abundant juice of *Costus*—more plentiful than that of *Oxalis*—is almost entirely in the cane.) In his 1530 edition, however, Enciso says nothing about the top of the plant being taken off; he states only that the herb was 'crushed, the juice extracted, and the copper washed with it.' (An herb, '*q ay en aqlla tierra: la qual majada y sacado el cumo y lauado el cobre conella*'; again, some diacritical marks on Spanish words are not reproducible.) Perhaps the 1518 or 1519 edition of *Suma de Geographia* says something different: the two copies of the 1530 version in the Bancroft

would be common vinegar, always plentiful in these areas where wine and beer were made.

According to one study, immersion in

> an acid solution of a plant juice (oxalis, etc.), will result, after a month or two in the solution of the surface copper and the formation of a dark layer of finely divided gold. If the object is now heated in the absence of air, or in a reducing atmosphere, to prevent oxidation of the remaining copper, the gold will consolidate to form a gold layer. The process then may be repeated. This method can be used for tumbagas of any composition, and is the only one which is suitable for copper-rich alloys. (Root 1961, 252)

Like the 16th century Peruvians, the Indian inhabitants of northern Colombia used weighing and measuring devices (Castellanos 1857, 306: '*con peso y con medida'*). The vicinity of Antioquia was very rich in gold, and inhabited by Indians who '*use small scales* and weights for weighing the gold' (Cieza de León 1985, 107: '*Usan de romanas pequeñas y de pesos para pesar el oro*').

Not far from Lake Maracaibo, in northwestern Venezuela, scales were also used:

> The inhabitants of this village for the greater part work gold...They have some cunning balance-beam scales like those of the Romans (*romanas sotiles*) with which they weigh, and these are of a white bone which looks like ivory, and they have them also of black wood, like ebony. *They have grooved rods and points for increasing and diminishing the weight like our Roman scales of weighing.* They can weigh with their method from half a *castellano*, which would be forty-eight grains, up to a mark of fifty *castellanos* or eight ounces, and no more, because they are small scales. (Oviedo 1959a, 3:11-12; translated by Saville 1925, 271)

Farther eastward, metalworking knowledge and skills tapered off or disappeared: In the West Indies,

> Gold was the only metal procured and worked, casually and in small amount. The islanders were neither miners nor metallurgists. They distinguished between native gold and alloyed gold objects, knowing the latter only by trade and tradition, and having very little of either. Alloyed pieces, called *guañín* [tumbaga], were known *as having been brought out of the south*; they were not prized as gold but as rare and attractive ornaments. (Sauer 1966, 61)

Much the same was true of the tribes in the northwest Amazon—at least until recently: 'Their only method of working metal when obtained [in trade] is to heat and hammer it into various forms and shapes for ornaments' (Whiffen 1915, 94). The fact that metallurgy in South America is almost entirely confined to the western edge of the continent can hardly be explained simply by differences in physical environment. While the surface of Amazonia lacks the required minerals, they are present to the southeast and northeast, for instance in the Guiana and Brazilian highlands.

That much metal-working knowledge was communicated overland from southern Central America to Mexico by way of the Panamanian Isthmus seems unlikely: The peoples of northern Nicaragua through Honduras and Guatemala to southern Mexico had very limited metallurgical knowledge. For instance, among the Maya, 'All objects that show casting are of foreign origin. The only technique with which Maya goldsmiths were familiar was the hammering for repoussé work' (Morley and Brainerd 1983, 444). Unlike those of Peru and Mexico, Mayan metalworkers did not know of bronze making.

Western Mexico

At the time of the Spanish conquest there was considerable maritime trade along the coast of Peru and Ecuador where the inhabitants operated large sail-propelled rafts quite capable of making a journey to Mexico (Paulsen, 1977).

As they sailed along the north coast of Peru (near Tumbez), Francisco Pizarro's party 'saw a balsa approaching by sea, so large that it resembled a ship' (Cieza de León 1998, 104); the ship 'carried numerous metal objects including bells, bands, tweezers, tiaras, and crowns' (Hosler and Stresser-Pean 1992, 1216).

Perhaps as a result of trade like this, metallurgy developed between 600 and 800 A.D. in western Mexico (the area located between Nayarit and Guerrero, and extending inland to include Michoacán) with several complex processes already well developed. Pre-Columbian inhabitants appear to have learned much of their technology directly from South America or from southern Central America:

> West Mexican metallurgy had its roots in two major metalworking traditions of the ancient Americas: the casting technology of lower Central America and Colombia and the sheet metal tradition of the Central Andean area of Ecuador, Peru, and Bolivia. (Hosler and Stresser-Pean 1992, 1215-16)

In the Aztec empire of Mexico, metalworkers smelted copper very much like the ancient Peruvians: in clay crucibles, over charcoal fires, and fanned by blowpipes.

As in Peru, West Mexican smiths probably smelted sulfide ores, such as chalcopyrite (Hosler 1988, 338), and made artifacts not only of copper, but also of copper-tin and copper-arsenic alloys: both of these bronzes were used to make awls, eyed sewing needles, tweezers and bells. Cutting tools (including axes) were made of both

Library at the University of California were the only works of Enciso available to me.

Aztec metal-worker with his son, using a blowpipe (Codex Mendoza 1992, 4:145). On the right is the Aztec symbol for gold (Spinden 1928, 224). Note the encircled gold-symbol on the charcoal-burning brazier.

Figure 49. Aztec Metal-Worker Using Blowpipe

bronzes, and their cutting edges were cold-worked to increase their hardness. Copper-silver alloys were also commonly used.

The appearance of the lost-wax technique in Colombia predates its appearance 'in West Mexico by at least 500 years' (Hosler 1994, 99). The Aztecs, however, like the artisans of Colombia and southern Central America, did eventually learn this casting process, often combining it with gilding. The desired artifact was modeled in either wax or in a mixture of wax, clay and charcoal.[11] 'They especially esteem beeswax...For there is a model (in wax) of all they make, whether birds' wings, or flowers, or leaves of plants, or whatsoever beautiful design' (Sahagún 1950-1982, 10:77). Beeswax has a melting point between 62° and 65°C (at least the wax of European bees); thus it was the best available material for lost-wax casting, since it does not soften in the hands of the craftsman. Nevertheless, Aztec metalworkers, in making their models, hardened beeswax with 'white copal' (Sahagún 1950-1982, 10:74).

Gold was panned by the Aztecs from stream sands using pieces of tree gourds (*jícaras*). The gold dust and sand were 'washed in a tray or special pottery dish' (Guerra 1969, 45). The Aztecs extracted gold from placers as well as hard-rock mines: in some of the latter, archaeologists have found wedges and mattocks made of andesite or other very hard stones, with the mine walls still bearing the marks made by such tools.

The Aztecs had discovered the value of emery (a greyish black, impure variety of corundum, Al_2O_3, the hardest of the common rock minerals) as an abrasive. (Corundum is also much used as an abrasive in modern industry.) The Aztecs found their emery in:

> the mountains, the crags. It is white yellow, ashen, ruby, mixed black and green. It is a grinder; it is that which wears away, which thins things. (Sahagún 1950-82, 12:237)

They also used powdered flint sand (silicon dioxide) as an abrasive (as is the case with modern kitchen cleansers):

> It is crushed, pulverized, ground...ground very fine...like *pinole*; powdery. It is a medium for cleaning, for polishing... (Sahagún 1950-82, 12:238)

Huastec Area of Eastern Mexico

In the Huastec area, metallurgy appeared in the Late Post Classic period (beginning about 1300 A.D.)—centuries later than the West Mexican center. Here 'local metalsmiths were manufacturing Cu-As-Sn [copper-arsenic-tin] alloys from smelted ores' (Hosler and Stresser-Pean 1992, 1218).

> The bronze working component of Huastec metallurgy was transmitted from the metalworking regions of West Mexico, most likely through market systems that distributed Aztec goods. (Hosler and Stresser-Pean 1992, 1215)

---

[11] Sahagún's detailed description of the lost–wax process, as undertaken by an Aztec goldsmith using a clay and charcoal mold, is summarized below:
First the craftsman took charcoal and ground it to a fine powder. Mixing this powder with potter's clay, he kneaded the two into a paste, and then set the mixture out in the sun to dry. When the mixture was dry and hard, a piece was 'carved, sculptured with a small metal blade [probably of copper]' into a model shaped exactly as the artisan wished his ornament to be.
Then melted beeswax 'was mixed with white copal' to give it firmness, and the mixture was filtered to remove all foreign matter. When cooled, the mixture was rolled out thin 'on a very smooth, flat stone with a round piece of wood...When it was flattened, just like a cobweb, nowhere of uneven thickness, then it was placed over' the sculptured clay–and–charcoal model; 'cautiously little pieces (of wax) were cut off or pared away. By this means a little (wax) entered hollows, covered eminences, filled depressions' where the clay–and–charcoal core had been carved away.
The whole wax–covered model was overlaid with a paste of finely powdered charcoal, followed by an overall covering of charcoal mixed with clay. A 'mouth,' or pouring tube, made of beeswax coated with clay, was affixed.
Then the whole enveloped model was placed in a brazier to melt out the wax through the affixed pouring tube, and molten gold was poured in through the same opening.
When the gold inside had cooled and solidified, the envelope was broken away and the exposed artifact was both 'burnished with a pebble' and treated with a wash of ground alum. It was then heated again in the fire and 'at once washed, rubbed with what was called 'gold medicine,' the ingredients of which, though uncertain, are described as 'just like yellow earth [a clay ?] mixed with a little salt' which gave it a good color (Sahagún 1950–82, 10:73–75).
'The treatment with clay and salt...will remove some of the silver, if present in the gold as an impurity, and make the surface more yellow' (Root 1961, 251–52).
Sahagún's account 'is an astonishingly accurate job of technical reporting. With very few changes it might have been written yesterday or today' (Easby 1956, 404).
Though in the above account only gold–or silver–casting is described, it is believed that copper was cast using basically the same technique: the whole process has been reproduced experimentally in every detail using copper instead of gold (Long, 1964).

The favorite metal artifacts in the Huastec area seem to have been bells; the number found there exceeds many times over that of tools, such as axes.

Little is known about the diffusion of intricate technological knowledge between widely-spaced areas when there is lack of written documents. It may be that direct personal communication is needed for the transmission of some metallurgical skills; that is to say, the 'transfer of a complex technology demands long-term, face-to-face contacts' (Hosler and Stresser-Pean 1992, 1219). This possibility should also be kept in mind when considering the diffusion of metallurgical technology between South America and West Mexico.

When the Europeans first arrived, maritime trade between the regional metallurgical centers was not limited to the Pacific coast. It is well known that boats capable of long ocean journeys existed in pre-Columbian times both on the Pacific and Atlantic coasts (on the latter, see Oviedo 1959b, 92). Thus, Columbus's ships, pausing near an island off the north coast of Honduras, saw a great dugout canoe carrying trade goods coming from the northwest. The Spanish confiscated the goods which included implements for smelting:

> ...there came at that time a great canoe as great as a galley, eight feet wide, all of a single trunk...[The canoe] came loaded with merchandise from western parts, from the side of New Spain [that is, from Mexico]...[This merchandise included] hatchets to cut wood, like those of stone used by other Indians, save for the fact that *these were of good copper*, of which metal they also had bells *and crucibles for smelting.* (Sauer 1966, 128-citing and translating the account of Ferdinand Columbus)

To which people the canoe's crew belonged is not stated, but it is unlikely that they were Maya. As mentioned before, the Maya—like the people of Honduras and Nicaragua—knew little about smelting. The copper bells and axes suggest that the Huastec area of eastern Mexico was the boat's source. In any case, we have (in the quoted passage above) an eyewitness account of smelting knowledge being spread by means of oceanic commerce. Early documents refer to a settlement of Nahuatl-speaking Indians in the Chiriquí area of northern Panama (Gordon 1957, 39). Perhaps the canoe was carrying goods from their homeland.

As noted before, between Panama and Mexico metallurgywas non-existent or poorly developed. In northern Mexico, beyond the Huastec area, knowledge of metallurgy again drops off sharply; for instance, the inhabitants of the American Southwest never learned the art of smelting ores or casting molten metals. To the east, among the Mound-Builders (various prehistoric and early historic peoples of the southeastern United States), no smelting was done: there, native copper was hammered into artifacts, and fine repoussé work was done on sheets of that metal. Farther north among non-agricultural peoples, meteoric iron (largely iron-nickel compounds) was beaten into artifacts in several localities.

In summary, the pre-Columbian artisans of the New World, working with the simplest of tools, but with sophisticated processes, produced some of the finest goldwork in the world. Auctions in 16th century Aztec markets displayed

> marvellous works of silver or engraved metals [probably repoussé work] or of cast bronze: hexagonal dishes which contain three parts of gold alternating with as much of silver, stuck one to another, not glued in any manner—but fused, consolidated and soldered in the same union; little amphorae of bronze with detachable handles ['*con asas sueltas*']; *fishes with one scale of gold and another of silver*; parakeets which have the tongue, the head, feet and wings movable; monkeys with the head and feet flexible and twisting a spindle, as if spinning—and others holding an apple, or some other fruit, seeming to eat it. All of which our [Spanish] craftsmen could not possibly emulate... (Hernández 1986, 104)

With regard to the gold and silver fish spoken of in the above passage, Benvenuto Cellini (one of the most celebrated goldsmiths of the Italian Renaissance) 'is said to have tried *in vain* to discover exactly how a jeweler of ancient Mexico had made a *fish of silver with a delicate inlay of gold*' (Easby 1956, 401).

An exhibition of manufactured articles, including goldwork (Easby 1956, 401), from the New World was shown at Brussels in 1520, only a year after Hernán Cortés had begun the conquest of the Aztec empire. Albrecht Dürer, the German Renaissance master, saw the exhibition—and, despite the quite different artistic idiom current in Europe, he was able to appreciate the beauty of the artifacts immediately. This is Dürer's remarkable appraisal, as translated from his diary:

> I saw the things which they brought the King [of Spain] from the new golden land [i.e., from the Spanish Main and Mesoamerica]...two chambers full of...*wonderful things for numerous uses*, more pleasing to view than marvels...*And all my life I have seen nothing which has so pleased my heart as these things.* For *I saw in them wondrous skillful things and marveled at the subtle ingenuity of the people in foreign lands.* (Dürer 1982, 65)

Concluding Remarks

Although the fifteenth-century chemical arts possessed by inhabitants of the Americas were substantial as compared to their conquerors, some European skills were no doubt superior in several important respects. For example, transparent glass artifacts were not made by pre-Columbian Americans: this deficiency, almost certainly, would have slowed development of chemical knowledge. Moreover, it can be said with little exaggeration that lack of two chemical skills—namely, the ability to make iron and gunpowder—lost for peoples of the New World political control of their hemisphere. It is only fair to say, however, that those skills were not the invention of Europeans themselves: the knowledge was introduced from elsewhere.

Epilogue

The main purpose of this discussion has been to describe the chemical arts and technologies of indigenous America, not to discuss opinions as to their ultimate origin. Nevertheless, occasional comment is made on the latter topic, mostly in response to the quoted literature.

For many, the roughly-parallel cultural progression that took place in the Old World and the New is puzzling, and still not adequately explained.

In the last few decades, the prevailing opinion has been that there was little cultural interchange between the Old and the New World from the beginning of Neolithic until the European conquest—a period of more than ten thousand years. Assuming that this is so, the similar development of the chemical arts between the eastern and western hemispheres is indeed surprising. Despite this long cultural separation (and despite the unique chemical inventions made in both hemispheres), major related accomplishments were attained in both the Old World and the New.

In making fabrics, species of wild cotton (genus *Gossypium*) were domesticated; fibers on their seeds were selected for length and spun. In both hemispheres, a loom with a heddle (a rather complex device) was invented to weave cotton threads into fabrics. Plants of the genus *Indigofera* were domesticated; 'vat-dyeing' was practiced, using indigo to dye cotton thread and fabrics; similar mordants were used to fix dyes, the chief mordant being alum. Pottery was invented in both hemispheres, as well as book making (i.e., literacy), and so on. Lacquering was discovered in both Old and New Worlds. In architecture, lime mortar and true and dry fresco murals were made in both hemispheres.

In metallurgy even more more striking parallels are found in both the hemispheres: a number of the same ores were smelted, with the deliberate use of fluxes (even the proposed smelting of sulfides); in addition, such relatively complex processes as the sintering of gold-platinum alloys; depletion gilding; in casting, the lost-wax method was employed; separating lead from silver through cupellation; both bronzes (copper-arsenic and copper-tin) were made in both the Old and New Worlds. For some, that all of this was discovered independently in both hemispheres strains credibility.

In opposition to the notion of cultural diffusion over long oceanic routes (for example, across the Pacific Ocean), others—seeking to explain the parallel cultural progression—have proposed that these similarities are mainly the result of independent inventions. The archaeological finds at Mina Perdida, Peru, tend to support this point of view: they suggest a sequence of metallurgical development over a long period, rather than trait introduction (Quiller 1998, 1058; Burger and Gordon 1998).

Some have proposed that 'cultural evolution,' according to Herbert Spencer's model, has produced these parallels (Sahlins and Service 1960, 23-24). Spencer, a contemporary of Charles Darwin, proposed a theory that human social progress is *inherent*—that human society 'is a growth and not a manufacture, and has its laws of evolution' (Spencer 1885, 3:321); that is to say, human artifacts, states, laws, morals all evolve and become more perfect. (Darwin also believed in progression, at least with regard to biological species.)

Considering metallurgy, this evolutionary viewpoint has not been accepted by all authorities—even amongst those who feel that metallurgy was independently developed in the New World. For example,

> While the great metallurgical traditions did, of course, proceed generally from the simple to the complex in their technologies, current views of that progression seem to me *overly evolutionary in spirit.* (Lechtman 1975, 24)

In any case, if the possibility of significant diffusional influences across the oceans were to be definitely eliminated, according to Spencer's progressive version of evolution, one might see a rough (but seemingly necessary) sequential order in the growth of the chemical arts.

WORKS CITED

Early Sources, Including Biographical Notes

Acosta, José de (2002) *Natural and Moral History of the Indies*. Translated by F. M. López-Morillas and edited by J. E. Mangan with an introduction and commentary by W. D. Mignolo. Durham and London: Duke University Press.
José de Acosta, a Jesuit, arrived in Peru in 1572.

Andagoya, Pascual de (1967) *Narrative of the Proceedings of Pedrarias de Avila*. Translated and edited by C. R. Markham. Hakluyt Society, ser. 1, no. 34, 1865; New York: Burt Franklin.
Pascual de Andagoya came to Panama with Pedrarias de Avila in 1514.

Anghiera, Pietro Martire d' (1964-65) *Decadas del Nuevo Mundo. Por Pedro Mártir de Angleria, primer cronista de indias*. 2 vols. Mexico: J. Porrua.
Italian-born Pietro Martire d'Anghiera was known to the Spanish as Pedro Mártir, and to the English as Peter Martyr. Although he was never in the Americas, Mártir was present at the Spanish court when Columbus returned from his voyages. Thus he was able to inspect the first accounts, and some of the first materials, received from the New World; his observations, written between 1493 and 1525, are among the most accurate of early accounts.

Barba, Alvaro Alonso (1923) *El arte de los metales*. Translated from the Spanish by R. E. Douglass and E. P. Mathewson. New York: J. Wiley.
Alvaro Alonso Barba was a Jesuit who lived in Peru. His work, first published in 1640, has been described as "the great classic of early Andean metallurgy."

Benzoni, Girolamo (1985) *La Historia del Mondo Nuovo*. Traducida por primera vez en Lengua Castellana por Carlos Radicati di Primeglio. Guayaquil: Museo Antropológico y Pinacoteca, Banco Central del Ecuador.
Girolamo Benzoni was in South America between 1547 and 1550.

Betanzos, Juan de (1996) *Narrative of the Incas*. Translated and edited by R. Hamilton and D. Buchanan from the Palma de Mallorca manuscript. Austin: University of Texas Press.
Juan de Betanzos, who finished his Narrative of the Incas in 1557, had worked in Peru as an interpreter and translator of Quechua since shortly after Pizarro and his soldiers had conquered the area.

Capoche, Luis (1959) *Relaciones histórico-literarias de la América meridional: Relacion General de Villa Imperial de Potosi [por] Luis Capoche*. Biblioteca de Autores Españoles. Madrid: Ediciones Atlas.
Luis Capoche, a silver miner at Potosí, wrote around 1585.

Casas, Bartolomé de las (1951) *Historia de las Indias*. Vols. 1-3. Mexico: Fondo de Cultura Economica.
Fray Bartolomé de las Casas arrived in Hispaniola in 1502. He became a Dominican priest and the first Bishop of Chiapas.

Castellanos, Juan de (1857) *Elegías de Varones Ilustres de Indias*. Biblioteca de Autores Españoles, vol. 4. Madrid: M. Rivadeneyra.
Juan de Castellanos arrived in America in 1541; in the form of a long poem, his Elegías... *is a primary source of information on indigenous north Colombia.*

Cervantes de Salazar, Francisco (1966) Cronicas de la Nueva España. In *Relacion de las cosas de Yucatan*, by Diego de Landa, a translation, edited with notes by Alfred M. Tozzer. Papers of the Peabody Museum of American Archaeology and Ethnology, Harvard University, vol. 18, appendix D, pp. 233-39. Cambridge: The Museum, 1941; New York: Kraus Reprint Corp.
Francisco Cervantes de Salazar came to Mexico in 1550, and became one of the first professors at the University of Mexico.

Cieza de León, Pedro de (1959) *The Incas*. Translated by H. de Onis, and edited with an introduction by V. W. von Hagen. Norman: University of Oklahoma Press.
Pedro Cieza de León, a soldier in the Spanish armies, was in Peru between 1532-1550 during the early days of the Spanish conquest.

—(1985) *La Crónica del Peru*. Madrid: Edición de Ballesteros.

—(1998) *The Discovery and Conquest of Peru: Chronicles of the New World Encounter*. Edited and translated by A. P. Cook and N. D. Cook. Durham: Duke University Press.

Cobo, Bernabé (1956) *Obras [Historia del Nuevo Mundo]. Estudio preliminar y edición del P. Francisco Mateos*. 2 vols. Biblioteca de Autores Españoles. Madrid: Atlas.
Father Bernabé Cobo, who was in America from 1599 to 1657, wrote mainly about Peru where he traveled widely. He also spent a year in the West Indies and visited Mexico.

—(1979) *History of the Inca Empire: An account of the Indians' customs and their origin, together with a treatise on Inca legends, history and social institutions*. Translated and edited by Roland Hamilton. Austin: University of Texas Press.

—(1990) *Inca Religion and Customs*. Translated and edited by Roland Hamilton, with a foreword by John H. Rowe. Austin: University of Texas Press.

Codex Borbonicus (*Códice Borbónico*): *manuscrito mexicano de la Biblioteca del Palais Bourbon* (1979) Colección America Nuestra. Mexico City: Siglo

Veintiuno.
This codex was probably created after the Spanish conquest of Mexico.

Codex Fejérváry-Mayer; a manuscrit mexicain des Free Public museums de Liverpool (1901) Paris: Imprime par P. Renouard.
A pre-Conquest document.

Codex Mendoza (1992) Edited by F. F. Berdan and P. R. Anawalt. Vols. 1-4. Berkeley: University of California Press.
The codex "probably can be dated to 1547."

Cortés, Hernán (1908) *Letters of Cortés.* Edited and translated by F. A. MacNutt. 2 vols. New York and London: G. P. Putnam's sons.
Famed as Spanish conqueror of Montezuma's Aztec empire, Hernán Cortés landed on the Yucatan Peninsula in 1519.

Cruz, M. de la (1940) *The Badianus Manuscript: An Aztec Herbal of 1552.* Translated by Juannes Badianus into Latin from Aztec, and edited by E. W. Emmart. Baltimore: Johns Hopkins Press.
Martin de la Cruz and his translator Juan Badiano, Aztec students at the College of Santa Cruz in Tlatilulco, compiled an illustrated list of medicinal herbs used by the Aztecs. Written in both Nahuatl and Latin, it was sent as a present to Charles V of Spain.

Díaz del Castillo, Bernal (1960) *Historia Verdadera de la Conquista de la Nueva España.* 2 vols. Mexico: Editorial Porrúa, S. A.
Bernal Díaz del Castillo arrived at Darién in 1514 and later served as a soldier with Hernán Cortés, the conqueror of Mexico.

Dürer, Albrecht (1982) *Schriften und Briefe.* Leipzig: Ernst Ullman.
Albrecht Dürer, whose father was a goldsmith, was trained as a painter and printmaker. He also made copper engravings and woodcuts.

Durán, Diego (1994) *The Aztecs: The History of the Indies of New Spain.* Translated by Doris Heyden. Norman: University of Oklahoma Press.
Fray Diego Durán, a Dominican who grew up in Mexico and spoke Nahuatl, wrote in 1581.

Enciso, Martín Fernández de (1530) *Suma de Geographia.* Sevilla: J. Cromberger.
Martín Fernández de Enciso came to the West Indies in 1500. In 1510, he founded the settlement of Darién on the west shore of the Gulf of Urabá in Panama. His Suma de Geografía was among the first published accounts of Spanish discoveries in the New World.

Fernández de Oviedo y Valdés, Gonzalo (1959a) *Historia General y Natural de las Indias.* Biblioteca de Autores Españoles. Madrid: Ediciones Atlas.
Gonzalo Fernández de Oviedo y Valdés, a Spanish chronicler, arrived in Darién in 1514, and later became governor of Cartagena.

—(1959b) *Natural History of the West Indies.* Translated and edited by S. A. Stoudmire. Studies in Romance Languages and Literatures, no. 32. Charlotte, North Carolina: University of North Carolina Press.

Garcilaso de la Vega: See Vega, Garcilaso de la.

Hernández, Francisco (1986) *Antigüedades de la Nueva España.* Edited by A. H. de León-Portilla. Cronicas de America, historia 16, vol. 28. Madrid: Hermanos Garcia Noblejas.
Francisco Hernández, 1517-1587, a physician and botanist, obtained his information from Aztec assistants and interpreters.

—(1888) *Cuatro Libros de la Naturaleza y Virtudes Medicinales de las Plantas y Animales de la Nueva España.* Morelia: La Escuela de Artes.
Between 1570 and 1577, Francisco Hernández produced a flora and fauna of Mexico. This Spanish edition--the only one available to me--has antiquated spellings.

—(1991) *El Preguntador del rey: Francisco Hernández.* Edited by Xavier Lozoya. Mexico: Pangea Editores.

Landa, Diego de (1966) *Relacion de las cosas de Yucatan.* A translation, edited with notes by Alfred M. Tozzer. Papers of the Peabody Museum of American Archaeology and Ethnology, Harvard University, vol. 18. Cambridge: The Museum, 1941; New York: Kraus Reprint Corp.
Diego de Landa, a Franciscan priest, wrote extensively on the culture of the Maya during the last half of the 16th century. In 1572, he was elected Bishop of Yucatan.

Las Casas, Bartolomé de: See Casas, Bartolomé de las.

Mártir de Angleria, Pedro: See Anghiera, Pietro Martire d'.

Martyr, Peter: See Anghiera, Pietro Martire d'.

Montesinos, Fernando (1967) *Memorias antiguas historiales del Peru.* Translated and edited by Philip Ainsworth Means. Hakluyt Society, ser. 2, vol. 48, 1920; Nendeln/Liechtenstein: Kraus Reprint Limited.
Fernando Montesinos was in Peru in the first half of the 17th century.

Motolinía, Fray Toribio de (1950) *History of the Indians of New Spain.* Translated and edited by E. A. Foster. Berkeley: The Cortés Society.
Toribio de Motolinía, a Franciscan friar, came to Mexico in 1524, only a few years after Hernán Cortés.

Narrative of some things of New Spain and of the great city of Temestitan, Mexico, written by the anonymous conqueror, a companion of Hernan Cortés (1917) Translated into English and annotated by Marshall H. Saville. New York: The Cortés Society.

Oviedo y Valdés, Gonzalo Fernández de: See Fernández de Oviedo y Valdés, Gonzalo.

Raleigh, Walter (1848) *The Discoverie of the Large, Rich and Bewtiful Empire of Guiana.* Edited from the edition of 1596 by R. H. Schomburgk, 1st Series, London: Hakluyt Society; New York: Burt Franklin.
In 1595 Sir Walter Raleigh traveled up the Orinoco, through modern Venezuela and into Colombia.

Sahagún, Bernardino de (1950-1982) *General History of the things of New Spain: Florentine Codex.* Translated from the Aztec into English by A. J. O.

Anderson and C. E. Dibble. Vols. 1-13. Salt Lake City: University of Utah.

Bernardino de Sahagún, a Franciscan friar, arrived in Mexico only a few years after the Spanish defeat of the Aztec armies. He soon became proficient in Nahuatl, the Aztecan language.

—(1956) *Historia General de las Cosas de Nueva España.* Edited by Angel Maria Garibay K. Vols. 1-3. Mexico: Editorial Porrua, S.A.

—(1997) *Primeros Memoriales.* Paleography of Nahuatl text and English translation by Thelma D. Sullivan. Norman: University of Oklahoma Press.

Salazar, Francisco Cervantes de: See Cervantes de Salazar, Francisco.

Stade(n), Hans (1874) *The Captivity of Hans Stade of Hesse, in A.D. 1547-1555, among the Wild Tribes of Eastern Brazil.* Hakluyt Society, 1st ser., no. 51; New York: Burt Franklin.

Stade, a mercenary soldier, was taken prisoner by the Tupinamba Indians along the south coast of Brazil in the mid-16th century, and lived for some time as their captive.

Vega, Garcilaso de la (1966) *Royal Commentaries of the Incas.* Translated by H. V. Livermore. Vol. 1. Austin: University of Texas Press.

Garcilaso de la Vega was born in Cuzco, Peru, in 1539 about a decade after the Spanish conquest. He was the son of a Spanish conqueror and an Inca princess.

Modern References

Abrams, E. M. and A. Freter (1996) A Late Classic Lime-Plaster Kiln from the Maya Centre of Copan, Honduras. *Antiquity* 70:422-428.

Ackerknecht, E. H. (1963) Medical Practices. In *Handbook of South American Indians*, vol. 5, pp. 621-643. Smithsonian Institution, Bureau of American Ethnology Bulletin 143, series editor J. H. Steward. New York: Cooper Square Publishers, Inc.

Aikens, C. M. (1995) First in the World: The Jomon Pottery of Early Japan. In *The Emergence of Pottery: Technology and Innovation in Ancient Societies*, edited by W. K. Barnett and J. W. Hoopes, pp. 11-21. Washington, D.C.: Smithsonian Institution Press.

Aikens, C. M. and T. Higuchi (1982) *Prehistory of Japan.* New York: Academic Press.

Amsden, C. A. (1964) *Navaho Weaving: Its Technic and History.* Chicago: The Rio Grande Press.

Andrews, A. P. (1983) *Maya Salt Production and Trade.* Tucson: The University of Arizona Press.

Arango, Luis C. (1941) *Recuerdos de la Guaquería en el Quindío.* 2nd ed. 2 vols. Bogotá.

Archer, A. (1934) The Dental Plant of the Citará Indians in Colombia. *Journal of the Washington Academy of Sciences* 24:402-404.

Arnold, D. E. (1975) Ceramic Ecology of the Ayacucho Basin, Peru: Implications for Prehistory. *Current Anthropology* 16:183-205.

—(1985) *Ceramic Theory and Cultural Process.* Cambridge: Cambridge University Press.

Arnold, D. E. and B. F. Bohor (1975) Attapulgite and Maya Blue. *Archaeology* 28:23-29.

Arriaza, B. T. (1995) Chinchorro Bioarchaeology: Chronology and Mummy Seriation. *Latin American Antiquity* 6(1):35-55.

Baessler, Arthur (1902-1903) *Ancient Peruvian Art....* Translated by A. H. Kean. Vol. 4. Berlin and New York: Dodd, Mead.

Bailey, F. L. (1940) Navaho Foods and Cooking Methods. *American Anthropologist* 42:270-290.

Bakewell, Peter (1984) *Miners of the Red Mountain: Indian Labor in Potosí, 1545-1650.* Albuquerque: University of New Mexico Press.

Balfour-Paul, Jenny (1997) *Indigo in the Arab World.* Richmond, Surrey: Curzon.

—(1998) *Indigo.* London: British Museum Press.

Balls, E. K. (1972) *Early Uses of California Plants.* Berkeley: University of California Press.

Barham, Peter (2001) *The Science of Cooking.* Berlin and New York: Springer.

Barnett, W. K. and J. W. Hoopes, eds. (1995) *The Emergence of Pottery: Technology and Innovation in Ancient Societies.* Washington, D.C.: Smithsonian Institution Press.

Barrett, S. A. (1925) *The Cayapa Indians of Ecuador.* Museum of the American Indian, Indian Notes and Monographs, no. 40, part 1. Heye Foundation, New York.

Barrows, D. P. (1931) Desert Plant Foods of the Coahuilla. In *Source Book in Anthropology*, edited by A. L. Kroeber and T. T. Waterman, pp. 202-210. New York: Harcourt, Brace and Company.

Bates, H. W. (1962) *The Naturalist on the River Amazon.* Foreword by R. L. Usinger. Berkeley: University of California Press.

Batra, L. R. and P. D. Millner (1974) Some Asian Fermented Foods and Beverages, and Associated Fungi. *Mycologia* 66:942-950.

Beechey, F. W. (1831) *Narrative of a Voyage to the Pacific and Bering's Strait.* Vol. 2. London: Henry Colburn & Richard Bentley.

Belitz, H.-D. and W. Grosch (1987) *Food Chemistry.* Translated from the 2nd German edition by D. Hadziyev. Berlin: Springer Verlag.

Bender, Max (1947) Colors for Textiles, Ancient and Modern. *Journal of Chemical Education* 24:2-12.

Bennett, W. C. and R. M. Zingg (1976) *The Tarahumara: An Indian Tribe of Northern Mexico.* Glorietta, New Mexico: Rio Grande Press.

Bergsøe, Paul (1937) *The Metallurgy of Gold and Platinum among the Pre-Columbian Indians.* Translated by F. C. Reynolds. Ingeniørvidenskabelige Skrifter No. A-44. Copenhagen: Danmarks Naturvidenskabelige Samfund.

—(1938) *The Gilding Process and the Metallurgy of Copper and Lead among the Pre-Columbian Indians.* Translated by F. C. Reynolds. Ingeniørvidenskabelige

Skrifter, No. A-46. Copenhagen: Danmarks Naturvidenskabelige Samfund.

Berkowitz, Mark (1996) World's Oldest Wine. A report on the research of P. E. McGovern and his team. *Archaeology* 49: 26.

Bingham, Hiram (1948) *Lost City of the Incas: The Story of Machu Picchu and Its Builders*. Westport, Connecticut: Greenwood Press, Publishers.

—(1962) Types of Macchu Picchu Pottery. *American Anthropologist* 17:259-271.

Bird, Junius (1979) Legacy of the stingless bee. *Natural History* 88(9):49-51.

Bishop, R. L. (1994) Pre-Columbian Pottery in the Maya Region. In *Archaeometry of Pre-Columbian Sites and Artifacts*, edited by D. A. Scott and P. Meyers, pp. 15-66. Los Angeles: The Getty Conservation Institute.

Bodman, Gösta (1924) Untersuchung eines Balsamierungsmittels aus Gräbern in Arica (Chile). *Meddelanden från Geografiska Föreningen i Göteborg* 3:79-86. Göteborg: Wettergren & Kerber.

Boman, Eric (1908) *Antiquités de la région andine de la république argentine et du désert d'Atacama.* 2 vols. Paris: H. Le Soudier.

Bonavia, Duccio (1985) *Mural Paintings in Ancient Peru.* Translated by P. J. Lyon. Bloomington: Indiana University Press.

Boynton, R. S. (1980) *Chemistry and Technology of Lime and Limestone.* 2nd ed. New York: John Wiley & Sons, Inc.

Brand, D. D. (1951) *Quiroga, a Mexican Municipio.* Smithsonian Institution, Institute of Social Anthropology, Publication 11.

Bray, W. and C. Dollery (1983) Coca Chewing and High-Altitude Stress: A Spurious Correlation. *Current Anthropology* 24(3):269-274.

Bressani, R., R. Paz y Paz, and N. S. Scrimshaw (1958) Chemical Changes in Corn during Preparation of Tortillas. *Journal of Agricultural and Food Chemistry* 6:770-774.

Brooks, J. E. (1937) Introductory essay, glossary and bibliographic notes in *Tobacco, Its History illustrated by the Books, Manuscripts and Engravings in the Library of George Arents, Jr.*, by George Arents, vol. 1. New York: The Rosenbach Company.

Browman, D. L. (1974) Pastoral Nomadism in the Andes. *Current Anthropology* 15(2):186-196.

Bruhns, K. O. (1994) *Ancient South America.* Cambridge: Cambridge University Press.

Bruman, H. J.(1940) Aboriginal Drink Areas in New Spain. Ph.D. dissertation, University of California, Berkeley.

—(1948) The Culture History of Mexican Vanilla. *Hispanic American Historical Review* 28:360-376.

Brunhes, J. and P. Deffontaines (1952) *Human Geography.* Translated by E. F. Row. New York: Rand McNally.

Burger, R. L. and R. B. Gordon (1998) Early Central Andean Metalworking from Mina Perdida, Peru. *Science* 282:1108-1111.

Bühler, A. (1948) Dyeing among Primitive Peoples. *Ciba Review* 68:2478-2512.

Cahill, J. P. (2003) Ethnobotany of Chia, *Salvia hispanica* L. (Lamiaceae). *Economic Botany*, 54(4):604-618.

Caley, E. R. (1973) Chemical Composition of Ancient Copper Objects of South America. In *Application of Science in Examination of Works of Art; proceedings of the Seminar, June 15-19, 1970, conducted by the Research Laboratory*, edited by William J. Young, pp. 53-61. Boston: Museum of Fine Arts.

Caley, E. R. and D. T. Easby Jr. (1959) The Smelting of Sulfide Ores of Copper in Preconquest Peru. *American Antiquity* 25:59-65.

—(1964) A Lead Lip Plug from Western Mexico. *American Antiquity* 29(4):497-500.

Case, R. J., A. O. Tucker, M. J. Maciarello, and K. A. Wheeler (2003) Chemistry and Ethnobotany of Commercial Incense Copals, Copal Blanco, Copal Oro, and Copal Negro, of North America. *Economic Botany* 57(2):189-202.

Christensen, Bodil (1963) Bark Paper and Witchcraft in Indian Mexico. *Economic Botany* 117:361-372.

Churchill, Awnsham (1746) The Mosqueto Indian and His Golden River. In *A Collection of Voyages and Travels*, vol. 6, pp. 297-312. London: Printed by assignment from Messrs. Churchill, for H. Lintot.

Clark, J. E. and D. Gosser (1995) Reinventing Mesoamerica's First Pottery. In *The Emergence of Pottery: Technology and Innovation in Ancient Societies,* edited by W. K. Barnett and J. W. Hoopes, pp. 209-221. Washington, D.C.: Smithsonian Institution Press.

Coe, F. G. and G. J. Anderson (1999) Ethnobotany of the Sumu (Ulwa) of Southeastern Nicaragua and Comparisons with Miskitu Plant Lore. *Economic Botany* 53(4):363-386.

Coe, W. R. (1965) Tikal, Guatemala, and Emergent Maya Civilization. *Science* 147:1401-1419.

Coe, M. D. and J. Kerr (1997) *The Art of the Maya Scribe.* London: Thames and Hudson.

Colton, M. F. (1938) The Arts and Crafts of the Hopi Indians. *Museum Notes of the Museum of Northern Arizona* 11(1):1-12.

Conzemius, E. (1932) *Ethnographic Survey of the Miskito and Sumu Indians of Honduras and Nicaragua.* Smithsonian Institution, Bureau of Ethnography Bulletin 106. Washington, D.C.: Smithsonian Institution Press.

Cook, O. F. (1903) Food Plants of Ancient America. *Smithsonian Institution Annual Report*, pp. 481-97. Washington, D.C.: Smithsonian Institution Press.

Coon, C. S. (1971) *The Hunting Peoples.* Boston: Little Brown and Company.

Cooper, J. M. (1963) Stimulants and Narcotics. In *Handbook of South American Indians*, vol. 5, pp. 525-558. Smithsonian Institution, Bureau of American Ethnology Bulletin 143, series editor J. H. Steward. New York: Cooper Square Publishers, Inc.

Cordan, Wolfgang (1962) *Das Buch des Popol Vuh.* Düsseldorf-Köln: Eugen Diederichs Verlag.

Coultate, T. P. (2002) *Food--The Chemistry of Its Components.* 3rd ed. Cambridge: The Royal Society of Chemistry.

Covarrubias, Miguel (1954) *The Eagle, the Jaguar and the Serpent: Indian Art of the Americas.* New York: Knopf.

Cox, P. A. (1980) Two Samoan Technologies for Breadfruit and Banana Preservation. *Economic Botany* 34:181-5.

Crawford, M. D. C. (1945) Introduction to *Ancient Peruvian Textiles.* Washington, D.C.: The Textile Museum of the District of Columbia.

Cuatrecasas, José (1964) *Cacao and Its Allies; A Taxonomic Revision of the Genus "Theobroma."* Contributions from the United States National Herbarium, vol. 35, part 6. Washington, D.C.: Smithsonian Institution.

Cushing, F. H. (1920) *Zuñi Breadstuff.* Indian Notes and Monographs, vol. 8. New York: Museum of the American Indian, Heye Foundation.

Cutler, H. C. and M. Cardenas (1947) Chicha, A Native South American Beer. *Harvard University Botanical Museum Leaflets* 13(3):33-60.

Dahlin, B. H. and W. J. Litzinger (1986) Old Bottle, New Wine: The Function of Chultuns in the Maya Lowlands. *American Antiquity*, 51(4):721-736.

Daly, J. W., J. Caceres, R. W. Moni, F. Gusovsky, M. Moos Jr., K. B. Seamon, K. Milton, and C. W. Myers (1992) Frog Secretions and Hunting Magic in the Upper Amazon: Identification of a Peptide that Interacts with an Adenosine Receptor. *Proceedings of the National Academy of Science* 89:10960-10963.

Danzel, Theodor-Wilhelm (1922) *Mexiko.* Vols. 1-3. Hagen i.W und Darmstadt: Folkwang-Verlag G.M.B.H.

Darwin, Charles (1897) *The Variation of Animals and Plants under Domestication.* 2 vols. New York: D. Appleton And Company.

DeBoer, W. R. and D. Lathrap (1979) The Making and Breaking of Shipibo-Conibo Ceramics. In *Ethnoarchaeology: Implications of Ethnography for Archaeology,* edited by C. Kramer, pp. 102-138. New York: Columbia University Press.

De Booy, Theodor (1962) Pottery from Certain Caves in Eastern Santo Domingo, West Indies. *American Anthropologist* 17:69-97.

D'Harcourt, Raoul (1962) *Textiles of Ancient Peru and Their Techniques.* Translated by Sadie Brown, edited by G. G. Denny and C. M. Osborne. Seattle: University of Washington Press.

De la Fuente, B. and L. Staines Cicero (2001) Mural Painting. In *Archaeology of ancient Mexico and Central America: an encyclopedia,* edited by S. T. Evans and D. L. Webster, pp. 491-494. New York: Garland Publisher.

Dibble, C. E. (1971) Writing in Central Mexico. In *Handbook of Middle American Indians*, general editor R. Wauchope, vol. 10, pp. 322-331. University of Texas Press, Austin.

Dole, G. E. (1978) The Use of Manioc among the Kuikuru: Some Interpretations. *Anthropological Papers (Museum of Anthropology, University of Michigan)* 67:217-247.

Donnan, C. B. (1972) Moche-Huari Murals from Northern Peru. *Archaeology* 25:85-95.

—(1973) A Precolumbian Smelter from Northern Peru. *Archaeology* 26:289-97.

Doyon-Bernard, S. J.(1990) From Twining to Triple Cloth: Experimentation and Innovation in Ancient Peruvian Weaving. *American Antiquity* 55(1):68-87.

Driver, H. E. (1953) The Acorn in North American Indian Diet. *Indiana Academy of Science* 62:56-62.

Driver, H. E. and W. C. Massey (1957) Comparative Studies of North American Indians. *Transactions of the American Philosophical Society* 47:165-456.

Duff, A. I. (2002) *Western Pueblo Identities.* Tucson: The University of Arizona Press.

Easby, Dudley T., Jr. (1956) Ancient American Goldsmiths. *Natural History* 65(8):401-9.

—(1961) Fine Metalwork in Pre-Conquest Mexico. In *Essays in Pre-Columbian Art and Archaeology*, by Samuel K. Lothrop and others, pp. 35-42. Cambridge: Harvard University Press.

—(1962) Two "South American" Metal Techniques Found Recently in Western Mexico. *American Antiquity* 28:19-24.

Eckel, E. C. (1928) *Cements, Limes and Plasters.* New York: John Wiley & Sons, Inc.

Ekholm, G. F.(1942) *Excavations at Guasave, Sinaloa, Mexico.* Anthropological Papers of the American Museum of Natural History, vol. 38, pt. 2, pp. 25-139. New York: American Museum of Natural History.

—(1944) *Excavations at Tampico and Panuco in the Huasteca, Mexico.* Anthropological Papers of the American Museum of Natural History, vol. 38, pt. 5, pp. 25-149. New York: American Museum of Natural History.

Emmerich, André (1965) *Sweat of the Sun and Tears of the Moon: Gold and Silver in Pre-Columbian Art.* Seattle: University of Washington Press.

Euler, R. C. and V. H. Jones (1956) Hermetic Sealing as a Technique of Food Preservation Among Indians of the American Southwest. *Proceedings of the American Philosophical Society* 100:87-99.

Falcón, Tatiana (1999) Prehispanic Mural Paintings: Materials and Techniques. In *The Pre-Columbian Painting: Murals of the Mesoamerica*, coordination Beatriz de la Fuente, pp.33-39. Milano: Jaca Book.

Farabee, W. C. (1915) Conebo Pottery. *The Museum Journal, University of Pennsylvania* 6:94-98.

—(1922) *Indian Tribes of Eastern Peru.* Papers of the Peabody Museum of American Archaeology and Ethnology, Harvard University, vol. 10. Cambridge: The Museum.

Fernández, León (1881-1907) *Coleccion de Documentos Para la Historia de Costa Rica.* 10 tomos. San José, Paris, Barcelona.

Fester, G. A. (1953) Einige Farbstoffe südamerikanischer Kulturvolker. *Isis* 44:13-16.

Findlay, Alexander (1934) *The Spirit of Chemistry.* London: Longmans, Green and Co.

Forbes, R. J. (1967) Metallurgy. In *A History of Technology,* edited by C. Singer, E. J. Holmyard, and A. R. Hill, vol. 1, pp. 572-598. London: Oxford University Press.

French, Dexter (1973) Chemical and Physical Properties of Starch. *Journal of Animal Science* 37(4):1048-1061.

Friederici, Georg (1960) *Amerikanistisches Wörterbuch und Hilfswörterbuch für den Amerikanisten.* Hamburg: Cram, de Gruyter.

Fritz, Samuel (1922) *Journal of the travels and labours of Father Samuel Fritz in the river of the Amazon between 1686 and 1723.* Translated by George Edmundson. London: Hakluyt Society.

Furst, P. T. (1976) *Hallucinogens and Culture.* San Francisco: Chandler and Sharp.

Gadsby, Patricia (2002) Endangered Chocolate. *Discover* (August):64-70.

Gage, Thomas (1958) *Travels in the New World.* Edited by J. E. S. Thompson. Norman: University of Oklahoma Press.

Gerhard, P. (1964) Shellfish Dye in America. In *Los Centros Coordinadores indigenistas: sobretiro de la edición conmemorativa en ocasión del XXXV Congreso Internacional de Americanistas, México, 1962*, vol. 3, pp. 177-191. México: El Instituto.

Gettens, R. J. (1962) Maya Blue: An Unsolved Problem in Ancient Pigments. *American Antiquity* 27(4):557-564.

Gibaja O., A. M. (2004) Dos Ofrendas al Agua de Ollantaytambo. *Ñawpa Pacha* 25-27:177-188.

Gifford, E. W. (1936) Californian Balanophagy. In *Essays in Anthropology, presented to A. L. Kroeber in celebration of his sixtieth birthday, June 11, 1936,* edited by R. H. Lowie, pp. 87-98. Berkeley: University of California Press.

Gifford, E. W. and A. L. Kroeber (1937) *Pomo.* University of California Publications in American Archaeology and Ethnology, vol. 37, no. 4. Berkeley: University of California Press.

Gilmore, R. M. (1963) Fauna and Ethnozoology of South America. In *Handbook of South American Indians*, vol. 6, pp. 345-464. Smithsonian Institution, Bureau of American Ethnology Bulletin 143, series editor J. H. Steward. New York: Cooper Square Publishers, Inc.

Glass, J. B. (1975) A Survey and Census of Native Middle American Pictorial Manuscripts. In *Handbook of Middle American Indians*, general editor R. Wauchope, vol. 14, pp. 3-81. University of Texas Press, Austin.

Goffer, Zvi (1980) *Archaeological Chemistry.* New York and Toronto: John Wiley & Sons.

Goodspeed, T. H. (1953) *Species Origins and Relationships in the Genus Nicotiana.* University of California Publications in Botany, vol. 26, no. 5. Berkeley: University of California Press.

Gordon, B. L. (1957a) *Human Geography and Ecology in the Sinú Country of Colombia.* Ibero-Americana, vol. 39. Berkeley: University of California Press.

—(1957b) A Domesticated Wax-Producing Scale Insect Kept by the Guaymí Indians of Panama. *Ethnos* (Stockholm) 22(1-2):36-49.

—(1962) Notes on Shell Mounds Near the Caribbean Coast of Western Panama. *Panama Archaeologist* (Panama City) 5:1-9.

—(1970) Heroes and Ethos of the Jicarilla Apache. *The Masterkey* (Southwest Museum, Los Angeles) 44(2):54-62.

—(1982) *A Panama Forest and Shore: Natural History and Amerindian Culture in Bocas del Toro.* Pacific Grove, CA: The Boxwood Press.

—(1996) *Monterey Bay Area: Natural History and Cultural Imprints.* 3rd ed. Pacific Grove, CA: The Boxwood Press.

Gordon, G. B. (1898) *Researches in the Uloa Valley, Honduras: report on explorations by the Museum.* Memoirs of the Peabody Museum of American Archaeology and Ethnology, Harvard University, vol. 1, no. 4, pp. 95-148. Cambridge: The Museum.

Gordon, R. B. and J. W. Rutledge (1984) Bismuth Bronze from Machu Picchu, Peru. *Science* 223:585-86.

Gordus, A. A., C. E. Henderson and I. Shimada (1996) Electron Microprobe and Neutron Activation Analysis of Gold Artifacts from a 1000 A.D. Peruvian Gravesite. In *Archaeological Chemistry: organic, inorganic, and biochemical analysis*, edited by Mary Virginia Orna. Washington, D.C.: American Chemical Society.

Gottlieb, O. R. (1985) The Chemical Uses and Chemical Geography of Amazon Plants. In *Amazonia*, edited by G. T. Prance and T. E. Lovejoy, pp. 218-238. Oxford: Pergamon Press.

Gourdin, W. H. and W. D. Kingery (1975) The Beginnings of Pyrotechnology: Neolithic and Egyptian Lime Plaster. *Journal of Field Archaeology* 2:133-150.

Graffam, G., M. Rivera, and A. Carevič (1996) Ancient Metallurgy in the Atacama: Evidence for Copper Smelting during Chile's Early Ceramic Period. *Latin American Antiquity* 7(2):101-113.

Grossman, J. W. (1972) An Ancient Goldworker's Toolkit. *Archaeology* 25:270-275.

Guerra, Francisco (1969) Aztec Science and Technology. *History of Science* 8:32-72.

Hall, G. D., S. M. Tarka Jr., W. J. Hurst, D. Stuart and R. E. W. Adams (1999) Cacao Residues in Ancient Maya Vessels from Rio Azul, Guatemala. *American Antiquity* 55:138-143.

Hamilton, D. W. (1957) The Use of Alkaline Admixtures with Narcotic Plants. Manuscript, Library of Economic Botany, Harvard Botanical Museum. Cambridge.

Hartman, C. V. (1910) Le Calabassier de l'Amérique

tropicale (*Crescentia*). Étude d'Ethnobotanique. *Journal de la Société des Américanistes de Paris*, n.s., vol. 7. Paris.

Haslam, Edwin (1988) *Plant Polyphenols: Vegetable tannins revisited*. Cambridge: Cambridge University Press.

Haury, E. W. (1937) A Pre-Spanish Rubber Ball from Arizona. *American Antiquity* 2:282-88.

—(1976) *The Hohokam, desert farmers and craftsmen: excavations at Snaketown, 1964-1965*. Tucson: University of Arizona Press.

Hawley, F. M. (1929) Prehistoric Pottery Pigments in the Southwest. *American Anthropologist* 31:731-754.

Heinen, H. D. and K. Ruddle (1974) Ecology, Ritual and Economic Organization in the Distribution of Palm Starch among the Warao of the Orinoco Delta. *Journal of Anthropological Research* 30:116-138.

Heizer, R. F. (1943) *Aconite Poison Whaling in Asia and America: An Aleutian Transfer to the New World*. Anthropological Papers, Smithsonian Institution, Bureau of American Ethnology no. 24. Washington, D.C.: U.S. Government Printing Office.

Helfritz, Hans (1963) Vorkolumbische Salzgewinnung in Guatemala. *Kosmos. Gesellschaft der Naturfreunde* (Stuttgart) 7:300-305.

Hesseltine, C. W. (1966) A Millennium of Fungi, Food and Fermentation. *Mycologia* 57:149-197.

Hill, W. D., M. Blake and J. E. Clark (1998) Ball Court Design Dates Back 3,400 Years. *Nature* 392:878-879.

Hill, W. W. (1937) *Navajo Pottery Manufacture*. University of New Mexico, Bulletin. Anthropological Series, vol. 2, no. 3, pp. 7-19. Albuquerque: University of New Mexico Press.

Holmes, W. H. (1908) Book Review. *American Anthropologist* 10:128-133.

—(1931) Ancient Maya Ruins. In *Source Book in Anthropology*, edited by A. L. Kroeber and T. T. Waterman, pp. 188-201. New York: Harcourt, Brace and Company.

Hoopes, J. W. and W. K. Barnett (1995) The Shape of Early Pottery Studies. In *The Emergence of Pottery: Technology and Innovation in Ancient Societies*, edited by W. K. Barnett and J. W. Hoopes, pp. 1-7. Smithsonian Series in Archaeological Inquiry. Washington, D.C.: Smithsonian Institution Press.

Hosler, Dorothy (1988) The Metallurgy of Ancient West Mexico. In *The Beginning of the Use of Metals and Alloys*, edited by R. Maddin, pp. 328-343. Cambridge: The MIT Press.

—(1994) *The Sounds and Colors of Power: The Sacred Metallurgical Technology of Ancient West Mexico*. Cambridge: The MIT Press.

Hosler, Dorothy and Guy Stresser-Pean (1992) The Huastec Region: A Second Locus for the Production of Bronze Alloys in Ancient Mesoamerica. *Science* 257:1215-1220.

Hosler, Dorothy, A. L. Burkett, and M. J. Tarkanian (1999) Prehistoric Polymers: Rubber Processing in Ancient Mesoamerica. *Science* 284:1988-1991.

Hough, Walter (1926) *Fire as an Agent in Human Culture*. Smithsonian Institution, United States National Museum, Bulletin 139. Washington, D.C.: U.S. Government Printing Office.

Howe, E. G. and U. Petersen (1994) Silver and Lead in the Late Prehistory of the Mantaro Valley, Peru. In *Archaeometry of Pre-Columbian Sites and Artifacts*, edited by D. A. Scott and P. Meyers, pp. 183-198. Los Angeles: The Getty Conservation Institute.

Hugh-Jones, C. (1978) Food for Thought. In *Sex and Age as Principles of Social Differentiation*, edited by J. S. LaFontaine, pp. 41-66. New York: Academic Press.

Humboldt, Alexander von (1966) *Political Essay on the Kingdom of New Spain*. Vols. 1-4. New York: AMS Press.

Humboldt, Alexander von and Aimé Bonpland (1972) *Personal Narrative of Travels to the Equinoctial Regions of the New Continent during the Years* 1799-1804. Translated by H. M. Williams. Vols. 1-7. Amsterdam and New York: Da Capo Press.

Hurst, W. J., S. M. Tarka Jr., T. G. Powis, F. Valdez Jr. and T. R. Hester (2002) Cacao Usage in the Earliest Maya Civilization. *Nature* 418:289.

Ikawa-Smith, F. (1976) On Ceramic Technology in East Asia. *Current Anthropology* 17:513-15.

Ixer, R. A. and S. Lunt (1991) The Petrography of Certain Pre-Spanish Pottery from Peru. In *Recent developments in ceramic petrology*, edited by A. Middleton and I. Freestone. British Museum, Occasional Paper 81, pp. 137-164. London: British Museum.

Izikowitz, Karl (1930) Calabashes with Star-Shaped Lids in South America and China. In *Comparative Ethnographical Studies*, edited by E. Nordenskiöld, vol. 9, pp. 130-153. Göteborg: Elanders Boktryckeri Aktiebolag.

Jenkins, K. D. (1967) Lacquer. In *Handbook of Middle American Indians*, general editor R. Wauchope, vol. 6, pp. 125-137. Austin: University of Texas Press.

Jijón y Caamaño, J. and C. M. Larrea (1918) Un Cementerio incasico en Quito. *Revista de la Sociedad Juridico-Literaria* (Quito).

Johns, Timothy (1986) Detoxification Function of Geophagy and Domestication of the Potato. *Journal of Chemical Ecology* 12(3):635-646.

Johns, T. and I. Kubo (1988) A Survey of Traditional Methods Employed for the Detoxification of Plant Foods. *Journal of Ethnobiology* 8:81-129.

Johnson, I. W. (1954) Chiptic Cave Textiles from Chiapas, Mexico. *Journal de la Société des Américanistes de Paris* 43:137-146.

—(1971) Baskets and Textiles. In *Handbook of Middle American Indians*, general editor R. Wauchope, vol. 10, pp. 297-321. Austin: University of Texas Press.

Johnsson, Mick (1986) *Food and Culture among Bolivian Aymara: Symbolic Expressions of Social Relations*. Stockholm: Almqvist & Wiksell International.

José-Yacamán, M., L. Rendón, J. Arenas, and M. C. Serra Puche (1996) Maya Blue Paint: An Ancient

Nanostructured Material. *Science* 273:223-226.

Joyce, Thomas A. (1916) *Central American and West Indian Archaeology*. New York: Putnam's Sons.

—(1970) *Mexican Archaeology*. Hacker Art Books, New York.

Julien, C. J. (2004) Las Tumbas de Sacsahuaman y el Estilo Cuzco-Inca. *Ñawpa Pacha* 25-27:1-126.

Kamen-Kaye, Dorothy (1975) Chimó—Why Not? A Primitive Form of Tobacco Still in Use in Venezuela. *Economic Botany* 29:47-69.

Karsten, Rafael (1926) *The Civilization of the South American Indians, with Special Reference to Magic and Religion*. London: Dawsons.

Kasha, M. (1948) Chemical Notes on the Coloring Matter of Chihuahua Textiles in Pre-Columbian Mexico. In *Textiles of Pre-Columbian Chihuahua*, by L. M. O'Neale, Contributions to American Anthropology and History, Carnegie Institution of Washington publication 574, vol. 9, pp. 151-157. Washington, D.C.

Katz, S. H., M. L. Hediger and L. A. Valleroy (1974) Traditional Maize Processing Techniques in the New World. *Science* 184:765-73.

Kelly, Isabel (1943) Notes on a West Coast Survival of the Ancient Mexican Ball Game. *Notes on Middle American Archaeology and Ethnology*. Division of Historical Research, Carnegie Institution of Washington, vol. 1, no. 26, pp. 163-174. Washington, D.C.

—(1947) *Excavations at Apatzingan, Michoacan*. Viking Fund Publications in Anthropology, no. 7, pp. 1-227. New York: Johnson Reprint.

Kidder, A. V. (1948) Introduction to *Textiles of Pre-Columbian Chihuahua*, by L. M. O'Neale. Contributions to American Anthropology and History, Carnegie Institution of Washington publication 574, vol. 9, pp. 99-100. Washington, D.C.

Kidder, A. V., J. D. Jennings and E. M. Shook (1977) *Excavations at Kaminaljuyu, Guatemala*. First published in 1946. University Park and London: The Pennsylvania State University Press.

Koch-Grünberg, Theodor (1909) *Zwei Jahre unter den Indianern. Reisen im Nordwest-Brasilien 1903/1905*. Berlin: Ernst Wasmuth.

Kozioł, M. J. and M. J. Macía (1998) Chemical Composition, Nutritional Evaluation, and Economic Prospects of *Spondias purpurea* (ANACARDIACEAE). *Economic Botany* 52(4):373-380.

Kroeber, A. L. (1923) *Anthropology*. New York: Harcourt, Brace and Company.

—(1976) *Handbook of the Indians of California*. New York: Dover Publications, Inc.

Krukoff, B. A. and A. C. Smith (1937) Rotenone-Yielding Plants of South America. *American Journal of Botany* 24:573-587.

La Barre, Weston (1938) Native American Beers. *American Anthropologist* 40:224-234.

La Niece, S. and N. Meeks (2000) Diversity of Goldsmithing Traditions in the Americas and the Old World. In *Precolumbian Gold: Technology, Style and Iconography*, edited by C. McEwan, pp. 220-239. London: British Museum Press.

Lancaster, P. A., J. S. Ingram, M. Y. Lim, and D. G. Coursey (1982) Traditional Cassava-Based Foods: Survey of Processing Techniques. *Economic Botany* 36:12-45.

Lathrap, D. W. (1975) *Ancient Ecuador-Culture, Clay and Creativity, 3000-300 B. C.* Chicago: Field Museum of Natural History.

Laufer, B. (1930) *Geophagy*. Field Museum of Natural History, Anthropological Series, vol. 18, no. 2, pp. 99-198. Chicago.

Lechtman, Heather (1975) Issues in Andean Metallurgy. In *Pre-Columbian Metallurgy of South America*, edited by E. P. Benson, pp. 1-52. Washington, D.C.: Dumbarton Oaks Research Library and Collections, Trustees for Harvard University.

—(1976) A Metallurgical Site Survey in the Peruvian Andes. *Journal of Field Archaeology* 3:1-42.

—(1980) The Central Andes: Metallurgy without Iron. In *The Coming of the Age of Iron*, edited by T. A. Wertime and J. D. Muhly, pp. 267-334. New Haven: Yale University Press.

—(1988) Traditions and Styles in Central Andean Metalworking. In *The Beginning of the Use of Metals and Alloys*, edited by R. Maddin, pp. 344-378. Cambridge: The MIT Press.

Lechtman, H., A. Erlij, and E. J. Barry Jr. (1982) New Perspectives on Moche Metallurgy: Techniques of Gilding Copper at Loma Negra, Northern Peru. *American Antiquity* 47(1):3-30.

Lee, F. A. (1983) *Basic Food Chemistry*. 2nd ed. Westport, Connecticut: The Avi Publishing Company, Inc.

Leechman, Douglas (1951) Bone Grease. *American Antiquity* 16:355-356.

Lenz, Hans (1973) *El Papel Indígena Mexicano: Historia y Supervivencia*. Mexico: SEP/SENTETAS.

Lévy-Strauss, Claude (1963a) The Nambicuara. In *Handbook of South American Indians*, vol. 3, pp. 361-369. Smithsonian Institution, Bureau of American Ethnology Bulletin 143, series editor J. H. Steward. New York: Cooper Square Publishers, Inc.

—(1963b) The Use of Wild Plants in Tropical South America. In *Handbook of South American Indians*, vol. 6, pp. 465-486. Smithsonian Institution, Bureau of American Ethnology Bulletin 143, series editor J. H. Steward. New York: Cooper Square Publishers, Inc.

—(1963c) The Tupi-Cawahib. In *Handbook of South American Indians*, vol. 6, pp. 299-306. Smithsonian Institution, Bureau of American Ethnology Bulletin 143, series editor J. H. Steward. New York: Cooper Square Publishers, Inc.

Lewin, Louis (1998) *Phantastica: A Classic Survey of the Use and Abuse of Mind-Altering Plants*. (The German edition of Lewin's book was published in 1964.) Rochester, VT: Park Street Press.

Li, Ch'iao-p'ing (1948) *The Chemical Arts of Old China.* Easton, PA: Journal of Chemical Education.

Linné, Sigvald (1925) *The Technique of South American Ceramics.* Kungl. Vetenskaps och Vitterhets-Samhälles Handlingar, Vol. 29. Göteborg: Elanders Boktryckeri Aktiebolag.

—1934) *Archaeological Researches at Teotihuacan, Mexico.* The Ethnographical Museum of Sweden, n.s., no. 1. Stockholm.

—(1949a) Kerus: Inca Wooden Cups. *Ethnos* 2-4:118-139.

—(1949b) Dental Decoration in Ancient Mexico. *Ethnos* 2-4:190-193.

—(1953) Prehistoric Peruvian Painting. *Ethnos* 18:110-123.

—(1957) Technical Secrets of American Indians. *Journal of the Royal Anthropological Institute of Great Britain and Ireland* 87:149-164.

—(1965) The Ethnologist and The American Indian Potter. In *Ceramics and Man,* edited by F. R. Matson, pp. 20-42. Chicago: Aldine Publishing Company.

Linton, Ralph (1944) North American Cooking Pots. *American Antiquity* 9:369-378.

Littmann, E. R. (1957) Ancient Mesoamerican Mortars, Plasters, and Stuccos: Comalcalco, Part I. *American Antiquity* 23:135-140.

—(1960) Ancient Mesoamerican Mortars, Plasters, and Stuccos: The Use of Bark Extracts in Lime Plasters. *American Antiquity* 25:593-7.

—(1973) The Physical Aspects of Some Teotihuacán Murals. Appendix 2 in *The Mural Painting of Teotihuacán,* by A. G. Miller, pp. 175-186. Washington, D.C.: Dumbarton Oaks.

—(1982) Maya Blue--Further Perspectives and the Possible Use of Indigo as the Colorant. *American Antiquity* 47:404-408.

Litto, Gertrude (1976) *South American Folk Pottery.* New York: Watson-Guptill Publications.

Long, Stanley (1964) Cireperdue Copper Casting in Pre-Columbian Mexico: An Experimental Approach. *American Antiquity* 30:189-92.

Lothrop, S. K. (1926) *Pottery of Costa Rica and Nicaragua.* Indian Notes and Monographs, vol. 2. New York: Museum of the American Indian, Heye Foundation.

—(1929) *Polychrome guanaco cloaks of Patagonia.* Indian Notes and Monographs, vol. 7. New York: Museum of the American Indian, Heye Foundation.

—(1930) *Notes on Indian Textiles of Central Chile.* Indian Notes and Monographs, vol. 7, pp. 324-335. New York: Museum of the American Indian, Heye Foundation.

—(1936) *Zacualpa: A Study of Ancient Quiche Artifacts.* Carnegie Institution of Washington.

—(1937) Gold and Silver from Southern Peru and Bolivia. *Journal of the Royal Anthropological Institute of Great Britain and Ireland* 67:305-325.

—(1950) Metal Working Tools from the Central Coast of Peru. *American Antiquity* 16:160-164.

—(1956) Peruvian Pacchas and Keros. *American Antiquity.* 21(3):233-293.

—(1963a) Indians of the Paraná Delta and La Plata Littoral. In *Handbook of South American Indians,* vol. 1, pp. 177-99. Smithsonian Institution, Bureau of American Ethnology Bulletin 143, series editor J. H. Steward. New York: Cooper Square Publishers, Inc.

—(1963b) *Archaeology of the Diquís Delta, Costa Rica.* Papers of the Peabody Museum of Archaeology and Ethnology, Harvard University, vol. 51. Cambridge: The Museum.

Lothrop, S. K. and Joy Mahler (1957) *A Chancay-Style Grave at Zapallan, Peru.* Papers of the Peabody Museum of Archaeology and Ethnology, Harvard University, vol. 50, no. 1. Cambridge: The Museum.

Lothrop, S. K. and others (1961) *Essays in pre-Columbian Art and Archaeology.* Cambridge: Harvard University Press.

Lowie, R. H. (1963) The Tropical Forests: An Introduction. In *Handbook of South American Indians,* vol. 3, pp. 1-56. Bureau of American Ethnology Bulletin 143, series editor J. H. Steward. New York: Cooper Square Publishers, Inc.

Lumholtz, Carl (1973) *Unknown Mexico.* 2 vols. New York: C. Scribner's sons, 1902; New York: AMS Press for Peabody Museum of Archaeology and Ethnology, Harvard University.

Luna Cavazos, M., M. Jiao and R. Bye (2000) Phenetic Analysis of *Datura* Section *Dutra* (Solanaceae) in Mexico. *Botanical Journal of the Linnean Society* 133:493-507.

Lundberg, D. E. and L. H. Kotschevar (1965) *Understanding Cooking.* Amherst: University of Massachusetts.

MacCauley, Clay (1887) *The Seminole Indians of Florida.* U.S. Bureau of American Ethnology, Fifth Annual Report, vol. 5. Washington, D.C.

MacNeish, R. S. and A. G. Cook (1972) Excavations in the Locality of the El Riego Oasis. In *The Prehistory of Tehuacan Valley,* edited by Douglas S. Byers, vol. 5, pp. 14-64. University of Texas Press, Austin.

Magaloni, Diana (1996) Técnicas de la pintura mural en Mesoamérica. *Arquelogía Mexicana* 3(16):16-23.

Marcos J. G. (2003) A Reassessment of the Ecuadorian Formative. In *Archaeology of Formative Ecuador,* edited by J. S. Raymond and R. L. Burger, pp. 7-32. Washington, D.C.: Dumbarton Oaks Research Library and Collections.

McBryde, F. W. (1945) *Cultural and Historical Geography of Southwest Guatemala.* Smithsonian Institution, Institute of Social Anthropology Publication No. 4. Washington, D.C.: U.S. Government Printing Office.

McGregor, J. C. (1931) Prehistoric Cotton Fabrics of Arizona. *Museum Notes of the Museum of Northern Arizona* 4(2).

Mead, C. W. (1907) *Techniques of Some South American Feather-Work.* Anthropological Papers of the Museum of Natural History, vol. 1, pt 1. New York:

American Museum of Natural History.

Means, P. A. (1964) *Ancient Civilizations of the Andes.* New York: Gordian Press.

Merkel, J. F., I. Shimada, C. P. Swann, and R. Doonan (1994) Pre-Hispanic Copper Alloy Production at Batán Grande, Peru: Interpretation of the Analytical Data for Ore Samples. In *Archaeometry of Pre-Columbian Sites and Artifacts*, edited by D. A. Scott and P. Meyers, pp. 183-198. Los Angeles: The Getty Conservation Institute.

Métraux, Alfred (1942) *The Native Tribes of Eastern Bolivia and Western Matto Grosso.* Smithsonian Institution, Bureau of American Ethnology Bulletin 134. Washington, D.C.: U.S. Government Printing Office.

—(1944) "Tapirage," a Biological Discovery of the South American Indians. *Journal of the Washington Academy of Sciences* 34:252-254.

—(1963a) The Guarani. In *Handbook of South American Indians*, vol. 3, pp. 69-94. Smithsonian Institution, Bureau of American Ethnology Bulletin 143, series editor J. H. Steward. New York: Cooper Square Publishers, Inc.

—(1963b) Bark Cloth. In *Handbook of South American Indians*, vol. 5, pp. 67-68. Smithsonian Institution, Bureau of American Ethnology Bulletin 143, series editor J. H. Steward. New York: Cooper Square Publishers, Inc.

—(1963c) Tribes of the Middle and Upper Amazon River. In *Handbook of South American Indians*, vol. 5, pp. 139-204. Smithsonian Institution, Bureau of American Ethnology Bulletin 143, series editor J. H. Steward. New York: Cooper Square Publishers, Inc.

—(1963d) Rubber. In *Handbook of South American Indians*, vol. 5, pp. 227-228. Smithsonian Institution, Bureau of American Ethnology Bulletin 143, series editor J. H. Steward. New York: Cooper Square Publishers, Inc.

Miller, G. R. and R. L. Burger (1995) Our Father the Cayman, Our Dinner the Llama: Animal Utilization at Chavin de Huantar, Peru. *American Antiquity* 60(3):421-458.

Moorey, P. R. S. (1998) Early Metallurgy in Mesopotamia. In *The Beginning of the Use of Metals and Alloys*, edited by R. Maddin, pp. 28-33. Cambridge: The MIT Press.

Morley, S. G. and G. W. Brainerd (1983) *The Ancient Maya.* 4th ed., revised by R. J. Sharer. Stanford: Stanford University Press.

Morris, E. H. (1980) *The Temple of the Warriors.* New York: Charles Scribner's Sons.

Muerer-Grimes, B., A. Berkov, and H. Beck (1998) Theobromine, Theophylline, and Caffeine in 42 Samples and Products of Guaraná (*Paullinia Cupana*, Sapindaceae). *Economic Botany* 52(3):293-301.

Murra, J. V. (1984) Andean Societies. *Annual Review of Anthropology* 13:119-141.

Myers, C. W. and J. W. Daly (1993) Tropical Poison Frogs. *Science* 262:1193.

Neff, H. and R. L. Bishop (1988) Plumbate Origins and Development. *American Antiquity* 53:505-522.

Nicholson, Edward (1960) Chicha Maize Types and Chicha Manufacture in Peru. *Economic Botany* 14:290-299.

Nimuendajú, Curt (1939) *The Apinayé.* Translated by R. H. Lowie. The Catholic University of America, Anthropological Series, no. 8. Washington, D.C.: The Catholic University of America Press.

Nordenskiöld, Erland (1920) *The Changes in Material Culture of Two Indian Tribes under the Influence of New Surroundings.* Comparative Ethnographical Studies, vol. 2. Göteborg: Elanders Boktryckeri Aktiebolag.

—(1924) *The Ethnography of South America Seen from Mojos in Bolivia.* Comparative Ethnographical Studies, vol. 3. Göteborg: Elanders Boktryckeri Aktiebolag.

—(1930a) Modifications in Indian Culture through Inventions and Loans. *Comparative Ethnographical Studies*, vol. 8, pp. 1-254. Göteborg: Elanders Boktryckeri Aktiebolag.

—(1930b) On the History and Geographical Distribution of Bee-Keeping by the American Indian. Appendix 2 to *Modifications in Indian Culture through Inventions and Loans.* Comparative Ethnographical Studies, vol. 8, pp. 196-218. Göteborg: Elanders Boktryckeri Aktiebolag.

—(1930c) The Ancient Peruvian System of Weights. *Man* 155:215-221.

—(1931a) The American Indian as an Inventor. In *Source Book in Anthropology*, edited by A. L. Kroeber and T. T. Waterman, pp. 489-504. Reprinted from *Journal of the Royal Anthropological Institute*, 1929, vol. 59, pp. 273-309. New York: Harcourt, Brace and Company.

—(1931b) *Origin of the Indian Civilizations in South America.* Comparative Ethnographical Studies, vol. 9. Göteborg: Elanders Boktryckeri Aktiebolag.

Nuttall, Zelia (1909) A Curious Survival in Mexico of the Use of the Purpura Shell-Fish for Dyeing. In *Anthropological Essays Presented to Frederic Ward Putnam in Honor of His Seventieth Birthday*, pp. 368-384. New York: G. E. Stechert & Co.

O'Neale, L. M. (1932) *Yurok-Karok Basket Weavers.* University of California Publications in Archaeology and Ethnology, vol. 32, no. 1. Berkeley: University of California Press.

—(1945) *Textiles of Highland Guatemala.* Carnegie Institution of Washington Publication 567. Washington, D.C.

—(1948) *Textiles of Pre-Columbian Chihuahua.* Contributions to American Anthropology and History, Carnegie Institution of Washington Publication 574, vol. 9, pp. 97-161. Washington, D.C.

—(1963) Weaving. In *Handbook of South American Indians*, vol. 5, pp. 97-138. Smithsonian Institution, Bureau of American Ethnology Bulletin 143, series editor J. H. Steward. New York: Cooper Square

Publishers, Inc.
O'Neill, John P. (1984) Feather Identification. In *Costumes and Featherwork of the Lords of Chimor: Textiles from Peru's North Coast*, edited by A. P. Rowe, pp. 145-150. Washington, D.C.: The Textile Museum.
Onion, D. K. (1964) Corn in the Culture of the Mohawk Iriquois. *Economic Botany* 18:60-66.
Osborne, L. de Jongh (1942) Materias Tintóreas Indígena. *America Indígena* (Mexico) 2:87-91.
—(1965) *Indian Crafts of Guatemala and El Salvador*. Norman: University of Oklahoma Press.
Parsons, E. C. (1945) *Peguche, A Study of Andean Indians*. Chicago: University of Chicago Press.
Parsons, M. H. (1975) The Distribution of Late Postclassic Spindle Whorls in the Valley of Mexico. *American Antiquity* 40:207-214.
Parsons, J. R. and M. H. Parsons (1990) *Maguey Utilization in Highland Central Mexico: An Archaeological Ethnography*. Anthropological Papers, Museum of Anthropology, University of Michigan, no. 82. Ann Arbor: University of Michigan.
Patterson, C. C. (1971) Native Copper, Silver, and Gold Accessible to Early Metallurgists. *American Antiquity* 36:286-321.
Paulsen, A. C. (1977) Patterns in Maritime Trade between South Coastal Ecuador and Western Mesoamerica, 1500 B.C.-A.D. 600. In *Dumbarton Oaks Conference on the Sea in the Pre-Columbian World*, pp. 141-160. Washington, D.C.: Dumbarton Oaks Research Library and Collections.
Petersen, Georg (1970) *Minería y metalurgía en el antiguo Perú*. Arqueológicas, no. 12. Lima: Museo Nacional de Antropología y Arqueología.
Pike, K. L. (1980) *A Mixtec Lime Oven*. Dallas: SIL Museum of Anthropology.
Piña Chan, Roman (1971) Preclassic or Formative Pottery and Minor Arts of the Valley of Mexico. In *Handbook of Middle American Indians*, general editor R. Wauchope, vol. 10, pp. 157-178. Austin: University of Texas Press.
Pio Corrêa, M. (1974) *Dicionário das Plantas Úteis do Brasil*. Vols. 1-5. Rio de Janeiro: Ministério de Agricultura.
Piso, Willem (1658) *De Indiae utriusque re naturali et medica: libri qvatuordecim...* Amstelaedami: Apud Ludovicum et Danielem Elzevirios.
Pittier, Henry (1908) *Ensayo sobre Las Plantas Usuales de Costa Rica*. Washington, D.C.: H. L. & J. B. McQueen Inc.
Plazas, C. and A. M. Falchetti de Sáenz (1979) Technology of Ancient Colombian Gold. *Natural History* 88(9):37-46.
Plowman, Timothy (1980) Chamairo: *Mussatia Hyacintha*—an Admixture to Coca from Amazonian Peru and Bolivia. *Harvard University Botanical Museum Leaflets* 28(3):253-259.
Pochettino, M. L., A. R. Cortella and M. Ruiz (1999) Hallucinogenic Snuff from Northwestern Argentina: Microscopical Identification of *Anadenanthera colubrina* var. *cebil* (Fabaceae) in Powdered Archaeological Material. *Economic Botany* 53(2):127-132.
Pohl, M. E. D., K. O. Pope and C. von Nagy (2002) Olmec Origins of Mesoamerican Writing. *Science* 298:1984-1987.
Pool, C. A. and G. M. Britt (2000) A Ceramic Perspective on the Formative to Classic Transition in Southern Vera Cruz, Mexico. *Latin American Antiquity* 11:139-161.
Popenoe, Wilson (1919) Batido and Other Guatemalan Beverages Prepared from Cacao. *American Anthropologist* 21:403-409.
Powis, T. G., F. Valdez Jr., T. R. Hester, W. J. Hurst and S. M. Tarka Jr. (2002) Spouted Vessels and Cacao Use among the Preclassic Maya. *Latin American Antiquity* 13(1):85-106.
Prance, G. T. (1972) Ethnobotanical Notes from Amazonian Brazil. *Economic Botany* 26:221-37.
—(1984) The Pejibaye, *Guilielma gasipaes* (HBK) Bailey, and the Papaya, *Carica papaya* L. In *Pre-Columbian Plant Migration*, edited by Doris Stone. Papers of the Peabody Museum of Archaeology and Ethnology, vol. 76, pp. 85-104. Cambridge: Harvard University Press.
Quiller, Jeffrey (1998) Metallic Reflections. *Science* 282:1058-1059.
Radicati di Primeglio, Carlos (1980) *El sistema contable de los Incas*. Lima: Libreria Studium S.A.
Read, John (1960) *A Direct Entry to Organic Chemistry*. New York: Harper & Brothers.
—(1995) *From Alchemy to Chemistry*. New York: Dover Publications, Inc.
Redfield, M. P. (1964) Notes on the Cookery of Tepoztlan, Morelos. *Journal of American Folklore* 42:167-196.
Redfield, Robert and Alfonso Villa Rojas. (1962) *Chan Kom, a Maya Village*. Abridged Edition. Chicago: University of Chicago Press.
Rehren, T. and M. Temme (1994) Pre-Columbian Gold Processing at Putushio, South Ecuador: The Archaeometallurgical Evidence. In *Archaeometry of Pre-Columbian Sites and Artifacts,* edited by D. A. Scott and P. Meyers, pp. 267-284. Los Angeles: The Getty Conservation Institute.
Reichel-Dolmatoff, G. (1975) *The Shaman and the Jaguar: A Study of Narcotic Drugs among the Indians of Colombia*. Philadelphia: Temple University Press.
Reina, R. E. and R. M. Hill (1978) *The Traditional Pottery of Guatemala.* Austin: University of Texas Press.
Rice, P. M. (1977) Whiteware Pottery Production in the Valley of Guatemala: Specialization and Resource Utilization. *Journal of Field Archaeology* 4:221-233.
Ries, Maurice (1932) Stamping: A Mass-Production Printing Method 2000 Years Old. In *Middle American Research Papers*, publ. no. 4, pp. 415-483. New

Orleans: The Tulane University of Lousiana.

Rivera, M. A. (1991) The Preceramic Chinchorro Mummy Complex of Northern Chile: Context, Style, and Purpose. In *Tombs for the Living: Andean Mortuary Practices*, edited by T. D. Dillehay, pp. 43-77. Washington, D.C.: Dumbarton Oaks Research Library and Collections.

Rivier, L. (1981) Analysis of Alkaloids in Leaves of Cultivated *Erythroxylum* and Characterization of Alkaline Substances Used during Coca Chewing. *Journal of Ethnopharmacology* 3:313-335.

Rivier, L. and J. Lindgren (1972) Ayahuasca, the South American Hallucinogenic Drink: An Ethnobotanical and Chemical Investigation. *Economic Botany* 26:101-129.

Robbins, W. W., J. P. Harrington and B. Freire-Marreco (1916) *Ethnobotany of the Tewa Indians.* Smithsonian Institution, Bureau of Ethnography Bulletin 55. Washington, D.C.: U.S. Government Printing Office.

Robertson, Donald (1975) Techialoyan Manuscripts and Paintings with a Catalog. In *Handbook of Middle American Indians*, general editor R. Wauchope, vol. 14, pp. 253-280. Austin: University of Texas Press.

Robicsek, Francis (1978) *The Smoking Gods: Tobacco in Maya Art, History, and Religion.* Norman: University of Oklahoma Press.

Robinson, Stuart (1969) *A History of Dyed Textiles.* Cambridge: The MIT Press.

Roosevelt, A. C. (1995) Early Pottery in the Amazon: Twenty Years of Scholarly Obscurity. In *The Emergence of Pottery: Technology and Innovation in Ancient Societies*, edited by W. K. Barnett and J. W. Hoopes, pp. 115-131. Smithsonian Series in Archaeological Inquiry. Washington, D.C.: Smithsonian Institution Press.

Roosevelt, A. C., M. Lima da Costa, G. Lopes Machado, M. Michab, N. Mercier, H. Valladas, J. Feathers, W. Barnett, M. Imazio da Silveira, A. Henderson, J. Sliva, B. Chernoff, D. S. Reese, J. A. Holman, N. Toth, K. Schick (1996) Paleoindian Cave Dwellers in the Amazon: The Peopling of the Americas. *Science* 272:373-384.

Root, W. C. (1961) Pre-Columbian Metalwork of Colombia and Its Neighbors. In *Essays in Pre-Columbian Art and Archaeology*, by Samuel K. Lothrop and others, pp. 242-257. Cambridge: Harvard University Press.

—(1963) Metallurgy. In *Handbook of South American Indians*, vol. 5, pp. 205-225. Smithsonian Institution, Bureau of American Ethnology Bulletin 143, series editor J. H. Steward. New York: Cooper Square Publishers, Inc.

Ross, G. N. (1986) The Bug in the Rug. *Natural History* 95(1-4):66-73.

Roth, W. E. (1929) *Additional Studies of the Arts, Crafts, and Customs of the Guiana Indians...* Smithsonian Institution, Bureau of American Ethnology no. 91. Washington, D.C.: U. S. Government Printing Office.

Rouse, Irving (1963) The Carib. In *Handbook of South American Indians*, vol. 4, pp. 547-609. Smithsonian Institution, Bureau of American Ethnology Bulletin 143, series editor J. H. Steward. New York: Cooper Square Publishers, Inc.

—(1972) Southwestern Archaeology Today. In *An Introduction to the Study of Southwestern Archaeology with a Preliminary Account of the Excavations at Pecos,* by A. V. Kidder, pp. 1-53. New Haven: Yale University Press.

Rowe, A. P., ed. (1984) *Costumes and Featherwork of the Lords of Chimor: Textiles from Peru's North Coast.* Washington, D.C.: The Textile Museum.

Rowe, J. H. (1961) The Chronology of Inca Wooden Cups. In *Essays in Pre-Columbian Art and Archaeology,* by Samuel K. Lothrop and others, pp. 317-341. Cambridge: Harvard University Press.

—(1963) Inca culture at the time of the Spanish Conquest. In *Handbook of South American Indians,* vol. 2, pp. 183-330. Smithsonian Institution, Bureau of American Ethnology Bulletin 143, series editor J. H. Steward. New York: Cooper Square Publishers, Inc.

Roys, R. L. (1931) *The Ethnobotany of the Maya.* Middle American Research Series 2. New Orleans: Tulane University.

Rutledge, J. W. and R. B. Gordon (1987) The Work of Metallurgical Artificers at Machu Picchu, Peru. *American Antiquity* 52(3):578-594.

Rye, O. S. (1981) *Pottery Technology: Principles and Reconstruction.* Washington, D.C.: Taraxacum.

Sahlins, M. D. and E. R. Service (1960) *Evolution and Culture.* Ann Arbor: University of Michigan Press.

Sakai, W. S., M. Hanson, and R. C. Jones (1972) Raphides with Barbs and Grooves in *Xanthosoma sagittifolium* (Araceae). *Science* 178:314-315.

Salaman, R. N. (1949) *The History and Social Influence of the Potato.* Cambridge: Cambridge University Press.

Salick, J., N. Cellinese and S. Knapp (1997) Indigenous Diversity of Cassava: Generation, Maintenance, Use and Loss among the Amuesha, Peruvian Upper Amazon. *Economic Botany* 51(1):6-19.

Salser, J. K., Jr. (1975) The Cubeo Gourd Blackening Process. *Economic Botany* 29:242-244.

Sancho, Pedro (1917) *An Account of the Conquest of Peru.* Translated into English and annotated by Philip Ainsworth Means. New York: The Cortés Society.

Santesson, C. G. (1931) An Arrow Poison with Cardiac Effect from the New World. In *Comparative Ethnographic Studies*, edited by E. Nordenskiöld, vol. 9, pp. 157-187. Göteborg: Elanders Boktryckeri Aktiebolag.

San Martín, R. and R. Briones (1999) Industrial Uses and Sustainable Supply of *Quillaja saponaria* (Rosaceae) Saponins. *Economic Botany* 53(3):302-311.

Sapper, Karl (1901) Speise und Trank der Kekchiindianer. *Globus* 80:259-263.

Sassaman, K. E. (1995) The Social Contradictions of

Traditional and Innovative Cooking in the Prehistoric American Southeast. In *The Emergence of Pottery: Technology and Innovation in Ancient Societies*, edited by W. K. Barnett and J. W. Hoopes, pp. 223-240. Smithsonian Series in Archaeological Inquiry. Washington, D.C.: Smithsonian Institution Press.

Sauer, Carl O. (1948) *Colima of New Spain in the Sixteenth Century*. Ibero-Americana, vol. 29. Berkeley: University of California Press.

—(1966) *The Early Spanish Main*. Berkeley: University of California Press.

Saville, M. H. (1917): See: *Narrative of some things of New Spain and of the great city of Temestitan, Mexico.*

—(1920) *The Goldsmith's Art in Ancient Mexico*. Indian Notes and Monographs, misc. no. 7. New York: Museum of the American Indian, Heye Foundation.

—(1925) *Balance-Beam Scales in Ancient Peru*. Indian Notes and Monographs, vol. 2, pp. 266-283. New York: Museum of the American Indian, Heye Foundation.

Sawyer, A. R. (1961) Paracas and Nazca Iconography. In *Essays in Pre-Columbian Art and Archaeology*, by Samuel K. Lothrop and others, pp. 269-298. Cambridge: Harvard University Press.

Schaub, H. P. (1947) Salzgewinnung im Hochland von Guatemala. *Acta Tropica (*Basel) 4:155-157.

Schmandt-Besserat, Denise (1980) Ocher in Prehistory: 300,000 Years of Iron Ores as Pigments. In *The Coming of the Age of Iron*, edited by T. A. Wertime and J. D. Muhly, pp. 127-150. New Haven: Yale University Press.

Schmeda-Hirschmann, Guillermo (1994) Tree Ash as an Ayoreo Salt Source in the Paraguayan Chaco. *Economic Botany* 48(2):159-162.

Schmieder, Oskar (1926) *The East Bolivian Andes South of the Rio Grande or Guapay.* University of California Publications in Geography, vol. 2, no. 5, pp. 85-165. Berkeley: University of California Press.

Schull, W. J. and F. Rothhammer, eds. (1990) *The Aymara: Strategies in Human Adaptation to a Rigorous Environment.* Dordrecht: Klower Academic Publishers.

Schultes, R. E. (1945) El uso del tabaco entre los Huitotos. *Agricultura Tropical* (Bogotá) 1(9):19-22.

—(1957a) A New Method of Coca Preparation in the Colombian Amazon. *Harvard University Botanical Museum Leaflets* 17(9):241-246.

—(1957b) The Genus *Quararibea* in Mexico and the Use of Its Flowers as a Spice for Chocolate. *Harvard University Botanical Museum Leaflets* 17(9):247-264.

—(1960) Native Narcotics of the New World. (The Pharmaceutical Sciences, Third Lecture Series 1960.) *Texas Journal of Pharmacology* 2:142–167.

—(1967) The Botanical Origin of South American Snuffs. In *Ethnopharmacologic Search for Psychoactive Drugs*, edited by D. H. Efron, pp. 291-305. Public Health Service Publication No. 1645. Washington, D.C.: U. S. Government Printing Office.

—(1984) Amazonian Cuktigens and Their Northward and Westward Migration in Pre-Columbian Times. In *Pre-Columbian Plant Migration*, edited by Doris Stone. Papers of the Peabody Museum of Archaeology and Ethnology, vol. 76, pp. 19-37. Cambridge: Harvard University Press.

Schultes, R. E. and B. Holmstedt (1968) De Plantis Toxicariis e Mundo Novo Tropicale Commentationes II: The Vegetal Ingredients of the Myristicaceous Snuffs of the Northwest Amazon. *Rhodora* 70:113-160.

Schultes, R. E. and A. Hofmann (1973) *The Botany and Chemistry of Hallucinogens*. Springfield, Illinois: Charles C. Thomas, Publisher.

Schwarz, Herbert F. (1948) Stingless Bees (Meliponidae) of the Western Hemisphere. *Bulletin of the American Museum of Natural History*, vol. 90. New York.

Schwede, Rudolf (1912) *Über das Papier der Maya-Codices und einiger altmexikanischer Bilderhandschriften*. Dresden.

—(1916) Ein weiterer Beitrag zur Geschichte des altamerikanischen Papiers. *Jahresbericht der Vereinigung für Angewandte Botanik* 13:35-55.

Scott, D. A. and W. Bray (1994) Pre-Hispanic Platinum Alloys: Their Composition and Use in Ecuador and Colombia. In *Archaeometry of Pre-Columbian Sites and Artifacts,* edited by D. A. Scott and P. Meyers, pp. 285-322. The Getty Conservation Institute, Los Angeles.

Seler-Sachs, Caecilie (1909) Mexikanische Küche. *Zeitschrift des Vereins für Volkskunde*, 19 Jahrgang, pp. 369-381. Berlin.

Sharer, R. J. (1994) *The Ancient Maya*. 5th ed. Stanford: Stanford University Press.

Sharon, D. G. and C. B. Donnan (1977) The Magic Cactus: Ethnoarchaeological Continuity in Peru. *Archaeology* 30: 374-381.

Sheldon, William (1820) *Brief Account of the Caraibs, Who Inhabited the Antilles.* Transactions and Collections of the American Antiquarian Society, vol. 1, pp. 365-433. Cambridge.

Shepard, A. O. (1942) *Rio Grande Glaze Paint Ware*. Contributions to American Anthropology and History 39, Carnegie Institution of Washington Publication 528. Washington, D.C.

—(1948) *Plumbate, A Mesoamerican Trade Ware*. Carnegie Institution of Washington Publication 573. Washington, D.C.

—(1965a) *Ceramics for the Archaeologist*. Carnegie Institution of Washington Publication 609. Washington, D.C.

—(1965b) Rio Grande Glaze-Paint Pottery: A Test of Petrographical Analysis. In *Ceramics and Man*, edited by F. R. Matson. Chicago: Aldine Publishing Company.

—(1977) Technological Notes. In *Excavations at Kaminaljuyu, Guatemala,* by A. V. Kidder, J. D. Jennings and E. M. Shook, pp. 261-277. University Park: Pennsylvania State University Press.

Shimada, Izumi (1994) *Pampa Grande and the Mochica Culture*. Austin: University of Texas Press.

—(1996-2000) Evolution of Andean Diversity: Regional Formations (500 B.C.E.-C.E.600). In *The Cambridge History of the Native Peoples of the Americas* 3(1):350-517. Cambridge: Cambridge University Press.

Shimada, I. and J. F. Merkel (1991) Copper-Alloy Metallurgy in Ancient Peru. *Scientific American* 265:80-86.

Simón, Pedro (1882-1892) *Noticias Historiales de las Conquistas de Tierra Firme en las Indias Occidentales*. 5 vols. Bogotá: M Rivas.

Sillar, Bill (2000) *Shaping Culture: Making Pots and Constructing Households, An Ethnoarchaeolgical Study of Pottery Production, Trade and Use in the Andes.* BAR International Series 883. Oxford: BAR Publishing.

Skibo, J. M. and M. B. Schiffer (1995) The Clay Cooking Pot: An Exploration of Women's Technology. In *Expanding Archaeology,* edited by J. M. Skibo, W. H. Walker, and A. E. Nielsen, pp. 80-91. Salt Lake City: University of Utah Press.

Solomons, T. W. G. (1996) *Organic Chemistry*. 6th ed. New York: John Wiley & Sons, Inc.

Spencer, Herbert (1885) *The Principles of Sociology.* 3 vols. New York: Appleton.

Spinden, H. J. (1928) *Ancient Civilizations of Mexico and Central America.* Handbook Series, no. 3, American Museum of Natural History. New York.

—(1931) The Origin and Distribution of Agriculture in America. In *Source Book in Anthropology*, edited by A. L. Kroeber and T. T. Waterman, pp. 227-233. New York: Harcourt, Brace and Company, Inc.

Standley, P. C. (1926) *Trees and Shrubs of Mexico.* Contributions from the United States National Herbarium, vol. 23. Washington, D.C.: Smithsonian Institution.

—(1928) *Flora of the Panama Canal Zone.* Contributions from the United States National Herbarium, vol. 27. Washington, D.C.: U.S Government Printing Office.

Standley, P. C. and Julian Steyermark (1946) *Flora of Guatemala.* Chicago Natural History Museum, Fieldiana, vol. 24. Chicago.

Starr, Frederick (1900-1902) *Notes upon the Ethnography of Southern Mexico.* Vols. 1-2. Davenport, Iowa: Putnam Memorial Publication Fund.

Steinkraus, K. H., ed. (1983) *Handbook of Indigenous Fermented Foods*. Microbiology Series, vol. 9. New York: Marcel Dekker, Inc.

Steward, J. H. (1963) The Native Population of South America. In *Handbook of South American Indians*, vol. 5, pp. 655-772. Smithsonian Institution, Bureau of American Ethnology Bulletin 143, series editor J. H. Steward. New York: Cooper Square Publishers, Inc.

Steward, J. H. and L. C. Faron (1959) *Native Peoples of South America.* New York: McGraw-Hill Book Company, Inc.

Stone, Doris (1949) *The Boruca of Costa Rica*. Papers of the Peabody Museum of American Archaeology and Ethnology, Harvard University, vol. 26, no. 2. Cambridge: The Museum.

—(1977) *Pre-Columbian Man in Costa Rica.* Cambridge: Peabody Museum Press.

Stone, Doris and Conchita Turnbull (1941/2) A Sula-Ulua Pottery Kiln. *American Antiquity* 7:39-47.

Stowe, B. B. (1959) Occurrence and Metabolism of Simple Indoles in Plants. *Fortschritte der Chemie organischer Naturstoffe* 17:248-297.

Stuart, David (1988) The Río Azul Cacao Pot: Epigraphic Observations on the Function of a Maya Ceramic Vessel. *Antiquity* 62:153-7.

Tastevin, Constant (1943) *The Middle Amazon; its people and geography, eleven articles.* Translated by the Strategic Index of the Americas. Washington, D.C.

Taube, K. A. (1989) The Maize Tamale in Classic Maya Diet, Epigraphy, and Art. *American Antiquity* 54:31-51.

Tessmann, Gunter (1930) *Die Indianer nordost-Perus, grundlegende Forschungen für eine systematische Kulturkunde.* Hamburg: Friederichsen.

Thomsen, H. H. (1960) Occurrence of Fired Bricks in Pre-Conquest Mexico. *Southwestern Journal of Anthropology* 16:429-441.

Thompson, D. W. (1947) *A Glossary of Greek Fishes.* London: Oxford University Press.

Thompson, E. H. (1897) *The Chultunes of Labná, Yucatan.* Memoirs of the Peabody Museum of American Archaeology and Ethnology, Harvard University, vol. 1, no. 3. Cambridge: The Museum.

—(1932) *People of the Serpent.* New York: Houghton Mifflin.

Thompson, F. C. (1958) The Early Metallurgy of Copper and Bronze. *Man* 58:1-7.

Thompson, J. E. S. (1934) *Sky Bearers, Colors and Directions in Maya and Mexican Religion.* Carnegie Institution of Washington Publication 436, pt. 10, pp. 211-215. New York: Johnson Reprint Corp.

—(1938) Sixteenth and Seventeenth Century Reports on the Chol Maya. *American Anthropologist* 40:584-604.

—(1972) *A Commentary on the Dresden Codex, A Maya Hieroglyphic Book.* Memoirs of the American Philosophical Society, vol. 93. Philadelphia: American Philosophical Society.

Tozzer, A. M. (1907) *A comparative study of the Mayas and the Lacandones.* New York: Published for the Archeological Institute of America by The Macmillan Company.

Trachsler, Walter (1965) Precursors of Polychrome Painted Pottery: Some Examples from the Prehistory of Switzerland. In *Ceramics and Man*, edited by F. R. Matson, pp. 152-160. Chicago: Aldine Publishing Company.

Troll, Carl (1958) *Las culturas superiores andinas y el medio geográfico.* Translated from German to Spanish by Carlos Nicholson. Publicaciones del Instituto de Geografía, Revista del Instituto de Geografía 5:3-55. Lima.

Tschopik, Harry (1963) The Aymara. In *Handbook of South American Indians*, vol. 2, pp. 501-573. Smithsonian Institution, Bureau of American Ethnology Bulletin 143, series editor J. H. Steward. New York: Cooper Square Publishers, Inc.

Tschudi, J. J. (1854) *Travels in Peru during the years 1838-1842...* Translated from the German by Thomasina Ross. New York: A. S. Barnes & Company.

Tylecote, R. F. (1980) Furnaces, Crucibles, and Slags. In *The Coming of the Age of Iron,* edited by T. A. Wertime and J. D. Muhly, pp. 183-228. New Haven: Yale University Press.

Uhle, Max (1889-90) *Kultur und Industrie südamerikanischer Völker*. 2 vols. Berlin: A. Ascher & Co.

Van de Velde, P. and H. R. Van de Velde (1939) *The Black Pottery of Coyotepec, Oaxaca, Mexico.* Southwest Museum Papers No. 13. Los Angeles: Southwest Museum.

Van Olphen, H. (1966) Maya Blue: A Clay-Organic Pigment? *Science* 154:645-46.

Velde, B. and I. C. Druc (1999) *Archaeological Ceramic Materials: Origin and Utilization.* Berlin: Springer.

Vezga, Florentino (1936) *Botánica Indigena.* Biblioteca Aldeana de Colombia, vol 47. Bogotá: Editorial Minerva.

Vogel, V. J. (1970) *American Indian Medicine.* Norman: University of Oklahoma Press.

von Hagen, V. W. (1939) *The Tsátchela Indians of Western Ecuador.* Indian Notes and Monographs, misc. no. 51. Heye Foundation. Museum of the American Indian, New York.

—(1977) *The Aztec and Maya Paper Makers.* With an introduction by Dard Hunter. New York: J. J. Augustin: 1944; New York: Hacker Art Books.

Wafer, Lionel (1934) *A New Voyage and Description of the Isthmus of America.* Edited by G. P. Winship. Cleveland: The Hakluyt Society.

Wagner, U., R. Gebhard, E. Murad, J. Riederer, I. Shimada, and F. Wagner (1994) Kiln Firing at Batán Grande: Today and in Formative Times. In *Archaeometry of Pre-Columbian Sites and Artifacts,* edited by D. A. Scott and P. Meyers, pp. 183-198. Los Angeles: The Getty Conservation Institute.

Wallace, A. R. (1890) *A Narrative of Travels on the Amazon and Rio Negro.* London: Ward, Lock and Co.

Wassén, S. H. (1935) *Notes on Southern Groups of Chocó Indians in Colombia.* Etnologiska Studier, pp. 35-182. Göteborg.

—(1957) *On Dendrobates-Frog-Poison Material Among Emperá (Chocó)-speaking Indians in Western Caldas, Colombia.* Etnografiska Museet, pp. 73-94. Göteborg.

—(1967) Anthropological Survey of the Use of South American Snuffs. In *Ethnopharmacologic Search for Psychoactive Drugs*, edited by D. H. Efron, pp. 233-289. Public Health Service Publication No. 1645. Washington, D.C.: U. S. Government Printing Office.

Wassén, S. H. and B. Holmstedt (1963) The Use of Paricá: An Ethnological and Pharmacological Review. *Ethnos* 28:5-45.

Waterman, T. T. (1931) The Architecture of the American Indians. In *Source Book in Anthropology,* edited by A. L. Kroeber and T. T. Waterman, pp. 512-524. New York: Harcourt, Brace and Company.

Webster, Steven (1975) Comments, in Ceramic Ecology of the Ayacucho Basin, Peru: Implications for Prehistory, by D. E. Arnold. *Current Anthropology* 16:201.

Wendorf, Fred (1953) *Archaeological Studies in the Petrified Forest National Monument.* Bulletin of the Museum of Northern Arizona, no. 27. Flagstaff, Arizona: Northern Arizona Society of Science and Art, Inc.

Werge, R. W. (1979) Potato Processing in the Central Highlands of Peru. *Ecology of Food and Nutrition* 7:229-234.

Wertime, T. A. (1973a) The Beginnings of Metallurgy: A New Look. *Science* 182:875-887.

—(1973b) Pyrotechnology: Man's First Industrial Uses of Fire. *American Scientist* 61(6):670-682.

Whiffen, Thomas (1915) *The North-West Amazons.* London: Constable and Company, Ltd.

Whitten, N. E. (1976) *Sacha Runa: Ethnicity and Adaptation of Ecuadorian Jungle Quichua.* Urbana: University of Illinois Press.

Wilbert, Johannes (1987) *Tobacco and Shamanism in South America.* New Haven and London: Yale University Press.

Willey, G. R. (1963) Ceramics. In *Handbook of South American Indians*, vol. 5, pp. 139-204. Smithsonian Institution, Bureau of American Ethnology Bulletin 143, series editor J. H. Steward. New York: Cooper Square Publishers, Inc.

—(1966) *An Introduction to American Archaeology.* Vol. 1, *North and Middle America.* Edgewood Cliffs: Prentice-Hall, Inc.

Willey, G. R. and P. Phillips (1944) Negative-Painted Pottery from Crystal River, Florida. *American Antiquity* 10:173-185.

Wisdom, Charles (1940) *The Chorti Indians of Guatemala.* Chicago: University of Chicago Press.

Wissler, Clark (1957) *The American Indian.* London: Oxford University Press.

Woodroffe, J. F. (1915) *The Rubber Industry of the Amazon, and how Its Supremacy can be Maintained.* London: John Bale, Sons & Danielsson, Ltd.

www.ingramcontent.com/pod-product-compliance
Lightning Source LLC
LaVergne TN
LVHW071211210926
848482LV00015B/783
9781407303901